Basic Concepts
of Probability
and Statistics

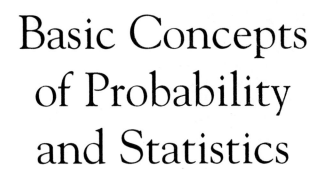

SIAM's Classics in Applied Mathematics series consists of books that were previously allowed to go out of print. These books are republished by SIAM as a professional service because they continue to be important resources for mathematical scientists.

Classics in Applied Mathematics

C. C. Lin and L. A. Segel, *Mathematics Applied to Deterministic Problems in the Natural Sciences*

Johan G. F. Belinfante and Bernard Kolman, *A Survey of Lie Groups and Lie Algebras with Applications and Computational Methods*

James M. Ortega, *Numerical Analysis: A Second Course*

Anthony V. Fiacco and Garth P. McCormick, *Nonlinear Programming: Sequential Unconstrained Minimization Techniques*

F. H. Clarke, *Optimization and Nonsmooth Analysis*

George F. Carrier and Carl E. Pearson, *Ordinary Differential Equations*

Leo Breiman, *Probability*

R. Bellman and G. M. Wing, *An Introduction to Invariant Imbedding*

Abraham Berman and Robert J. Plemmons, *Nonnegative Matrices in the Mathematical Sciences*

Olvi L. Mangasarian, *Nonlinear Programming*

*Carl Friedrich Gauss, *Theory of the Combination of Observations Least Subject to Errors: Part One, Part Two, Supplement*. Translated by G. W. Stewart

Richard Bellman, *Introduction to Matrix Analysis*

U. M. Ascher, R. M. M. Mattheij, and R. D. Russell, *Numerical Solution of Boundary Value Problems for Ordinary Differential Equations*

K. E. Brenan, S. L. Campbell, and L. R. Petzold, *Numerical Solution of Initial-Value Problems in Differential-Algebraic Equations*

Charles L. Lawson and Richard J. Hanson, *Solving Least Squares Problems*

J. E. Dennis, Jr. and Robert B. Schnabel, *Numerical Methods for Unconstrained Optimization and Nonlinear Equations*

Richard E. Barlow and Frank Proschan, *Mathematical Theory of Reliability*

Cornelius Lanczos, *Linear Differential Operators*

Richard Bellman, *Introduction to Matrix Analysis, Second Edition*

Beresford N. Parlett, *The Symmetric Eigenvalue Problem*

*First time in print.

Classics in Applied Mathematics (continued)

Basic Concepts
of Probability
and Statistics

Second Edition

J. L. Hodges, Jr.
and E. L. Lehmann

Society for Industrial and Applied Mathematics
Philadelphia

 is a registered trademark.

To J. Neyman

Contents

PART I. PROBABILITY

1. PROBABILITY MODELS

2. SAMPLING

3. PRODUCT MODELS

4. CONDITIONAL PROBABILITY

5. RANDOM VARIABLES

6. SPECIAL DISTRIBUTIONS

7. MULTIVARIATE DISTRIBUTIONS

PART II. STATISTICS

INTRODUCTION TO STATISTICS

8. ESTIMATION

9. ESTIMATION IN MEASUREMENT AND SAMPLING MODELS

Preface to the Classics Edition

Basic Concepts was originally published in 1964; a slightly enlarged second edition appeared in 1970. It filled the need for an introduction to the fundamental ideas of modern statistics that was mathematically rigorous but did not require calculus. This was achieved by restricting attention to discrete situations. The book was translated into Italian, Hebrew, Danish, and, more recently, Farsi. It went out of print in 1991 when its publisher, Holden-Day, went out of business.

Despite its age, the book in many ways is modern in outlook. This is particularly true for its emphasis on models and model-building but also by its coverage of such topics as survey sampling (both simple and stratified), experimental design (with a proof of the superiority of factorial design over varying one factor at a time), its presentation of nonparametric tests such as the Wilcoxon, and its discussion of power (including the Neyman–Pearson lemma). The book is very much in the spirit of texts on discrete mathematics and could well be used to supplement high school and college courses on this subject.

Although the book contains a large number of examples from a great variety of fields of application, it does not base these on real data. When used as a textbook, this drawback could be remedied by adding a laboratory in which actual situations are discussed.

Basic Concepts was written jointly with my friend and colleague Joseph L. Hodges, Jr. Each section, nearly each sentence, was vigorously debated, drafted, and then subjected to additional debate. We greatly enjoyed this collaboration. My delight in seeing the book reissued by SIAM is tempered by the fact that, after a long period of poor health, Joe died on March 1, 2000, of a heart attack. Seeing *Basic Concepts* brought to life again would have given him great pleasure.

E. L. LEHMANN

Preface to the Second Edition

The second edition differs from its predecessor primarily through the addition of new material. Stimulated by suggestions of users of the book and by our own teaching experience, we have added more than 300 problems, most of them elementary. We have also provided answers to selected problems at the end of the book. The present volume contains four sections not in the first edition. Of these, Sections 6.9 and 6.10 were introduced in the separately published first part, *Elements of Finite Probability*, in 1965. Section 12.7 discusses the problem of ties for the rank tests treated in Chapter 12, thereby adding to the usefulness of these tests. A final Section 13.6 presents Student's t-test, which many instructors believe should be included in an introductory course, and the Wilcoxon one-sample test. This permits us to compare the t-test with the nonparametric approach emphasized in the other testing material. (Without the calculus the exact distribution of the t-statistic cannot be derived, but we give an approximation for the significance probability.)

We are very grateful to the many readers who have taken the trouble to let us have their reactions and suggestions. These were most helpful as we tried to improve the presentation. In addition to making many minor changes, we rewrote Section 4.1 to correct an error in the earlier version; we should like to thank the several readers who pointed this out to us. We also took a new approach in Section 12.4, in response to a criticism for which we are indebted to Ellen Sherman. Our thanks are due to Howard D'Abrera for checking the problems and providing the answers to the new ones. An answer book can be obtained by instructors from the publishers.

J. L. HODGES, JR.
E. L. LEHMANN

Berkeley
January, 1970

Preface to the First Edition

Statistics has come to play an increasingly important role in such diverse fields as agriculture, criminology, economics, engineering, linguistics, medicine, psychology and sociology. Many of the statistical methods used in these and other areas are quite complicated, and cannot be fully understood without considerable mathematical background. However, the basic concepts underlying the methods require no advanced mathematics. It is the purpose of this book to explain these concepts, assuming only a knowledge of high-school algebra, and to illustrate them on a number of simple but important statistical techniques. We believe that this material is both more interesting and more useful than complicated methods learned by rote and without full understanding of their limitations.

A satisfactory statistical treatment of observational or experimental data requires assumptions about the origin and nature of the data being analyzed. Because of random elements in the data, these assumptions are often of a probabilistic nature. For this reason, statistics rests inherently on probability, and an adequate introduction to statistics, even at the most elementary level, does not seem possible without first developing the essentials of probability theory.

The first part of the book is therefore devoted to the basic concepts of probability. It is centered on the notion of a probability model, which is the mathematical representation of the random aspects of the observations. We attempt not only to develop the essentials of the mathematical theory of probability but also to give a feeling for the relations between the model and the reality which it represents.

Since both probability theory and statistics involve many new concepts which need time for their comprehension, we feel that an introductory course to the subject ideally should extend over more time than the usual one semester or quarter. The 66 sections of the book provide enough material for a year course. However, we have repeatedly used the material in teaching an introduction to probability and statistics in a one-semester course meeting three hours a week. This can be accomplished

xvii

by taking up only those topics in probability of which essential use is made in the statistics part, omitting Chapter 4 (conditional probability), Chapter 7 (multivariate distributions), as well as certain special topics (Sections 2.4, 3.4, and 6.6–6.8). With this arrangement, the second half of the semester is available for specifically statistical material; here the emphasis may, if desired, be placed either on estimation or on hypothesis testing since the two topics are presented independently of each other.

The conceptual interdependence of different parts of the book, and the resulting possible different course plans, are indicated on the diagram. It should be remarked that certain problems, illustrations and technicalities involve cross-reference outside this schema. However, any of the course plans suggested by the diagram can be used by omitting an occasional illustration or problem, and by accepting a few formulas without proof.

A careful treatment of probability without calculus is possible only if attention is restricted to finite probability models, i.e., to models representing experiments with finitely many outcomes. Fortunately, the basic notions of probability, such as probability model, random variable, expectation and variance, are in fact most easily introduced with finite models. Furthermore, many interesting and important applications lead to finite models, such as the binomial, hypergeometric and sampling models.

In the statistical part of the book, random experimental designs are seen to lead to finite models which, together with the binomial and sampling models and a model for measurement, provide the background for a simple and natural introduction of the basic concepts of estimation and hypothesis testing. Without calculus, it is of course not possible to give a satisfactory treatment of such classical procedures as the one- and two-sample t-tests. Instead, we discuss the corresponding Wilcoxon tests, which we believe in any case to be superior.

While basically we have included only concepts that can be defined, and results that can be proved, without calculus, there is one exception: we give approximations to certain probabilities, whose calculation we have explained in principle, but whose actual computation would be too laborious. Thus, in particular, we discuss the normal approximation to the binomial, hypergeometric and Wilcoxon distributions; the Poisson approximation to the binomial and Poisson-binomial distributions, and the chi-square approximation to the goodness-of-fit criterion. Since the limit theorems which underlie these approximations require advanced analytical techniques and are hence not available to us, we give instead numerical illustrations to develop some feeling for their accuracy.

An unusual feature of the book is its careful treatment of the independence assumption in building probability models. Many complex experiments are made up of simpler parts, and it is frequently possible to build models for such experiments by first building simple models for the parts, and then combining them into a "product model" for the experiment as a whole. Particular attention is devoted to the realism of this procedure, and in general to the empirical meaning of the independence assumption, which so often is applied invalidly.

A decimal numbering system is used for sections, formulas, examples, tables, problems, etc. Thus, Section 5.4 is the fourth section of Chapter 5. To simplify the writing, within Chapter 5 itself we omit the chapter and refer to this section simply as Section 4. Similarly Example 5.4.2 is the second example in Section 5.4. However, within Section 5.4 we omit the chapter and section and refer to the example as Example 2; in other sections of Chapter 5 we omit the chapter and refer to the example as Example 4.2. Eleven of the more important tables, including tables for the two Wilcoxon tests, are collected at the end of the book as Tables A–K.

We are grateful to the following colleagues, who have used or read parts of a preliminary edition of the book, for many corrections and suggestions: F. J. Anscombe, M. Atiqullah, D. R. Cox, Tore Dalenius, William Kruskal, L. J. Savage, Henry Scheffé, Rosedith Sitgreaves, Curt Stern, Herbert Solomon and David L. Wallace. Peter Nuesch and Stephen Stigler worked through the problems, and we are indebted to them for numerous corrections. Our thanks are due to Mrs. Julia Rubalcava and Mrs. Carol Rule Roth for patient and expert typing. Finally, we wish to express our appreciation to our publishers, Holden-Day, for their efficiency and helpfulness during all stages of publication.

J. L. HODGES, JR.
E. L. LEHMANN

Berkeley
January, 1964

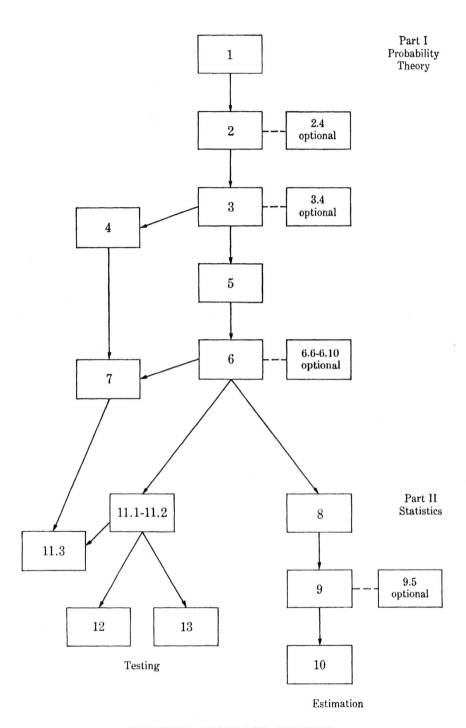

RELATIONS AMONG THE CHAPTERS

PART I

PROBABILITY

CHAPTER 1

PROBABILITY MODELS

1.1 RANDOM EXPERIMENTS

The theories of probability and statistics are mathematical disciplines, which have found important applications in many different fields of human activity. They have extended the scope of scientific method, making it applicable to experiments whose results are not completely determined by the experimental conditions.

The agreement among scientists regarding the validity of most scientific theories rests to a considerable extent on the fact that the experiments on which the theories are based will yield essentially the same results when they are repeated. When a scientist announces a discovery, other scientists in different parts of the world can verify his findings for themselves. Sometimes the results of two workers appear to disagree, but this usually turns out to mean that the experimental conditions were not quite the same in the two cases. If the same results are obtained when an experiment is repeated under the same conditions, we may say that the result is determined by the conditions, or that the experiment is *deterministic*. It is the deterministic nature of science that permits the use of scientific theory for predicting what will be observed under specified conditions.

However, there are also experiments whose results vary, in spite of all efforts to keep the experimental conditions constant. Familiar examples are provided by gambling games: throwing dice, tossing pennies, dealing from a shuffled deck of cards can all be thought of as "experiments" with unpredictable results. More important and interesting instances occur in many fields. For example, seeds that are apparently identical will produce plants of differing height, and repeated weighings of the same object with a chemical balance will show slight variations. A machine which sews together two pieces of material, occasionally—for no apparent reason—will miss a stitch. If we are willing to stretch our idea of "experiment," length of life may be considered a variable experimental result, since people living under similar conditions will die at different and unpredictable ages. We shall refer to experiments that are not deterministic, and thus do not always yield the same result when repeated under the same conditions, as *random experiments*. Probability theory and statistics are the branches of mathematics that have been developed to deal with random experiments.

Let us now consider two random experiments in more detail. As the first example, we take the experiment of throwing a die. This is one of the simplest random experiments and one with which most people are personally acquainted. In fact, probability theory had its beginnings in the study of dice games.

EXAMPLE 1. *Throwing a die.* Suppose we take a die, shake it vigorously in a dice cup, and throw it against a vertical board so that it bounces onto a table. When the die comes to rest, we observe as the experimental result the number, say X, of points on the upper face. The experiment is not deterministic: the result X may be any of the six numbers 1, 2, 3, 4, 5, or 6, and no one can predict which of the values will be obtained on any particular performance of the experiment. We may make every effort to control or standardize the experimental conditions, by always placing the die in the cup in the same position, always shaking the cup the same number of times, always throwing it against the same spot on the backboard, and so on. In spite of all such efforts, the result will remain variable and unpredictable.

EXAMPLE 2. *Measuring a distance.* It is desired to determine the distance between two points a few miles apart. A surveying party measures the distance by use of a surveyor's chain. It is found in practice that the measured distance will not be exactly the same if two parties do the job, or even if the same party does the job on consecutive days. In spite of the best efforts to measure precisely, small differences will accumulate and the final measurement will vary from one performance of the experiment to another.

How is it possible for a scientific theory to be based on indeterminacy? The paradox is resolved by an empirical observation: while the result on any particular performance of such an experiment cannot be predicted, a long sequence of performances, taken together, reveals a stability that can serve as the basis for quite precise predictions. The property of *long-run stability* lies at the root of the ideas of probability and statistics, and we shall examine it in more detail in the next section.

1.2 EMPIRICAL BASIS OF PROBABILITY

To obtain some idea about the behavior of results in a sequence of repetitions of a random experiment, we shall now consider some specific examples.

EXAMPLE 1. *Throwing a die.* Figure 1 shows certain results of 5000 throws of a die. In this experiment, we were not interested in the actual value of the number X of points showing, but only in whether X was less

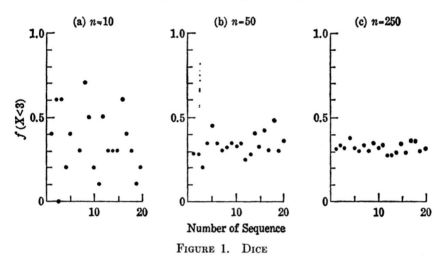

FIGURE 1. DICE

than three $(X < 3)$, or on the contrary was greater than or equal to three $(X \geq 3)$. The values of X on the first ten repetitions or *trials* of the experiment are as follows:

Trial number	1	2	3	4	5	6	7	8	9	10
Value of X	6	3	2	1	5	6	1	3	5	2

The result $(X < 3)$ occurred on trials 3, 4, 7, and 10 and did not occur on trials 1, 2, 5, 6, 8, and 9. Thus, the number of occurrences of the result $(X < 3)$ on the first ten trials was 4: we denote this for brevity by the formula $\#(X < 3) = 4$. Dividing this by the number 10 of trials, we obtain the fraction .4, which will be called the *frequency* of the result $(X < 3)$. This frequency will be denoted by $f(X < 3)$, so that $f(X < 3) = .4$ in the first 10 trials. In general, if a result R occurs $\#(R)$ times in n trials of an experiment, the frequency $f(R)$ of the result is given by

$$f(R) = \frac{\#(R)}{n}.$$

We have been discussing the first sequence of ten trials. In a similar way, 19 other sequences of ten trials were carried out. On the 20 sequences, various values of $f(X < 3)$ were observed, as follows:

Sequence	1	2	3	4	5	6	7	8	9	10	11	12	13	14	15	16	17	18	19	20
$f(X < 3)$	.4	.6	.6	.2	.4	0	.3	.7	.5	.2	.1	.5	.3	.3	.3	.6	.4	.3	.1	.2

Thus, on the second sequence of ten trials, the result $(X < 3)$ occurred six times out of ten, or $f(X < 3) = .6$. The values observed for $f(X < 3)$ varied from 0 (Sequence 6) to .7 (Sequence 8). These observations are displayed in Figure 1a, where each point corresponds to one sequence of ten trials; on this diagram, the horizontal scale shows the number of the sequence, while the vertical scale shows the value of $f(X < 3)$ observed on that sequence.

An examination of Figure 1a shows that while $f(X < 3)$ varies from one sequence to another, it never exceeded .7, and we might perhaps predict that $f(X < 3)$ would not exceed .7 in another sequence of ten trials. Of course $f(X < 3)$ *could* be greater than .7, or even as great as 1, since it is possible that X will be either 1 or 2 every time the die is rolled. However, our experience of 20 sequences suggests that a value of $f(X < 3)$ larger than .7 will not occur very often.

Would the behavior of $f(X < 3)$ be less erratic if a longer sequence of trials were used? If we denote by n the number of trials in the sequence, Figure 1b shows the values of $f(X < 3)$ in 20 sequences of $n = 50$ trials each. These values are still variable, but clearly less so than for $n = 10$, since $f(X < 3)$ now ranges from .20 to .48 instead of from 0 to .7. It appears that, with longer sequences, the frequency of the result $X < 3$ is less variable and hence more predictable than when the sequences are shorter. After studying Figure 1b, even a cautious person might predict that in another sequence of $n = 50$ trials, $f(X < 3)$ is not likely to fall below .15 or above .55. Thus encouraged, we finally turn to Figure 1c, which shows the values of $f(X < 3)$ in 20 sequences of $n = 250$ trials each, based on 5000 throws of the die in all. Again the variation is reduced: now $f(X < 3)$ ranges only from .276 to .372.

EXAMPLE 2. *Random digits.* A rather parallel illustration is provided by a random digit generator. This device, which has been called an "electronic roulette wheel," is designed to produce at random the digits from 0 to 9. (The precise meaning and usefulness of such "random digits" will be made clear in Section 3.4.) In the present experiment we were not interested in the actual value of the digit produced but only in whether it was even, that is, was one of the digits 0, 2, 4, 6, or 8. Figure 2 is analogous to Figure 1. It shows the frequency $f(\text{even})$ with which an even digit occurred, in 20 sequences of $n = 10$, $n = 50$, and $n = 250$ digits. Qualitatively the features of Figure 2 agree with those of Figure 1: it appears that f(even) also stabilizes as n is increased, although the values are placed somewhat higher on the vertical scale.

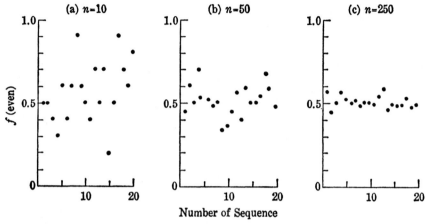

FIGURE 2.   RANDOM DIGITS

EXAMPLE 3.   *Male births.*   The general features of our preceding examples, indeterminacy combined with stability as the sequences get longer, may also appear with sequences of occurrences which at first would not seem to constitute repetitions of an experiment under constant conditions.   Figure

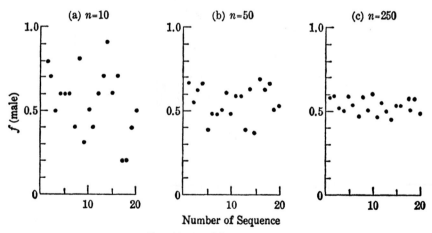

FIGURE 3.   MALE BIRTHS

3 relates to the frequency of males in a sequence of human births, showing the frequency f(male) for 20 sequences of $n = 10$, 50, and 250 births.*

---

* We are indebted to our colleague J. Yerushalmy for the record of sex in 5000 births represented here.

Since the parents in the 5000 cases were of course different, how can we say—even if we agree to regard the sex of a baby as the result of an experiment—that the conditions are constant?   Yet, in appearance Figures 2 and 3 are so similar as to suggest that a theory built to deal with one would also work for the other—as indeed it does.

EXAMPLE *4*.   *Two-letter words*.   The data for Figure 4 were obtained by inspecting 5000 lines in a book, and determining for each line whether or not it contained at least one two-letter word.   Figure 4 shows the fre-

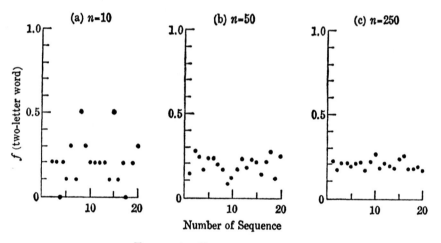

FIGURE 4.   TWO-LETTER WORDS

quency f(two-letter word) of getting at least one two-letter word, for 20 sequences of $n = 10$, 50, and 250 lines.   Again, we would hardly think of an author, when he sits down to write, as performing an experiment which may result in the appearance of a two-letter word in a line of type, but operationally the sequence of results seem to portray features quite like those of the other cases.

A careful examination of Figures 1–4, or of similar records in other cases, will suggest that the tendency towards stability proceeds at a fairly definite rate as the number n of trials is increased.   The table below gives, for each of our four experiments and for each of the three values of n, the difference between the maximum and minimum values observed for f in the 20 sequences.   This difference is known as the *range* of f.   We see that when the sequences are made five times as long, the ranges become about half as large.

TABLE 1. RANGE OF f

	Dice	Random digits	Male births	Two-letter words
$n = 10$	.7	.7	.7	.5
$n = 50$	.28	.36	.32	.20
$n = 250$	.096	.140	.144	.096

These facts lead to the speculation that we could make f as nearly constant as we please, by taking sufficiently large values for n. In some cases it is possible to see what happens when n is made very large. For example, there has been published a table* giving the number of times each of the ten digits was produced in 20 sequences of $n = 50,000$ trials of a random digit generator. If we consider the frequency with which even digits occurred in these sequences, it is found that the lowest value observed among the twenty sequences was .4971 and the highest was .5054, giving a range of only .0083. As a second illustration, let us consider the frequency of males among registered births, which are available from publications of the U.S. Public Health Service. In the twenty years from 1937 to 1956, f(male) for live births in the state of New York varied between .5126 and .5162, giving a range of .0036. (On the average, the number n of these births was about 230,000 per year.)

Data of this kind, gathered from many sources over a long period of time, indicate the following *stability property of frequencies:* for sequences of sufficient length the value of f will be practically constant; that is, if we observed f in several such sequences, we would find it to have practically the same value in each of them. This *long-run frequency* may of course depend on the conditions under which the experiment is performed. The frequency of an ace in a long sequence of throws of a die, for example, may depend on the die that is used, on how it is thrown, on who does the throwing, and so forth. Similarly, the frequency of males in human births has been found to depend, to a greater or lesser extent, on such "conditions of the experiment" as the nationality of the parents and the order of the child in the family. In fact, an active field of research in human biology is concerned with determining just what factors do influence the frequency of males.

It is essential for the stability of long-run frequencies that the conditions of the experiment be kept constant. Consider for example the machine mentioned earlier, which sews together two pieces of material. If the

---

* The RAND Corporation, "A Million Random Digits with 100,000 Normal Deviates," Free Press, Glencoe, Illinois, 1955.

needle that does the stitching gradually begins to blunt, the frequency with which it will miss stitches might increase instead of settling down, in which case a plot of the frequencies would look quite different from Figure 1. Actually, in reality it is of course never possible to keep the conditions of the experiment exactly constant. There is in fact a circularity in the argument here: we consider that the conditions are *essentially* constant as long as the frequency is observed to be stable. As suggested above in connection with the frequency of males, it may be important and useful to study what aspects of the experimental conditions influence the frequency. Thus, special techniques (known as quality control) have been developed to help isolate those conditions of a production process, under which the frequency of defective items is high, in order to change them and thereby improve the quality of the product.

The stability property of frequencies, which we have discussed in this section, is not a consequence of logical deduction. It is quite possible to conceive of a world in which frequencies would not stabilize as the number of repetitions of the experiment becomes large. That frequencies actually do possess this property is an empirical or observational fact based on literally millions of observations. This fact is the experimental basis for the concept of probability with which we are concerned in this book.

PROBLEMS

1. Perform the random experiment of tossing a penny 200 times, and make a plot like Figure 2a. To standardize the experimental conditions, place the penny on the thumb heads up and toss it so that it rotates several times and lands on a hard flat surface. (By combining the results of five such sequences, a figure like 2b can be produced.) Keep this data for use in Problem 11.3.8.

2. In a long sequence of throws of a die as described in Example 1, let $f(1)$, $f(2), \ldots, f(6)$ denote the frequencies with which the upper face shows 1 point, 2 points, $\ldots$, 6 points. Approximately what values would you expect $f(1)$, $f(2)$, $\ldots$ to have if the die is carefully made so that in shape it approximates a perfect cube and if the material is uniform throughout?

3. State qualitatively what changes you would expect in the answer to the preceding problem if the die were loaded by placing a weight in the center of the face with 6 points.

4. State qualitatively what changes you would expect in the answer to Problem 2 if two of the sides of the cube had been shaved so as to bring the sides with 1 point and 6 points closer together.

1.3　SIMPLE EVENTS AND THEIR PROBABILITIES

In Section 1, probability theory and statistics were stated to be mathematical disciplines dealing with random experiments. Such experiments were illustrated, and some of their basic features described, in Section 2. But how does one deal with them mathematically? Observational phenomena are made accessible to mathematical treatment by constructing *mathematical models* for them. Such a model is an abstraction of the real situation, compared with which it has two important advantages.

(i) Properties and relationships, which in reality hold only approximately, in the model can be postulated to be mathematically precise, and their consequences can then be worked out rigorously (that is, as mathematical theorems). This is for example the case with geometry, where the mathematical theory deals with idealized points, straight lines, etc., satisfying certain postulates. The results which are rigorously deduced from these postulates (for example the theorem of Pythagoras) are then applied to such approximately straight lines as those laid out by a surveyor.

(ii) Each actual situation is in some way unique; results worked out directly in concrete terms would therefore have only very limited applicability. In the model, on the other hand, the situation is stripped of all that is special and irrelevant. The results obtained in the model thus are applicable to all real situations having the same general structure.

It is seen that the construction of a mathematical model involves a twofold simplification: only the important and relevant features of the actual situation are portrayed in the model; and in this portrayal they are idealized, the "imperfections" of the actual situation being smoothed away. How well a model works depends on how realistic it is in spite of these simplifications. One of the most important tasks in the application of mathematical models is to see that the model employed represents the real situation sufficiently well so that the results deduced from it will have the desired accuracy. For example, consider the model which represents the earth as a sphere. For a rough calculation of the surface in terms of the diameter this may be adequate. For slightly more refined results it may be necessary to take account of the flattening at the poles and to represent the earth as a spheroid.

We now turn to the problem of building a mathematical model for a random experiment. In the experiment, we are interested in the frequencies with which various results occur in the long run. To serve our purpose, the model will have to have features corresponding to these results, and to the frequencies with which they occur. Let us first consider how to represent the results of an experiment.

In any trial of a random experiment, the record of the result could be almost indefinitely detailed. When a die is thrown, we could record how many times it bounces, where on the table it comes to rest, the angle its edge makes with the edge of the table, how many decibels of noise it produces, and so forth. However, if we are concerned with the use of the die in gaming, none of these results is of interest: we just want to know the number of points on the upper face.

It is important, before conducting any experiment, to decide just which features are worth recording, or one may find when the experiment is over that it is not possible to answer the questions of interest. If for example in an investigation of human births the day but not the hour of each birth is recorded, it will be possible to study the distribution of births throughout the year, but not to determine whether more births occur at night than during the day.

Having decided what to observe, we can (at least in principle) make a list of all possible results of our observations in such a way that on each trial of the experiment one and only one of the results on the list will occur. We shall refer to the results on such a list as the *simple results* of the experiment. If in the investigation of human births, for example, we have decided to record just the month of birth because we are only interested in certain seasonal trends, the simple results would be the twelve months of the year. If instead we had decided that both the month of birth and the sex of the child are relevant to the investigation, there would be 24 simple results: (Jan., male), (Jan., female), (Feb., male), . . . , (Dec., female). In an experiment with a die, if we have decided to observe only the number of points showing on the die, the six simple results will be 1 point, 2 points, . . . , 6 points.

According to the above definition, the list must divide the possible outcomes of the experiment into categories, which between them cover all possibilities and which leave no ambiguity as to the category to which the outcome of an experiment belongs. Suppose for example that we are interested in the weather, and are considering a list which classifies each day as sunny, cloudy, rainy. This would not be an adequate list of simple results. It is not complete, since it does not for example include the possibility of it snowing. Also, the list is ambiguous since a day may be rainy in the morning but sunny in the afternoon.

In the model, we shall represent each simple result of the experiment by an element to be called a *simple event.* For example, in the experiment of throwing a die, we may represent the simple result "1 point" by the integer 1, "2 points" by the integer 2, and so forth. The integers 1, . . . , 6 would then constitute the simple events of the model. Equally well, we could use as simple events the words One, . . . , Six. Again, when a penny is tossed we may be interested in whether it falls heads or tails. There are two simple results, and the model correspondingly needs two simple events, which may be labeled H (for heads) and T (for tails).

We shall also want our model to possess features that correspond to the long-run frequencies with which the various simple results occur. To see how these should be specified, consider any particular simple result, say r. Suppose that in a long sequence of n performances or trials of the experiment, the result r occurred $\#(r)$ times. Its frequency is then $f(r) = \#(r)/n$, and this is a number satisfying the two inequalities

$$0 \leq f(r) \leq 1,$$

since $\#(r)$ cannot be less than 0 or greater than n. If e is the simple event corresponding to r, we shall associate with e a number called the *probability of e*, to be denoted by $P(e)$. This number represents in the model the long run frequency $f(r)$ of the corresponding simple result r. Accordingly we shall require that

$$0 \leq P(e) \leq 1$$

for every simple event e.

The probability model for a random experiment will thus contain simple events (corresponding to the simple results of the experiment) to which are assigned probabilities (corresponding to the long-run frequencies of the simple results). We illustrate the construction of such models with two examples.

EXAMPLE 1. *Throwing a die.* As we have suggested above, for the experiment of throwing a die the integers from 1 to 6 may be used as the simple events. If the die is symmetric, one would expect the six faces to occur about equally often in a long sequence of throws. Experience usually bears out this expectation, and the result "1 point" for instance is found to occur in about $\frac{1}{6}$ of the trials; that is, $f(1 \text{ point})$ is usually found to be near $\frac{1}{6}$. This suggests that in the model one should set $P(1) = \frac{1}{6}$, and similarly $P(2) = \frac{1}{6}, \ldots, P(6) = \frac{1}{6}$. The customary probability model for throwing a die thus consists of six simple events, to each of which is assigned the probability $\frac{1}{6}$.

EXAMPLE 2. *Throwing a loaded die.* An amusement store offers for sale dice that are loaded to favor showing the face with one point. The preceding model would not be reasonable for use with such a die, since $f(1 \text{ point})$ will tend to be larger than $\frac{1}{6}$ because of the loading. Suppose that the die is thrown $n = 1000$ times, and that the number of occurrences of each face is as follows:

$\#(1 \text{ point}) = 214,$ $\#(2 \text{ points}) = 152,$ $\#(3 \text{ points}) = 178$
$\#(4 \text{ points}) = 188,$ $\#(5 \text{ points}) = 163,$ $\#(6 \text{ points}) = 105.$

For the probabilities in the model we might reasonably take the corresponding frequencies, putting $P(1) = .214$ and so forth. As we shall see later, the chance fluctuations in the 1000 throws are so great that there is little or no validity in the third decimal, and the simpler probabilities

$$P(1) = .21, \qquad P(2) = .15, \qquad P(3) = .18,$$
$$P(4) = .19, \qquad P(5) = .16, \qquad P(6) = .11,$$

would serve as well.

The two examples illustrate the two principal methods of determining probabilities in practice. Considerations of symmetry, backed up by general experience, will in many problems suggest that certain frequencies should be nearly equal, and one may then reasonably use a model in which the corresponding probabilities are equal. In other problems one may rely directly on frequencies observed in a long sequence of repeated experiments. (This corresponds to everyday usage. When the probability of twins for example is stated to be 1.3%, it means that this was the observed frequency of twins in an extensive series of birth records.) In more complicated problems, a mixture of observed frequencies and symmetry considerations may be appropriate. Suppose for example that a die is loaded by placing a weight in the center of the face with six points. This will cause the die to tend to fall with that face down, and hence with the opposite face (which has one point) up. However, the remaining four faces are still symmetric, so that we might want to impose the condition $P(2) = P(3) = P(4) = P(5)$ on the model. If the observed frequencies are those given in Example 2, we would retain the values of $P(1)$ and $P(6)$ assumed there but might replace the probabilities .15, .18, .19, .16 by their average .17, so that the six simple events would be assigned the probabilities

(1)
$$P(1) = .21, \qquad P(2) = .17, \qquad P(3) = .17,$$
$$P(4) = .17, \qquad P(5) = .17, \qquad P(6) = .11.$$

EXAMPLE 3. *Tossing a penny.* Experience and symmetry suggest that in tossing a penny, heads and tails will tend to occur about equally often, so that in a long series of tosses f(heads) and f(tails) would both be near $\frac{1}{2}$. One might then reasonably form a model consisting of two simple events H and T, with $P(H) = \frac{1}{2}$ and $P(T) = \frac{1}{2}$.

EXAMPLE 4. *Boy or girl?* What is the probability that a newborn baby will be a boy? If we have no detailed knowledge of the frequency of male births, we might feel that there should be about equal numbers of male and female births (this is also suggested by a consideration of the genetic mechanism by which the sex of a child is determined), and hence assign to the two simple events M (for male) and F (for female) the probabilities

$$(2) \qquad P(M) = \tfrac{1}{2}, \qquad P(F) = \tfrac{1}{2}.$$

Actually, examination of any sufficiently extensive sequence of birth records shows the frequency of male births to be slightly higher than $\tfrac{1}{2}$. The observations for the state of New York quoted in Section 1, for example, suggest putting

$$(3) \qquad P(M) = .514, \qquad P(F) = .486.$$

This example shows again that theoretical considerations are not necessarily completely reliable guides to a correct model. In the present case, of course, the difference between the two models turns out to be rather slight. For many purposes, the simpler model (2) will be quite adequate, and we shall in fact later in the book occasionally use it because of its simplicity.

The various simple events taken together form a collection that will be called the *event set* and denoted by $\mathcal{E}$. Thus in both Examples 1 and 2, $\mathcal{E}$ consists of the first six positive integers; in Example 3 it consists of the two letters H and T; and in Example 4 of the two letters M and F.

PROBLEMS

1. In an investigation concerned with age, decide for each of the following whether it would be a possible list of simple results.

(i) r_1 = young : less than 25 years
r_2 = middle : 20–55 years
r_3 = old   : more than 55 years

(ii) r_1 = young : less than 20 years
r_2 = middle : 20–55 years
r_3 = old   : more than 55 years

(iii) r_1 = young : less than 20 years
r_2 = middle : 20–55 years
r_3 = old   : 56–80 years

2. In a study of three-child families, decide for each of the following whether it would be a possible list of simple results.

(i) r_1 = 1st child is a boy
r_2 = 2nd child is a boy
r_3 = 3rd child is a boy

(ii) r_1 = 1st child is a boy
r_2 = 1st child is a girl

(iii) r_1 = all three children are boys
   r_2 = all three children are girls
(iv) r_1 = exactly one child is a boy
   r_2 = exactly two children are boys
   r_3 = all three children are boys

3. In a tasting experiment, subjects are asked to state which of three brands A, B and C they like best. What outcomes would have to be added to "A is best," "B is best," "C is best" if a subject is permitted to remain undecided between two or more brands?

4. In a sociological study, a variable of interest is marriage status. Make a suitable list of simple results.

5. For a study of the effects of smoking, make a list of simple results which could be used to summarize the smoking habits of the subjects.

6. An event set ε consists of the three simple events e_1, e_2, and e_3. Explain for each of the following why it cannot be a probability model:
   (i) $P(e_1) = .3$, $P(e_2) = .6$, $P(e_3) = 1.2$
   (ii) $P(e_1) = .6$, $P(e_2) = .9$, $P(e_3) = -.5$.

7. A die is thrown twice; if you are interested in the numbers of points on both the first and on the second throws, what would you take as your list of simple results? How many simple results are there in this list?

8. If a penny and nickel are tossed, make a list of simple results, in such a way that all of them might be expected to occur with approximately equal frequency.

9. In a study of the effects of birth order, each subject is asked to list his birth order (i.e., first-born, second-born, etc.). Discuss some complications which might prevent this from being an adequate list.

10. A box contains three marbles, one blue, one red, and one white. Make a list of simple results for an experiment consisting of a selection of two of these marbles.

11. In the preceding problem make a list of simple results if the two marbles are drawn out one after the other (without replacing the first marble before the second one is drawn) and if the order in which the two marbles are obtained is relevant to the experiment.

12. Make a list if in the preceding problem the two marbles are drawn out with replacement, that is, if the first marble is put back into the box before the second marble is drawn.

13. A box contains six marbles, two white and four red. If an experiment consists of the selection of three of these marbles, make a list of simple results
   (i) in terms of the number of white marbles;
   (ii) in terms of the number of red marbles.

14. A pyramid has as its base an equilateral triangle ABC, and its three sides ABD, BCD, CAD are congruent isosceles triangles. Assuming that in 1000 throws the pyramid fell on its base 182 times, use the symmetry with respect to the three sides to build a model for the four possible results of a throw.

15. The base of a prism is an equilateral triangle. In 2000 throws, the prism fell 1506 times on its rectangular sides, and the remaining 494 times on its upper and lower ends. Use the symmetries of the prism to build a model for the five possible results of a throw.

16. In an amusement park one can shoot 100 times at a circular target with a bull's-eye. If for a certain person the target shows 31 hits of which 8 are bull's-eyes, build a probability model with the three simple events: bull's-eye, hit (but not a bull's eye), miss.

1.4 DEFINITION OF PROBABILITY MODEL

As is easily checked in all examples considered in the preceding section, the probabilities of the simple events in the event set add up to 1. To see why this should be true in general, suppose that the simple results of an experiment are labeled $r_1, r_2, \ldots$.   On each performance of the experiment exactly one simple result occurs. In a sequence of n performances we must therefore have

$$\#(r_1) + \#(r_2) + \ldots = n.$$

On dividing both sides by n, this equation becomes

$$f(r_1) + f(r_2) + \ldots = 1.$$

Let $e_1, e_2, \ldots$ denote the simple events that in the model correspond to the simple results $r_1, r_2, \ldots$.   Since the probabilities $P(e_1), P(e_2), \ldots$ are to represent the frequencies $f(r_1), f(r_2), \ldots$ when n is large, it seems reasonable to require that the probabilities in the model should satisfy

$$P(e_1) + P(e_2) + \ldots = 1.$$

The model would not give a satisfactory picture of the experimental situation if the probabilities were defined so as not to add up to 1.

We can now give a formal definition of a probability model, and describe its identification with the random experiment.

Definition. By a *probability model* we mean the following:

(i)   there is an *event set* $\mathcal{E}$, which is a set of elements called *simple events*;

(ii)   to each simple event e is attached a number $P(e)$ called the *probability* of the event e; the probabilities must satisfy $0 \leqq P(e) \leqq 1$ for each e, and their sum must be equal to 1.

In using the model to represent a random experiment, we identify the simple events of the model with the simple results of the experiment, and the probability of a simple event with the frequency of the corresponding result in a long sequence of trials of the experiment. This relationship may be summarized in the following "dictionary" for translating from the real world to the model and back:

Real World	Model
random experiment	probability model
simple result	simple event
list of simple results	event set
long-run frequency	probability

A probability model can be exhibited by listing its simple events together with their probabilities, for example in the form

$$e_1, \quad e_2, \ldots$$

$$P(e_1), P(e_2), \ldots$$

In Example 3.2, for instance, the probability model may be specified by

1	2	3	4	5	6
.21	.15	.18	.19	.16	.11.

The simple results of an experiment are essentially the "atoms" from which other results can be built up. Any result consisting of more than one simple result is said to be *composite*. Consider for instance the situation mentioned in the preceding section, in which the simple results are the month of birth and sex of a child. The following are some examples of composite results together with the simple results of which they are composed.

(a) The composite result "The child is a male born during the first three months of the year" is composed of the simple results (Jan., male), (Feb., male), (Mar., male).

(b) The composite result "The child is a male" is composed of the simple results (Jan., male), (Feb., male), . . . , (Dec., male).

(c) The composite result "The child is born in January" is composed of the simple results (Jan., male), (Jan., female).

Suppose now that R is the composite result composed, say, of the two simple results r_1 and r_2. This may be indicated by writing $R = \{r_1, r_2\}$. Since on each trial of the experiment exactly one simple result occurs, we must have $\#(R) = \#(r_1) + \#(r_2)$, and hence

$$f(R) = f(r_1) + f(r_2).$$

In general the frequency of a composite result is the sum of the frequencies of the simple results that make it up.

To a composite result R of the experiment, there corresponds in the model a *composite event*, say E. This consists of the simple events that correspond to the simple results making up R. For example, if

$R = \{r_2, r_6, r_8\}$ consists of the simple results r_2, r_6, and r_8, then the composite event E corresponding to R in the model consists of the simple events e_2, e_6, and e_8. Since the frequency of R would then be given by $f(R) = f(r_2) + f(r_6) + f(r_8)$, we should have in the model $P(E) = P(e_2) + P(e_6) + P(e_8)$. Generalization of this example motivates the following definition of the probability of a composite event.

Definition. The probability of a composite event E is the sum of the probabilities of the simple events of which E is composed.

As an illustration, consider the experiment of throwing a die, and the composite result that the number of points on the upper face is even. This composite result consists of the three simple results: 2 points, 4 points, 6 points. In the models of Examples 3.1 and 3.2, the corresponding composite event (even) will then consist of the integers 2, 4, and 6, so that

$$P(\text{even}) = P(2) + P(4) + P(6).$$

In the model of Example 3.1 we thus have

$$P(\text{even}) = \tfrac{1}{6} + \tfrac{1}{6} + \tfrac{1}{6} = \tfrac{1}{2} = .5,$$

while in the model of Example 3.2

$$P(\text{even}) = .15 + .19 + .11 = .45.$$

Since the probabilities of all simple events are nonnegative and add up to 1, it is clear that for any event E,

$$0 \leqq P(E) \leqq 1.$$

In presenting probability models, we have so far been very careful to distinguish between the results of real experiments and the events that represent them in the model. However, it is sometimes convenient to be less careful and to employ the same terms for the results and for the corresponding events in the model. Thus, we may for example write "P (die falls with six points showing)" rather than "$P(6)$, where 6 denotes the event representing the result that the die falls with six points showing," as we did in Examples 3.1 and 3.2. When discussing applicational examples, we shall usually adopt this simpler terminology.

We shall conclude this section with an important example, in which the values of the probabilities are obtained mainly on the basis of extensive experience.

EXAMPLE 1. *Life table.* Consider the "experiment" that consists in observing the length of a human life. It is usual to think of life span as a continuous variable, but for our present purpose it will suffice to observe merely the decade A in which the individual dies, so that the simple results are "dies before 10th birthday" ($A = 1$), "survives 10th birthday but dies before 20th" ($A = 2$), etc. Can we regard such an experiment as random?

It is clearly not repeatable unless we agree that the identity of the individual is not essential to the constancy of the conditions. If we observe the life spans of a number of similar individuals, we might perhaps regard them as trials of the same experiment. The entire life insurance industry is founded on the observation that the property of stability does indeed hold in this case: the fraction of a large number of similar individuals who die in a specified decade of their lives will be nearly constant and can therefore be predicted on the basis of past experience.

The probability model for life span is known as a *life table*, the earliest example of which was given by John Graunt in 1662. The simple events correspond to the age of the individual at death. The probability assigned to one of these events corresponds to the frequency with which such individuals die in the decade in question. An example of a life table is shown below, where for simplicity we assume there is no chance of reaching age 100.

TABLE 1. LIFE TABLE

Decade A	Probability of death	Decade A	Probability of death
1	.064	6	.127
2	.013	7	.211
3	.022	8	.269
4	.032	9	.170
5	.064	10	.028

The meaning of the entry .032 opposite decade 4 is that among a large number of individuals, of the kind for whom the table is appropriate, approximately 3.2% will die between their 30th and 40th birthdays.

According to the above table, what is the probability of reaching the age of 60? This is a composite result, consisting of the simple results of dying in the 7th, 8th, 9th, or 10th decade. The probability of reaching 60 is therefore .211 + .269 + .170 + .028 = .678.

It is important when using a life table to be sure that the table represents the life spans of individuals of the kind to whom it is to be applied. For example, women in our culture live considerably longer than men. The table above is based on experience of white males living in the United States at the present time—it should not be used for a woman, a Negro, a resident of Pakistan, or someone in the twenty-first century.

PROBLEMS

1. An event set ε consists of the four simple events e_1, e_2, e_3, e_4. Determine for each of the following whether it satisfies the conditions of a probability model:

(i) $P(e_1) = .01$, $P(e_2) = .03$, $P(e_3) = .11$, $P(e_4) = .35$

(ii) $P(e_1) = .3$, $P(e_2) = .6$, $P(e_3) = .9$, $P(e_4) = -.8$

(iii) $P(e_1) = .01$, $P(e_2) = .04$, $P(e_3) = .06$, $P(e_4) = .89$

(iv) $P(e_1) = .2$, $P(e_2) = .3$, $P(e_3) = 0$, $P(e_4) = .5$.

2. An event set ε consists of five simple events $e_1, \ldots, e_5$.

(i) If $P(e_1) = .01$, $P(e_2) = .02$, $P(e_3) = .03$, $P(e_4) = .04$, determine $P(e_5)$.

(ii) If $P(e_1) = .01$, $P(e_2) = .02$, $P(e_3) = .04$, what are the largest and smallest possible values for $P(e_4)$?

(iii) In part (ii), what is the value of $P(e_4)$ if it is known that $P(e_4) = P(e_5)$?

3. An event set ε consists of ten simple events $e_1, \ldots, e_{10}$. Determine the probabilities of these events if their probabilities are all equal.

4. With the model of Example 3.1, find the probability of the following events.

(i) The number of points is at least three.

(ii) The number of points is at most two.

(iii) The number of points is more than four.

5. Find the probabilities of the events (i)–(iii) of the preceding problem for model (1) of Example 3.2.

6. With model (1) of Example 3.2 would you (at even money) prefer to bet for or against the die showing an even number of points?

7. A person driving to work every day on a route with four traffic lights has observed the following to be a suitable probability model for the number R of red lights encountered on a trip:

$$P(0 \text{ red lights}) = .05$$
$$P(1 \text{ red light}) = .25$$
$$P(2 \text{ red lights}) = .36$$
$$P(3 \text{ red lights}) = .26$$
$$P(4 \text{ red lights}) = .08.$$

Find the probabilities of his encountering

(i) at least 2 red lights;

(ii) more than 2 red lights;

(iii) at most 2 red lights.

8. In grading certain boards sold by a sawmill, each board is given 0, 1 or 2 points on each of two characteristics, namely A: the number of knotholes, and B: the pattern of the grain. Suppose the following probability model is suitable.

A \\ B	0	1	2
0	.24	.19	.10
1	.16	.12	.06
2	.09	.03	.01

Here the entry .19 in the first row and second column means that the probability is .19 of a board receiving 1 point on B and 0 on A.

(i) Check that the table defines a probability model.

Find the probability that a board will receive

(ii) a total number of 4 points;

(iii) a total number of 3 points or more;

(iv) a total number of at most three points;

(v) at least one point on both A and B;

(vi) the same number of points on A as on B.

9. In the preceding problem suppose that 0 to 3 points are given for B but only 0 or 1 point for A and that the probabilities are as follows:

B\A	0	1	2	3
0	.25	.19	.12	.07
1	.16	.13	.06	.02

(i) Check that the table defines a probability model.
Find the probability that a board will receive
  (ii) a total of at least 2 points;
  (iii) a total of at most 2 points;
  (iv) the highest number of points for at least one of the categories;
  (v) at least one point on both A and B;
  (vi) the same number of points on both A and B.

10. In a study of the relationship of the performance in course 1 and 2, a department finds that the following model suitably describes the distribution of grade points for students completing both courses:

2\1	0	1	2	3	4
0	.00	.01	.00	.00	.00
1	.03	.05	.04	.00	.00
2	.01	.04	.26	.05	.00
3	.00	.03	.11	.15	.03
4	.00	.00	.03	.07	.09

(i) Check that the table defines a probability model.
Find the probability that a student will
  (ii) do better in course 2 than in course 1;
  (iii) get the same grade in both courses;
  (iv) change his grade by more than one grade point from course 1 to course 2;
  (v) get a total of at least six grade points;
  (vi) get a total of exactly six grade points;
  (vii) get a total of not more than 3 grade points.

11. In a population for which Table 1 is appropriate, find approximately the *median* age of death; that is, the age such that the probability is $\frac{1}{2}$ of dying before that age and $\frac{1}{2}$ of surviving beyond it.

12. In a long sequence of observations on length of life of male rats, it was found that 98 percent still survived 200 days after birth, 83 percent survived 400 days, 40 percent survived 600 days, 8 percent survived 800 days, and that there were no survivors after 1000 days. On the basis of this information, construct a probability model with the simple events "death within the first 200 days," "death between 200 and 400 days," etc.

1.5 UNIFORM PROBABILITY MODELS

A reasonable probability model for throwing a die or tossing a coin is frequently obtained by assigning equal probability to all simple events. In general, if all simple events are assigned the same probability, we say that the probability model is *uniform*. For the case of a die or a coin we have discussed this model in Examples 3.1 and 3.3. Surprisingly, such models, which are of course realistic only if the simple results of the experiment tend to occur about equally often, are appropriate in a great many interesting and important cases.

We shall in the next chapter take up applications of the uniform model to sampling theory, and in later chapters to genetics, quality control, and the design of scientific experiments. For the moment we mention only that gambling devices other than dice or coins also provide simple illustrations of the uniform model. Thus, when drawing a card from a standard deck containing 52 cards, it is customary to assign equal probability $\frac{1}{52}$ to each card of the deck. Similarly, for a roulette wheel one would assign equal probability to the simple events that correspond to the ball coming to rest in any particular slot. In fact, the use of such devices for gaming depends on the equal frequency of the outcomes. If, for example, a roulette wheel favored one slot over the others, gamblers would soon discover this and break the bank.

The calculation of probabilities is particularly simple when the model is uniform. Suppose the event set $\mathcal{E}$ consists of $\#(\mathcal{E})$ simple events. If each simple event is to have the same probability, and if the sum of the probabilities is to be 1, then the probability of each simple event must be $1/\#(\mathcal{E})$. Consider now a composite event E consisting of $\#(E)$ simple events. Then, by the definition of the probability of a composite event, the probability of E is the sum of $\#(E)$ terms, each equal to $1/\#(\mathcal{E})$, so that

$$(1) \qquad\qquad P(E) = \frac{\#(E)}{\#(\mathcal{E})}.$$

To compute the probability, it is therefore only necessary to count how many simple events there are in E and $\mathcal{E}$, and to take the ratio.

EXAMPLE *1*. *Drawing a card.* A standard bridge deck contains 13 cards of each of four different suits: two red suits (hearts and diamonds) and two black suits (clubs and spades). The 13 cards of each suit are Ace, King, Queen, Jack and cards numbered from 2 to 10. The deck is shuffled and a card is drawn. What is the probability that it will be a spade, if the uniform model is assumed? In the present case, there are 52 simple events corresponding to the 52 cards, and hence $\#(\mathcal{E}) = 52$. Of these 52 events, 13 (namely those corresponding to the 13 spades) make up the event E = spade. Hence $P(E) = \#(E)/\#(\mathcal{E}) = \frac{13}{52} = \frac{1}{4}$.

High school algebra books often include a section on probability, which is usually devoted to the uniform model. The simple events are called "equally likely cases," and the simple events that make up the composite event whose probability is sought are called "favorable cases." The probability of the event is defined as the ratio of the number of favorable cases to the number of equally likely cases. This of course is in agreement with formula (1).

The uniform model was used much more freely in the past than it is today, the simple results of an experiment being often quite uncritically treated as equally likely. The practice was justified by citing the Principle of Insufficient Reason, which declared that cases were equally likely unless reasons to the contrary were known. Unfortunately, the results were often quite unrealistic, since events might occur with quite different frequencies even though the model builder was unaware of the reasons for the difference. In modern practice, the burden of proof is shifted: it is up to the model builder to advance reasons for using the uniform model. Sometimes the model is justified on grounds of the symmetry of the results, as in the case of the die. The final test, however, is empirical: do the results occur in practice with nearly equal frequency?

The inadequacy of the Principle of Insufficient Reason is made clear by the fact that the list of simple results of an experiment may be drawn up in many different ways. When the uniform model is used with the different lists, different answers are obtained all of which according to the Principle of Insufficient Reason are equally valid. We illustrate the point with a simple example.

EXAMPLE 2. *Probability of a boy.* What is the chance that a two-child family will have at least one boy? If the number of boys is denoted by B, there are three possible results: $B = 0$, $B = 1$, $B = 2$. Of these three cases, the last two are favorable to the family having at least one boy, so that the use of the uniform model gives the value $\frac{2}{3}$ for the desired probability.

But we could list the results differently, recording the sex of the first and second children separately. Then there would be four cases: MM, MF, FM, FF; where for example MF means "first child male, second child female." Of the four cases, the first three imply that there is at least one boy, so that we get $\frac{3}{4}$ for the desired probability if the uniform model is used with this list.

Which answer is right, $\frac{2}{3}$ or $\frac{3}{4}$? Both are correctly deduced from the premises, and no amount of theoretical argument could prove or disprove either. To find out which model is the more realistic, it is necessary to observe the sex distributions actually occurring in two-child families. When this is done, it is found that the second model is the more realistic, although even moderately large samples would convince us that it too does not give an adequate description. For example, MM occurs somewhat more frequently than FF, reflecting the fact that boys constitute about 51.4% of all births. That boys are more frequent than girls could not have been predicted by theoretical arguments, and the reason for it is still not fully understood.

To drive home the point that the test of a model must in the last analysis be empirical, we will mention two circumstances in which the second model would be quite wrong. Identical twins are always of the same sex, which tends to increase the frequency of MM and FF families. The frequency of identical twins is in fact low, but if they constituted about $\frac{1}{3}$ of the two-child families, then the first model would fit much better than the second. Again, in some cultures it is thought essential that a family have a son. Suppose that in such a society, families were terminated as soon as a boy was born. This would mean that only FM and FF could occur in two-child families, so that both models would be quite wrong.

We conclude the section by giving two further examples of the use of uniform models.

EXAMPLE 3. *Sum of points on two dice.* In many dice games, the player throws two dice, and the outcome of the game depends on the total number T of points showing. As in Example 2, there are two rather natural ways to make out the list of simple results. We might observe that T may take on the 11 different values 2, 3, . . . , 12, and take these as our equally likely cases. This would, for example, imply $P(T = 7) = \frac{1}{11}$. Alternatively, we might record the results of the two dice separately, writing for example (4, 3) if the first die showed 4 points and the second 3 points. This record has the following 36 possible values:

$$
\begin{array}{cccccc}
(1, 1) & (1, 2) & (1, 3) & (1, 4) & (1, 5) & (1, 6) \\
(2, 1) & (2, 2) & (2, 3) & (2, 4) & (2, 5) & (2, 6) \\
(3, 1) & (3, 2) & (3, 3) & (3, 4) & (3, 5) & (3, 6) \\
(4, 1) & (4, 2) & (4, 3) & (4, 4) & (4, 5) & (4, 6) \\
(5, 1) & (5, 2) & (5, 3) & (5, 4) & (5, 5) & (5, 6) \\
(6, 1) & (6, 2) & (6, 3) & (6, 4) & (6, 5) & (6, 6)
\end{array}
$$

(2)

Of these the following six give $T = 7$: (1, 6), (2, 5), (3, 4), (4, 3), (5, 2) and (6, 1). With the second model, $P(T = 7) = \frac{6}{36} = \frac{1}{6}$, which is nearly twice as large as $\frac{1}{11}$. Which if either of these radically different results is correct, can be determined only by experience, though analogy with Example 2 may lead one to guess that the second model will work better. Extensive experience justifies attributing equal probability to the 36 simple events displayed above, provided the dice are sufficiently well made.

EXAMPLE *4. Floods.* When building a dam to protect a valley against the spring flood, the engineers consult the records of the past 50 years and make the dam just high enough to contain the worst flood recorded. How likely is it that the dam will hold for the next twenty-five years? As always, the answer depends on the model. If the climate and vegetation are not changing, then it seems plausible to assume that the worst flood is as likely to occur in one year as in another, of the 75 years consisting of the last 50 and the next 25. With this model, the probability that a larger flood will occur in the next 25 years than in the past 50, is $\frac{25}{75} = \frac{1}{3}$.

PROBLEMS

1. If a card is drawn under the assumptions of Example 1, find the probability that it is (i) red; (ii) an ace; (iii) a red ace; (iv) an ace or king.

2. A box contains 10 marbles, of which 2 are white, 3 red, and 5 black. A marble is drawn at random, that is, in such a way that each marble has the same chance of being selected.

   (i) Construct a probability model for this experiment.
   (ii) Find the probability that the marble drawn will be white.
   (iii) Find the probability that the marble drawn will be white or red.

3. In Example 3, find the probabilities of the following events.

   (i) The numbers of points on both dice are even.
   (ii) The number of points on at least one of the two dice is even.
   (iii) The sum of the points on the two dice is even.
   (iv) The number of points on at least one of the two dice is odd.
   (v) The sum of the numbers of points on the two dice is less than six.
   (vi) The sum of the numbers of points on the two dice is greater than six.
   (vii) The sum of the numbers of points on the two dice is less than three.
   (viii) The number of points on at least one of the two dice is less than three.
   (ix) The numbers of points on both dice are less than three.
   (x) The numbers of points on the two dice are equal.

4. In Example 3, find the probabilities
$$P(T = 2), P(T = 3), \ldots, P(T = 12).$$

5. In order to decide which of three persons will pay for a round of drinks, each tosses a penny. If the result of one toss differs from those of the other two (i.e., one head and two tails, or one tail and two heads), the "odd man" has to pay. Assuming the 8 possible results HHH, HHT, HTH, HTT, THH, THT, TTH, TTT to be equally likely, what is the probability that one toss will differ from the other two?

6. Assuming that in three-child families the 8 cases MMM, MMF, . . . , FFF are equally likely, find the probabilities of the following events:

   (i) at least one boy   (iv) exactly two boys
   (ii) at least two boys   (v) at most one boy
   (iii) exactly one boy   (vi) more boys than girls

(vii) at least one girl and one boy   (ix) the oldest a boy and the youngest

(viii) the oldest a boy   a girl

(x) no girl younger than a boy.

7. Twelve keys, of which only one fits, are tried one after another until a door opens. Let e_1 correspond to the result that the door opens on the first try; e_2 to the result that it opens on the second try; etc., and suppose that the 12 simple events $e_1, \ldots, e_{12}$ are equally likely. Find the probability that the door will be opened

(i) on the twelfth try,

(ii) on none of the first three tries,

(iii) on either the first or the twelfth try.

8. Suppose that a random digit generator produces two digits in such a way that all 100 cases $00, \ldots, 09; 10, \ldots, 19; \ldots; 90, \ldots, 99$ are equally likely. Interpreting 00 to be 0, 01 to be 1, $\ldots$, 09 to be 9, the machine produces a number between 0 and 99. What is the probability that this number is (i) positive and divisible by 11; (ii) less than 20; (iii) greater than 65; (iv) greater than 10 but less than 18?

9. In a surprise quiz, you are given two multiple choice questions, one with three possible answers a, b, c and the other with five possible answers A, B, C, D, E. Let (a, C) denote the event that you give answer a for the first and answer C for the second question, etc. Suppose that you have no idea as to the right answers and decide to write all possible combinations

$$(a, A), \ldots, (a, E); (b, A), \ldots, (b, E); (c, A), \ldots, (c, E)$$

on separate slips of paper of which you then draw one at random; that is, so that each has the same chance of being selected. Suppose the correct answers are b in the first and A in the second problem. What is the probability that you will give:

(i) the correct answer on both questions,

(ii) the incorrect answer on both questions,

(iii) at least one correct answer,

(iv) at most one correct answer,

(v) the correct answer on the first question,

(vi) the correct answer on the first but an incorrect answer on the second question?

10. In throws with three dice, the sum $T = 9$ can be produced in six ways, namely as $1 + 2 + 6, 1 + 3 + 5, 1 + 4 + 4, 2 + 2 + 5, 2 + 3 + 4, 3 + 3 + 3$. (i) Determine the number of ways in which the sum $T = 10$ can be produced. (ii) Would you conclude that $P(T = 9) = P(T = 10)$?

11. Assume that the $6^3 = 216$ different outcomes of throws with three dice: $(1, 1, 1), (1, 1, 2), \ldots, (6, 6, 6)$ are all equally likely. Under this assumption, compute $P(T = 9)$ and $P(T = 10)$, and discuss the difference between this result and that suggested by the preceding problem.

1.6 THE ALGEBRA OF EVENTS

Returning now to the consideration of probability models in general, we shall in the present section study certain relations that may exist between events, and certain operations that may be performed on them. These will enable us in the next section to derive desired probabilities from others that are given or assumed. We shall be working here within the mathematical model, with definitions as precise and proofs as rigorous as those of plane geometry or any other branch of pure mathematics, but we shall keep the random experiment in view as motivation for definitions and proofs.

Together with any result R that may occur on a trial of a random experiment, one may consider another result, namely "R does not occur." These two results are said to be *complements*, or complementary to one another. For example, the complement of the result that a three-child family has at least one boy is that all three children are girls. Again, if X is the number of points showing when a die is thrown, the results "X is even" and "X is odd" are complementary. We shall denote the complement of a result by placing a bar above it, so that $\overline{X \leq 2}$ means the same as $X > 2$. We shall employ the same terminology in the model.

Definition. The *complement* of an event E is the event $\overline{E}$ which consists just of those simple events in $\mathcal{E}$ that do not belong to E.

It follows from this definition that the sets E and $\overline{E}$ have no common member, and that between them they contain all simple events of the event set $\mathcal{E}$. The situation is illustrated in Figure 1, where E consists of

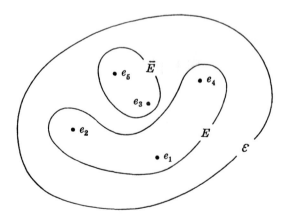

FIGURE 1. COMPLEMENTARY EVENTS

the simple events e_1, e_2, e_4, and $\overline{E}$ consists of the remaining simple events in $\mathcal{E}$, namely e_3, e_5.

An interesting special case is the complement of the event set ε itself. By definition, ε consists of all simple events of the model under discussion, so that $\bar{\varepsilon}$ is a set without members. This set is known as the *empty set*. If we think in terms of results, $\bar{\varepsilon}$ corresponds to a result that cannot occur, such as the result $X = 7$ when rolling a die.

Related to any two given results R, S is the result which occurs when both the result R and the result S occur, and which will be called the result "R and S." For example, if with two-child families R denotes the result that the first child is a boy and S that the second child is a boy, then "R and S" denotes the result that both children are boys. Similarly, if R indicates that a card drawn from a deck is a heart while S indicates that it is a face card, then the result "R and S" will occur if a heart face card is drawn.

Suppose the results R, S are represented in the model by the events E, F respectively. What event in the model then corresponds to the result "R and S"? To answer this question notice that the result "R and S" occurs whenever the experiment yields a simple result, which is a simple result both of R and of S. Therefore the event in the model corresponding to "R and S" will consist of those simple events that belong both to E and to F. The situation is illustrated in Figure 2a, where $E = \{e_1, e_3, e_4, e_6\}$

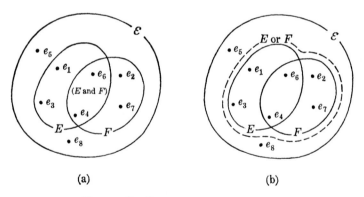

(a)   (b)

FIGURE 2. INTERSECTION AND UNION

corresponds to the result R consisting of the simple results r_1, r_3, r_4, r_6, and where $F = \{e_2, e_4, e_6, e_7\}$ corresponds to the result S consisting of the simple results r_2, r_4, r_6, r_7. Here "R and S" will occur whenever the experiment yields one of the simple results r_4 or r_6. The event corresponding to the result "R and S" consists therefore of the simple events e_4, e_6. It is the common part or intersection of the two events E, F.

Definition. The *intersection* (E and F) of two events E, F is the event which consists of all simple events that belong both to E and to F.

It may of course happen that two events E, F have no simple events in common. The intersection (E and F) is then the empty set.

Definition. Two events E, F are said to be *exclusive* if no simple event belongs to both E and to F.

The results corresponding to two exclusive events will be such that not both of them can happen on the same trial of the random experiment. For example, when throwing a die, the results $X \leq 2$ and $X > 4$ are exclusive, but the results "X is even" and "X is divisible by 3" are not, since both occur when $X = 6$. Again, for two-child families, the results "first child is a boy" and "both children are girls" are exclusive, but the results "first child is a boy" and "second child is a boy" are not, since both children may be boys. The case of two exclusive events E and F is illustrated in Figure 3.

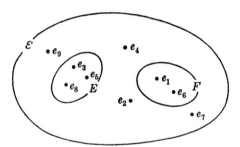

FIGURE 3.   EXCLUSIVE EVENTS

Another result related to the given results R, S is the result "R or S" which occurs when either the result R occurs or the result S occurs, or both occur. If for example R denotes the result that the first child of a two-child family is a boy while S denotes that the second child is a boy, then "R or S" will occur if at least one of the children is a boy. Similarly, if R denotes that a card drawn from a deck is a heart, while S denotes that it is a face card, then "R or S" will occur if the card is a heart or a face card, including the case when it is a heart face card.

What event in the model corresponds to the result "R or S"? To answer this question, notice that the result "R or S" occurs whenever the experiment yields one of the simple results of R or of S or both. Therefore the event in the model corresponding to "R or S" will consist of those simple events belonging to E or to F or both. The situation is illustrated in Figure 2b, where as before $E = \{e_1, e_3, e_4, e_6\}$ corresponds to the result R consisting of the simple results r_1, r_3, r_4, r_6, while $F = \{e_2, e_4, e_6, e_7\}$ corresponds to the result S consisting of the simple results r_2, r_4, r_6, r_7. Here "R or S" will occur whenever the experiment yields one of the simple results r_1, r_2, r_3, r_4, r_6, r_7. The composite event corresponding to "R or S" consists therefore of the simple events e_1, e_2, e_3, e_4, e_6, e_7. It is obtained by combining or uniting the simple events of E with those of F, producing the set (E or F) enclosed by the dashed line in Figure 2b.

Definition. The *union* (E or F) of two events E, F is the event which consists of all simple events that belong to E, or to F, or to both E and F.

EXAMPLE 1. *Two dice.* In the model for throwing two dice of Example 5.3, let E correspond to the result "first die shows one point," so that E consists of the first row of tableau (5.2). Similarly, let F correspond to "second die shows one point," consisting of the first column of (5.2). Then (E or F) consists of the eleven simple events in the top and left margins of (5.2), and corresponds to the result "at least one die shows one point." On the other hand, (E and F) consists only of the simple event (1, 1) and corresponds to the results "both dice show one point."

EXAMPLE 2. *Complementation.* Consider an event E and its complement $\overline{E}$. Since E and $\overline{E}$ have no simple event in common, they are exclusive. By the definition of complement, any simple event not in E belongs to $\overline{E}$, so that every simple event belongs either to E or to $\overline{E}$. It follows that the union of E and $\overline{E}$ is the entire event set $\mathcal{E}$: (E or $\overline{E}$) $= \mathcal{E}$.

The reader should be warned of a confusion that sometimes arises due to connotations of the word "and" in other contexts. From the use of "and" as a substitute for "plus," it might be thought that (E and F) should represent the set obtained by putting together or uniting the sets E and F. However, in the algebra of events (E and F) corresponds to the result "R and S" which occurs only if both R and S occur, and thus (E and F) is the common part or intersection of E and F.

In many books the notation ($E \cup F$) is used for (E or F) while ($E \cap F$) is used for (E and F).

The generalization of the concepts of intersection and union to three or more sets is obvious.

PROBLEMS

1. An experiment consists of ten tosses with a coin. Give a verbal description of the complement of each of the following events:
   (i) at least six heads
   (ii) at most six heads
   (iii) no heads.

2. In Problem 5.3, give verbal descriptions of the complements of the events described in (i), (iii), (v), (vii), (ix).

3. In Problem 5.6 give a verbal description of the complements of the events described in parts (i), (ii), (v), and (viii).

4. In Problem 5.9 give verbal descriptions of the complements of the events described in parts (i)–(v).

5. In Problem 5.6 give verbal descriptions of the intersections of each of the following pairs of events:

    (i) the events (i) and (ii)   (vi) the events (i) and (vi)
    (ii) the events (i) and (iii)   (vii) the events (ii) and (vi)
    (iii) the events (ii) and (iii)   (viii) the events (iii) and (vi)
    (iv) the events (i) and (iv)   (ix) the events (iv) and (vii)
    (v) the events (i) and (v)   (x) the events (vi) and (vii).

6. In Problem 5.3, determine the intersections of each of the following pairs of events:

    (i) the events (i) and (vi)   (iii) the events (v) and (viii)
    (ii) the events (ii) and (viii)   (iv) the events (i) and (iii).

7. In Problem 5.6 determine for each of the following pairs whether they are exclusive:

    (i) the events (i) and (iii)   (v) the events (iv) and (vi)
    (ii) the events (ii) and (iii)   (vi) the events (v) and (vi)
    (iii) the events (iii) and (v)   (vii) the events (v) and (vii)
    (iv) the events (iv) and (v)   (viii) the events (vi) and (vii).

8. In Problem 5.9 pick out three pairs from the events described in parts (i)–(vi) that are exclusive.

9. In Problem 5.3, determine for each of the events (i)–(x) whether it is exclusive of event (iv).

10. In Problem 5.3, determine the unions of each of the following pairs of events:

    (i) the events (v) and (vi)
    (ii) the events (iii) and (vi)
    (iii) the events (iv) and (vii).

11. Give verbal descriptions of the unions of each of the pairs of events of Problem 5.

12. In Problem 5.9, give verbal descriptions of the unions of each of the following pairs of events:

    (i) the events (i) and (iii)
    (ii) the events (ii) and (iii).

13. For each of the following statements determine whether it is correct for all E, F, and G.

(i) If E, F are exclusive, and F, G are exclusive, then E, G are exclusive.

(ii) If E, F are exclusive, and E, G are exclusive, then E, $(F$ or $G)$ are exclusive.

(iii) If E, $(F$ or $G)$ are exclusive, then E, F are exclusive.

14.   From a class of ten, an instructor selects a student to answer a certain question; if the first student cannot answer, he selects one of the remaining students; etc.   Let E denote the event that the first student knows the answer, and F the event that the first student does not know the answer but the second one does. Are E and F exclusive?

15.   Suppose that the event set ε is broken up into exclusive pieces E_1, E_2, ... , E_a. Is it true that then any event F can be written as

(1) $$F = (E_1 \text{ and } F) \text{ or } (E_2 \text{ and } F) \text{ or } \ldots (E_a \text{ and } F)?$$

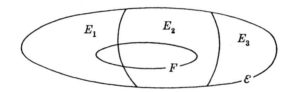

16.   Draw diagrams to illustrate the facts (known as De Morgan's Laws) that, for any sets E and F

(i) $(\overline{E \text{ or } F}) = \overline{E} \text{ and } \overline{F}$

(ii) $(\overline{E \text{ and } F}) = \overline{E} \text{ or } \overline{F}$.

1.7 SOME LAWS OF PROBABILITY

In the preceding section we have defined three operations on events: complementation, intersection, and union.   We shall now consider how probabilities behave under union and complementation.   The probabilities of intersections will be studied in Chapters 3 and 4.

A law connecting the probability of the union $(E$ or $F)$ with the probabilities of E and of F is obtained most easily in the special case that E and F are exclusive.   If R and S are two exclusive results, then not both can happen on the same trial of the experiment, and therefore in a sequence of trials

$$\#(R \text{ or } S) = \#(R) + \#(S).$$

For example, in an evening of poker, the number of red flushes dealt to a player will be the sum of the number of heart flushes and the number of diamond flushes that he receives. Dividing both sides of the displayed equation by the total number of trials we see that

$$f(R \text{ or } S) = f(R) + f(S) \qquad \text{if } R, S \text{ are exclusive.}$$

Since probability is supposed to represent long-run frequency, this suggests that in the model we should have the equation

(1) $\qquad P(E \text{ or } F) = P(E) + P(F) \qquad \text{if } E, F \text{ are exclusive.}$

The proof of this *addition law for exclusive events* may be illustrated on Figure 6.3. There $(E \text{ or } F)$ consists of the five simple events e_1, e_3, e_5, e_6, e_8, so by the definition of the probability of a composite event (Section 4),

$$P(E \text{ or } F) = P(e_1) + P(e_3) + P(e_5) + P(e_6) + P(e_8).$$

The right side may be written as

$$[P(e_3) + P(e_5) + P(e_8)] + [P(e_1) + P(e_6)]$$

which is just $P(E) + P(F)$. It is clear that the argument would hold in general.

EXAMPLE 1. *Two dice.* Suppose that in throwing two dice, the 36 simple events displayed in (5.2) are equally likely. If T denotes the sum of the points on the two dice, then, as shown in Example 5.3, $P(T = 7) = \frac{6}{36}$. The probabilities of the other values of T are as follows (Problem 5.4).

TABLE 1. PROBABILITIES OF VALUES OF T

t	2	3	4	5	6	7	8	9	10	11	12
$P(T = t)$	$\frac{1}{36}$	$\frac{2}{36}$	$\frac{3}{36}$	$\frac{4}{36}$	$\frac{5}{36}$	$\frac{6}{36}$	$\frac{5}{36}$	$\frac{4}{36}$	$\frac{3}{36}$	$\frac{2}{36}$	$\frac{1}{36}$

Let us now find the probability of the event G that the sum T will be divisible by 5. Since this occurs if and only if T is either equal to 5 or to 10, G is the union of the exclusive events $E: T = 5$ and $F: T = 10$. Hence

$$P(G) = P(E) + P(F) = \frac{4}{36} + \frac{3}{36} = \frac{7}{36}.$$

A special case of the addition law (1) is obtained by taking for F the complement $\overline{E}$ of E. Since E and $\overline{E}$ are exclusive and their union is the event set $\mathcal{E}$, it follows that

$$P(E) + P(\overline{E}) = P(E \text{ or } \overline{E}) = P(\mathcal{E}) = 1.$$

This establishes the important *law of complementation*

(2) $\qquad\qquad\qquad P(E) = 1 - P(\overline{E}).$

As we shall see later, the main usefulness of this relation resides in the fact that it is often easier to calculate the probability that something does not happen than that it does happen.

As an illustration of (2), consider once more the model of Example 1 and suppose we wish to determine the probability of the event E: the sum T of the points showing on the two dice is three or more. The complement of E is the event $\bar{E}$: $T = 2$, which has probability $\frac{1}{36}$. Hence

$$P(E) = 1 - \tfrac{1}{36} = \tfrac{35}{36}.$$

The concept of exclusive events extends readily to more than two events.

Definition. The events E_1, E_2, . . . are said to be *exclusive* if no simple event belongs to more than one of them.

The addition law also extends in the obvious manner:

(3) $\qquad P(E_1 \text{ or } E_2 \text{ or } \ldots) = P(E_1) + P(E_2) + \ldots$
$$\text{if } E_1, E_2, \ldots \text{ are exclusive.}$$

If E, F are not exclusive, law (1) will in general not be correct, as may be seen by examining Figure 6.2b. Here

$$P(E) = P(e_1) + P(e_3) + P(e_4) + P(e_6)$$
$$P(F) = \qquad\qquad\quad P(e_4) + P(e_6) + P(e_2) + P(e_7)$$

so that

$$P(E) + P(F) = P(e_1) + P(e_3) + 2P(e_4) + 2P(e_6) + P(e_2) + P(e_7)$$

while

$$P(E \text{ or } F) = P(e_1) + P(e_3) + P(e_4) + P(e_6) + P(e_2) + P(e_7).$$

Since $P(e_4)$ and $P(e_6)$ occur twice in $P(E) + P(F)$ but only once in $P(E \text{ or } F)$, it follows that $P(E) + P(F)$ is greater than $P(E \text{ or } F)$ (except in the trivial case when both $P(e_4)$ and $P(e_6)$ are equal to zero).

A correct law must allow for the double counting of the probabilities of those simple events which belong both to E and to F, that is to the intersection $(E \text{ and } F)$. In Figure 6.2a the event $(E \text{ and } F)$ consists of e_4 and e_6, so that

$$P(E \text{ and } F) = P(e_4) + P(e_6).$$

It follows that

(4) $\qquad P(E \text{ or } F) = P(E) + P(F) - P(E \text{ and } F).$

This addition law holds for all events E, F whether or not they are exclusive. If they happen to be exclusive (as in Figure 6.3), $(E \text{ and } F)$ will be empty and have probability 0, in which case (4) reduces to (1).

EXAMPLE 1. *Two dice (continued).* In Example 6.1 we considered for two dice the events E, F corresponding to the results "first die shows one point" and "second die shows one point" respectively. If the 36 simple events displayed in (5.2) are equally likely, it is seen that

$$P(E) = \tfrac{6}{36}, \quad P(F) = \tfrac{6}{36}, \quad \text{and} \quad P(E \text{ and } F) = \tfrac{1}{36}.$$

The probability of the event (E or F) corresponding to the result that at least one of the dice shows one point is then, by (4), equal to $\frac{6}{36} + \frac{6}{36} - \frac{1}{36} = \frac{11}{36}$. This can also be seen directly from the fact that the event (E or F) consists of 11 simple events each having probability $\frac{1}{36}$.

It is important to keep in mind that law (3) requires the assumption that the events are exclusive. When they are not, it is necessary to correct for multiple counting, as is done in law (4) for the case of two events. The formulas for nonexclusive events become rapidly more complicated as the number of events is increased (see Problem 12).

PROBLEMS

1. In Problem 5.3, find the probabilities of the complements of the events described in parts (ii), (iv), (vi), (viii), and (x).

2. In Problem 5.6, find the probabilities of the complements of the events described in parts (i), (ii), (v), and (vii).

3. In Problem 5.9, find the probabilities of the complements of the events described in parts (i)–(v).

4. In Problem 5.8, use the law of complementation to find the probabilities of the following events:
  (i) at least one of the two digits is greater than zero
  (ii) at least one of the two digits is even.

5. In Example 1, use the results of Table 1 and the addition law (3) to find the probabilities of the following events:
  (i) $T \geq 10$
  (ii) $T < 5$
  (iii) T is divisible by 3.

6. In Example 1, let M be the larger of the numbers of points showing on the two dice (or their common value if equal). Make a table similar to Table 1 showing the probabilities of M taking on its various possible values $1, 2, \ldots, 6$.

7. Use the table of the preceding problem and the addition law (3) to find the probabilities of the following events:
  (i) $M \leq 3$ (iv) $2 < M < 5$
  (ii) $M < 3$ (v) $2 \leq M < 4$.
  (iii) $M \geq 5$

8. In Problem 5.6 let B be the number of boys. Make a table similar to Table 1 showing the probabilities of B taking on its various possible values 0, 1, 2, and 3.

9. Use the table of the preceding problem and the addition law (3) to find the probabilities of the following events:
  (i) $B \geq 2$
  (ii) $B > 2$
  (iii) $B \leq 1$.

10. A manufacturer of fruit drinks wishes to know whether a new formula (say B) has greater consumer appeal than his old formula (say A). Each of four customers is presented with two glasses, one prepared from each formula, and is asked to state which he prefers. The results may be represented by a sequence of four letters; thus ABAB would mean that the first and third customers preferred A, while the second and fourth preferred B. Suppose that the $2^4 = 16$ possible outcomes of the experiment are equally likely, as would be reasonable if the two formulas are in fact equally attractive. If S denotes the number of consumers preferring the new formula, make a table similar to Table 1, showing the probabilities of S taking on its various possible values.

11. Use the table of the preceding problem and the addition law (3) to find the probabilities of the following events:
   (i) at least two of the customers prefer the new product
   (ii) at most two of the customers prefer the new product
   (iii) more than two of the customers prefer the new product.

12. Generalize Problem 10 to the case of five customers. If S again denotes the number of customers preferring the new formula, make a table similar to Table 1.

13. Use the table of the preceding problem and the addition law (3) to find the probabilities of the following events:
   (i) at least three of the customers prefer the new product
   (ii) at most two of the customers prefer the new product
   (iii) $2 < S < 4$
   (iv) $2 \leq S \leq 4$.

14. In Problem 4.10, make a table similar to Table 1 showing the probabilities of

   (i) the number U of grade points in course 1 taking on its various possible values 0,1,2,3,4;
   (ii) the number V of grade points in course 2 taking on its various possible values;
   (iii) the total number W of grade points in the two courses combined taking on its various possible values;
   (iv) the difference D between the grade points in course 2 and those in course 1 taking on its various possible values.

15. Use the table of part (iii) of the preceding problem and the addition law (3) to find the probabilities of the following events:
   (i) $W > 6$
   (ii) $5 \leq W \leq 7$
   (iii) $3 < W < 8$.

16. Use the table of part (iv) of Problem 14 and the addition law (3) to find the probabilities of the following events:
   (i) $D \geq 0$
   (ii) $-1 \leq D \leq 1$
   (iii) $D < -2$ or $D > 2$.

17. Let ε consist of the six simple events $e_1, \ldots, e_6$, and let $P(e_1) = P(e_2) = .1$, $P(e_3) = .2$, $P(e_4) = .25$, $P(e_5) = .3$, $P(e_6) = .05$. If

$$E = \{e_1, e_2\}, F = \{e_2, e_3, e_4\}, G = \{e_2, e_2, e_5\},$$

find

   (i) $P(\overline{E})$ (iv) $P(F$ or $G)$

   (ii) $P(\overline{F})$ (v) $P(E$ or F or $G)$.

   (iii) $P(E$ or $F)$

18. Let ε consist of the six simple events $e_1, \ldots, e_6$, and let $P(e_1) = .01$, $P(e_2) = .05$, $P(e_3) = .14$, $P(e_4) = .1$, $P(e_5) = .2$, $P(e_6) = .5$. If

$$E = \{e_1, e_4, e_6\}, F = \{e_3, e_4, e_5\}, G = \{e_2, e_5\},$$

check for each of the following equations whether it is true by computing both the right- and left-hand sides:

   (i) $P(E$ or $F) = P(E) + P(F)$

   (ii) $P(E$ or $G) = P(E) + P(G)$

   (iii) $P(F$ or $G) = P(F) + P(G)$.

Explain your results.

19. Check equation (4) for the events E and F of Problem **17.**

20. Check the following generalization of equation (4) by computing both sides when E, F, G are the events of Problem **17.**

$$P(E \text{ or } F \text{ or } G) = [P(E) + P(F) + P(G)]$$
$$- [P(E \text{ and } F) + P(E \text{ and } G) + P(F \text{ and } G)]$$
$$+ P(E \text{ and } F \text{ and } G)$$

CHAPTER 2

SAMPLING

2.1 A MODEL FOR SAMPLING

The notion that information about a population may be obtained by examining a *sample* drawn from the population has been made familiar to everyone in recent years.   Public opinion polling organizations predict how millions of people will vote, on the basis of interviews with a few thousands.   Advertisers decide to support or to drop television programs on the basis of popularity ratings that rest on the viewing practices of a sample of much less than one percent of the audience.   The quality of the product of large factories is controlled by inspection of a small fraction of the output.   Newspapers carry stories reporting on the cost of living, the number of unemployed, the acreage planted in certain crops, and many other economic variables which are estimated from the examination of a sample.   Like anything else sampling may be done well or badly, and it is desirable for everyone to understand the theoretical basis for an activity that has become so important to our society.   In its application, sampling involves many complications of detail, but the idea is essentially simple, and is in fact based on the uniform model.

Since the purpose behind taking a sample is to get information about the population, it seems natural to select a sample so that it will be representative of the population.   For instance, a fruit grower trying to estimate the harvest from a grove of lemon trees might pick out a few trees that he judges to be typical of the grove as a whole, count the fruit on the selected trees, and then assume that the average number of fruit per tree in the whole orchard is about equal to the average for the sampled trees. Again, in trying to forecast an election, we might select counties that have voted like the country as a whole in recent elections, and conduct interviews with voters in these counties.   While this method of "purposive" sampling is superficially attractive, it has not worked well in practice.

Biases of the selector creep in too easily when the items for the sample are selected purposively by the exercise of judgement. Also, items that are typical of the population in some respects need not be typical in others. For instance, counties that vote like the country on the last election may well not do so on the next, when the issues will be different.

Most sampling experts have come to the conclusion that they cannot rely on samples selected purposively, and that the only safe practice is to have the sample chosen by the operation of chance. Dice, pennies, and random number generators have the virtue of complete impartiality, and a sample selected by their operation is not subject to the bias of human judgement. Great ingenuity has been exercised in devising efficient ways to use chance mechanisms in picking the sample, some of which will be taken up in Chapter 10. We shall here consider only the simplest case, in which the sample is chosen so that all possible samples of the same size have the same probability of being chosen. Such a sample is known as a *random sample*. The concept can best be presented by means of an example.

EXAMPLE 1. *The delegation.* A city is governed by a council consisting of a mayor and six councilmen. It is necessary to send three members to testify in Washington. All seven want to go, and the mayor proposes that the three delegates be selected by lot. He places in a box seven marbles, similar in size, weight, etc., and labeled with the integers 1 through 7. A list of the seven men is made, and the names on the list are similarly numbered. The mayor's secretary is blindfolded, the marbles shaken up, and she draws three. The delegation will consist of the three men whose numbers are the same as those of the three marbles drawn.

The above procedure of selecting the sample by lot seems to be completely fair and impartial. Because the marbles are alike and are thoroughly mixed, and because the drawing is done blindfold, there appears to be no means whereby one possible three-man delegation could be favored over another.

We shall now build a probability model for this experiment. The simple results are the possible three-man delegations, or the possible sets of three marbles. They may be denoted by triples of integers, as follows:

	123	134	146	234	246	345	367
	124	135	147	235	247	346	456
(1)	125	136	156	236	256	347	457
	126	137	157	237	257	356	467
	127	145	167	245	267	357	567

It is seen that there are 35 possible samples of three marbles that the secretary might draw. Since the experiment is arranged so that none of the samples is favored over any other, it seems reasonable to regard all samples as equally likely.

Let us generalize the idea of the example. Suppose that it is desired to draw a sample of s items from a population of N items. We shall denote by $\binom{N}{s}$ the number of different samples of size s that can be formed from a population of N items. For example, $\binom{7}{3}$ is the number of different samples of three items that can be chosen from a given set of seven items; tableau (1) shows that $\binom{7}{3} = 35$.

We shall say that a sample of size s is drawn from a population of size N at random if the drawing does not favor any one sample over another, so that it is reasonable to expect each of the $\binom{N}{s}$ possible samples to be drawn about equally frequently in a long sequence of such draws. (In practice, it is often difficult to know whether a given method of drawing the sample has this property.)

In the model representing the drawing of a random sample, it is then natural to regard all $\binom{N}{s}$ samples as equally likely. Each will thus be assigned probability $1/\binom{N}{s}$, and the resulting model is a uniform probability model in the sense of Section 1.5.

Table 1 shows all values of $\binom{N}{s}$ for populations of size $N \leqq 10$. A more extensive table, which is discussed in the next section, is given as Table A at the end of the book. In principle, any entry of these tables could be computed by writing out a tableau similar to (1). However, much more convenient methods for obtaining the numerical values of $\binom{N}{s}$ will be presented in the next section.

EXAMPLE 1. *The delegation (continued).* Suppose that the mayor and three members of the council are Conservatives, while the other three are Liberals. It turns out that the delegation, although supposedly chosen

$$\text{TABLE 1. THE NUMBER } \binom{N}{s} \text{ OF SAMPLES}$$

$\dfrac{\ \ s}{N}$	1	2	3	4	5	6	7	8	9	10
1	1									
2	2	1								
3	3	3	1							
4	4	6	4	1						
5	5	10	10	5	1					
6	6	15	20	15	6	1				
7	7	21	35	35	21	7	1			
8	8	28	56	70	56	28	8	1		
9	9	36	84	126	126	84	36	9	1	
10	10	45	120	210	252	210	120	45	10	1

by lot, consists entirely of Conservatives. Is this so unlikely, under random sampling, that the Liberals have good reason for believing the drawing was rigged? Since there are four Conservatives on the Council, there are $\binom{4}{3}$ possible three-man delegations consisting entirely of Conservatives, and from Table 1 we see that $\binom{4}{3} = 4$. If the drawing is fair, so that the $\binom{7}{3} = 35$ delegations are equally likely, the probability of getting a delegation without Liberals is $\frac{4}{35} = .11$, or about one chance in nine. This is not small enough to make a strong case for rigging.

EXAMPLE 2. *Quality control.* A box of ten fuses contains three that will not work. If a random sample of four fuses is examined, what is the chance that no defective fuse will be found? From Table 1 we see that there are $\binom{10}{4} = 210$ possible samples of size $s = 4$. By assumption, the sample is random; that is, all 210 possible samples are equally likely. Since the box contains seven good fuses, there are $\binom{7}{4} = 35$ samples consisting entirely of good fuses. The probability that no defective fuse will be found in the sample is therefore $\frac{35}{210} = .167$.

EXAMPLE *3*. *Lunch counter*. Three persons occupy the seven seats at a small lunch counter. It is noticed that no two of them are sitting next to each other. Is this fact evidence that the customers tend to avoid taking a stool next to one already occupied? If the customers seat themselves without regard to whether the adjacent seat is occupied, we may think of them as choosing their places at random. Since three of the seven seats may be chosen in $\binom{7}{3} = 35$ different ways, we have only to count the number of choices which do not bring two customers together. There are 10 such arrangements:

OEOEOEE,   OEOEEOE,   OEOEEEO,   OEEOEOE,   OEEOEEO
OEEEOEO,   EOEOEOE,   EOEOEEO,   EOEEOEO,   EEOEOEO

The probability of the observed event is therefore $\frac{10}{35} = \frac{2}{7}$ if the customers seat themselves at random. This probability is too large for the observed seating to be very strong evidence in favor of the avoidance theory.

EXAMPLE *4*. *The diseased poplars*. Of a row of ten poplar trees, four adjacent trees are affected with a disease. Is there reason for thinking that the disease is spreading from one tree to another? This problem is much like the preceding. If the disease strikes at random, there are $\binom{10}{4} = 210$ possible sets of four that might be affected. Of these arrangements, only seven consist of four adjacent trees. Thus P(affected trees adjacent) $= \frac{7}{210} = \frac{1}{30}$, if the choice is random. This is rather a small probability, so there is some ground for thinking that a causal relation underlies the adjacency of the diseased trees.

The method of drawing a random sample by means of numbered marbles, described in Example 1, is feasible only if the population size N is sufficiently small. For larger N, a more convenient method utilizes a table of random numbers (see Section 3.4). An essential step in both these methods, or any others, is the construction of a numbered list of the items in the population. It is often difficult and expensive to make such a list. However, to draw a random sample one cannot do without it or some other equivalent method of numbering the items. Thus, a random sample of the houses in a block may be obtained by making a sketch map of the block and numbering the houses on this map. If light bulbs are packed in cartons in a regular pattern, one may devise a systematic numbering scheme that attaches to each bulb a different integer, thereby making it unnecessary actually to write down the list. Sampling experts have developed many devices to obtain the equivalent of a list at less expense and trouble.

PROBLEMS

1. In Example 1, find the probabilities of the following events.
   (i) The delegation consists of Liberals only.
   (ii) The delegation consists of two Conservatives and one Liberal.
   (iii) There are at least two Conservatives on the delegation.
   (iv) The mayor is a member of the delegation.

2. Suppose that the council of Example 1 consists of the mayor and eight council-men, and that the mayor and four of the council members are Conservatives while the other four are Liberals. Let a delegation of four be selected at random.
   (i) How many possible delegations are there?
   (ii) What is the probability that the delegation will consist entirely of Conservatives?
   (iii) What is the probability that the delegation will contain at least one Conservative?
   (iv) What is the probability that the delegation will contain at least one Liberal?

3. In the preceding problem, suppose that the mayor and three council members are Conservatives, three members are Liberals, and two are Independents. If a delegation of three is selected at random, find the answers to parts (i)–(iv) of the preceding problem.

4. Check the entries $\binom{5}{2}$ and $\binom{5}{3}$ in Table 1 by listing all possible samples of sizes two and three that can be drawn from the five persons A, B, C, D, E.

5. Suppose you know that a family has six children, two boys and four girls, but that you don't know their sexes according to age.
   (i) How many possible arrangements bbgggg (two boys, followed by four girls), bgbggg, . . . are there?
Assuming all possible arrangements of part (i) to be equally likely, find the probabilities of the following events.
   (ii) The oldest child is a girl.
   (iii) The two oldest children are both girls.
   (iv) The oldest child is a boy and the youngest is a girl.

6. Use the definition of $\binom{N}{8}$ to find the value of

$$\text{(i) } \binom{N}{N}, \quad \text{(ii) } \binom{N}{1}, \quad \text{(iii) } \binom{N}{N-1}.$$

7. Of a group of seven children, four are selected at random to receive instruction by method A, while method B will be used on the remaining three. Find the probability of the following events.
   (i) The four most intelligent children are all assigned to method A.
   (ii) The three most intelligent children are all assigned to method A.
   (iii) The most intelligent child is assigned to method A.
[Hint: In counting the numbers of samples having the properties in question, it may be helpful to label the children 1, 2, . . . , 7 in order of intelligence.]

8. Find the probabilities of the three events (i)–(iii) of the preceding problem assuming that five of the seven children are selected at random to be instructed by method A, the remaining two receiving instruction by method B.

9. Suppose that of the seven children in Problem 7, four are boys and three girls. Two of the four boys and two of the three girls are selected at random to be instructed by method A.

(i) How many possible samples of two boys and two girls are there?

(ii) Assuming all these samples to be equally likely, find the probability that the most intelligent boy and the most intelligent girl both receive instruction by method A.

10. A batch of ten items contains two defectives. If two items are drawn at random, what is the probability that both are nondefective?

11. In Example 4, what is the probability that no two diseased trees are adjacent?

12. By inspection of Table 1, formulate a rule for obtaining each entry as the sum of two entries in the row above it. Use the rule to add a new row to the table.

13. It appears in Table 1 that certain pairs of entries in each row are equal Formulate a rule that specifies when two entries are equal.

14. A square field is divided into nine square plots, arranged in three rows and three columns. Three plots are chosen at random. What is the probability that of the chosen plots

(i) all are in the same row;

(ii) one is in each column;

(iii) one is in each row and one is in each column?

15. Suppose the field of Problem 14 is divided into eight square blocks, arranged in two rows and four columns. Two plots are chosen at random. What is the probability that

(i) both are in the same row;

(ii) both are in the same column;

(iii) one is in each row?

16. A rectangular carton of 24 light bulbs consists of 2 layers, each with 3 rows and 4 columns. Three bulbs are selected at random. What is the probability that all of the selected bulbs are from corners of the package?

$$\left[\text{Hint:} \binom{24}{3} = 2024. \right]$$

17. In order to select a law firm at random, it is suggested to draw a lawyer from an available list of all N lawyers in the city and then select the firm to which he belongs. Suppose there are in the city 50 firms consisting of only one lawyer, ten firms consisting of two lawyers each, two firms consisting of three lawyers each, and one firm consisting of four lawyers. Find the probability that the selected firm consists of (i) four lawyers, (ii) one lawyer.

(iii) Has the law firm been selected at random?

18. One of the 63 law firms of the preceding problem is selected at random, and then a lawyer randomly chosen from the firm. Have we selected a lawyer at random from those of the city?

2.2 THE NUMBER OF SAMPLES

The number $\binom{N}{s}$ of different samples of size s that can be drawn from a population of size N is of great importance in probability and statistics, and also in other branches of mathematics.   It is known as the *number of combinations* of N things taken s at a time.   The numbers $\binom{N}{s}$ are also referred to as *binomial coefficients*.   We shall now point out several features of these quantities, and show how a table of values of $\binom{N}{s}$ may easily be computed.

An inspection of Table 1.1 shows that, at least in all cases covered by the table,

(1) $$\binom{N}{1} = N \quad \text{and} \quad \binom{N}{N} = 1.$$

These relations hold quite generally.   Suppose that a sample of size $s = 1$ is to be drawn.   Then each of the N items may serve as this sample, so that there are N possible samples of size 1.   At the other extreme, if the sample is to have size $s = N$, the "sample" must consist of the entire population, and this is possible in just one way.

To discover another property of these numbers, let us look at one of the rows of Table 1.1, for example the last row:

(2)

s	1	2	3	4	5	6	7	8	9	10
$\binom{10}{s}$	10	45	120	210	252	210	120	45	10	1

The entries are arranged symmetrically about the largest value, $\binom{10}{5} =$ 252.   Thus $\binom{10}{4} = 210 = \binom{10}{6}$, $\binom{10}{3} = 120 = \binom{10}{7}$, and so forth.   A similar symmetry holds in the other rows.   This phenomenon is easily explained.   For example, when a sample of size $s = 3$ is drawn from a population of size $N = 10$, there is left in the population a remnant of size $N - s = 10 - 3 = 7$.   We can think of these seven items as a sample of size 7.   Thus to each sample of size 3 that may be removed, there correresponds a sample of size 7 that is left.   Therefore the number of samples of size $s = 3$ equals the number of samples of size $N - s = 7$, or $\binom{10}{3} = \binom{10}{7}$ The same argument shows that quite generally

(3)
$$\binom{N}{N-s} = \binom{N}{s},$$

which is the formal expression of the observed symmetry.

Equation (3) makes it possible to cut in half the size of any table of the quantities $\binom{N}{s}$. A table utilizing this fact and giving all entries with $N \leq 26$ and $s \leq 13$ is given as Table A at the end of the book. If we wish to use this table for example to obtain the value of $\binom{24}{15}$, we look up instead the value of $\binom{24}{24-15}$ to find

$$\binom{24}{15} = \binom{24}{9} = 1{,}307{,}504.$$

For values beyond the range of the table, there exist a variety of formulas and approximations. One such formula will be derived in Section 4.

Equation (3) breaks down in one case, namely when s is equal to 0. The left side then becomes $\binom{N}{N}$, which by (1) has the value 1; but the right side becomes $\binom{N}{0}$, which is meaningless. Whenever a mathematical expression is meaningless because it is undefined, we are free to attach to it any meaning that we like. Usually, it is convenient to do this in such a way that certain formulas remain valid for the previously undefined case. This suggests that we *define*

(4)
$$\binom{N}{0} = 1,$$

so that equation (3) will continue to be valid when $s = 0$. With this definition it is possible to complete the symmetry of the rows of Table 1.1; by adding the entry for $s = 0$, the row for $N = 7$ for example will become:

s	0	1	2	3	4	5	6	7
$\binom{7}{s}$	1	7	21	35	35	21	7	1

Finally, we shall discuss a relation which permits much easier computation of the table than direct enumeration. Inspection of Table 1.1 shows that any entry is the sum of the entry above it and the entry to the left of the latter. The entry above $\binom{N}{s}$ is $\binom{N-1}{s}$, and the entry to the left of this is $\binom{N-1}{s-1}$. The suggested relation is therefore

(5)
$$\binom{N}{s} = \binom{N-1}{s} + \binom{N-1}{s-1},$$

and this does in fact hold quite generally.

To see how to prove equation (5), let us consider once more the problem of choosing a three-man delegation from a seven-man council (Example 1.1). The $\binom{7}{3} = 35$ possible delegations may be divided into two types: those to which the mayor belongs, and those to which he does not. There are just $\binom{6}{2} = 15$ of the former, since if the mayor belongs, there are still two members to be chosen from among the six other councilmen. On the other hand, the number of three-man delegations without the mayor is $\binom{6}{3} = 20$, since for these delegations all three must be chosen from the six ordinary council members. This proves (5) for the case $N = 7$, $s = 3$, and the argument can easily be extended to give a proof of the general case.

Relation (5) can be used to compute a table of $\binom{N}{s}$ by obtaining successively each row from the preceding one. To illustrate the method, let us get the value $\binom{8}{3}$ from the entries listed above for $N = 7$. The entry above $\binom{8}{3}$ is $\binom{7}{3} = 35$; the entry to the left of this is $\binom{7}{2} = 21$, and their sum gives $\binom{8}{3} = 35 + 21 = 56$. In this manner it is possible to build up the table quickly. The triangular scheme that is built up in this way is known as *Pascal's triangle*. Table 1.1 may easily be checked by this device.

The method outlined above has been used on an automatic computer to compute the values up to $N = 200$, and the resulting table has been published.* The values increase very rapidly; it may interest the reader to know that $\binom{200}{100} = 90,548,514,656,103,281,165,404,177,077,484,163,874,$ 504,589,675,413,336,841,320, which is a number of 59 digits! Values sometimes quoted are the number of possible hands at poker, $\binom{52}{5} = 2,598,960,$ and the number of possible hands at bridge, $\binom{52}{13} = 635,013,559,600.$

EXAMPLE 1. *Poker.* What is the chance that a poker hand will be a heart flush (i.e. consist only of hearts)? A poker hand consists of 5 cards dealt from a deck of 52 cards. Before they are dealt, the cards are shuffled, and if the shuffling is sufficiently thorough it is natural to assume that no one poker hand is more likely to be dealt than any other. This assumption cannot be directly checked, because the number of poker hands is so large

* *Table of Binomial Coefficients,* Cambridge University Press, 1954.

that it is not feasible to deal enough hands to show whether or not they all arise with equal frequency.   It is possible however to check certain consequences of the assumption of the uniform model, and experiments of this sort have been carried out.   The general conclusion is that much more thorough shuffling than is customary among card players would be needed before the uniform model could be safely used.

Suppose, however, that we are willing to assume that all hands are equally likely.   It is then easy to compute the probability of a heart flush. The number of favorable cases is the number of hands that can be chosen from the 13 hearts in the deck, and this is just $\binom{13}{5} = 1287$.   Therefore, $P(\text{heart flush}) = 1287/2,598,960 = .000495$, which is about 1 in 2000. The same argument shows this also to be the probability of each of the other types of flush.

EXAMPLE 2. *Inclusion of a specified item.* A term paper consists of 24 problems.   In order to save time, the instructor corrects only eight of them which he selects at random.   If in your assignment only one problem is wrong, what is the probability that it will be among those selected for correction?   The total number of possible samples of eight problems is $\binom{24}{8} = 735,471$.   The number of these including the one incorrect problem is $\binom{23}{7} = 245,157$, since this is the number of ways in which the remaining seven problems needed for the sample can be chosen from the 23 correct problems.   The desired probability is therefore

$$\frac{245,157}{735,471} = \frac{1}{3}.$$

This example illustrates a fact which holds quite generally: the probability that a random sample of size s includes any specified item of the population of N items from which the sample is taken is s/N.   (In the example $s = 8$, $N = 24$ so that $s/N = \frac{1}{3}$; for a general proof see Problem 17.)   It follows that the method of random sampling is "fair" in the sense that each item has the same probability (namely s/N) of being included in the sample.

PROBLEMS

1. Use Table A and relation (3) to find (i) $\binom{20}{16}$; (ii) $\binom{22}{14}$; (iii) $\binom{25}{17}$.

2. Use Table A and relation (5) to find (i) $\binom{27}{6}$; (ii) $\binom{27}{10}$; (iii) $\binom{27}{20}$.

3. Using Table A and making repeated use of relation (5), find (i) $\binom{28}{6}$; (ii) $\binom{28}{14}$; (iii) $\binom{29}{15}$; (iv) $\binom{29}{21}$.

4. A batch of 20 items contains four defectives. If three items are drawn at random, find the probability that
   (i) all three are nondefective;
   (ii) at least one is defective.

5. A batch of 25 items contains five defectives. If three items are drawn at random, what is the probability that
   (i) at least one is defective;
   (ii) at most two are defective?

6. A class of 24 students consists of 10 freshmen and 14 sophomores. The instructor decides to read a sample of six papers selected at random from an assignment handed in by all students. What is the probability that his sample contains
   (i) at least one paper by a freshman;
   (ii) at least one paper by a sophomore;
   (iii) at least one paper from each group?

7. Suppose that the class of the preceding problem consists of 7 freshmen, 8 sophomores and 9 juniors. Find the probability that the sample of six contains
   (i) at least one junior;
   (ii) no freshman.

8. During a flu epidemic, 18 patients have volunteered to be treated by an experimental drug. It is decided to give the drug to 12 of the 18 selected at random, and to use the other 6 as controls by giving them a more conventional treatment. Suppose that 7 of the patients would, without treatment, have had a serious case of the flu and the remaining 11 a light case. Find the probability that the control group contains at least one of the serious cases.

9. A group of 20, one of whom acts as secretary, selects at random a delegation of five. What is the probability that the secretary is included in the delegation?

10. In a poker hand, what is the probability of a "flush" (i.e., all five cards from the same suit)?

11. Find the probability that a poker hand contains
   (i) at least one black card;
   (ii) only Kings, Queens and Jacks.

12. A lot of 25 fuses contains five defectives. How large a sample must one draw so that the probability of getting at least one defective fuse in the sample will exceed (i) $\frac{1}{2}$; (ii) $\frac{4}{5}$; (iii) $\frac{9}{10}$?

13. In Problem 6, find the smallest sample size so that, with probability exceeding $\frac{1}{2}$, the sample will contain
   (i) at least one sophomore;
   (ii) at least one freshman;
   (iii) at least one member from each group.

14. A box contains four red and five white marbles. In how many ways can a sample be drawn consisting of

   (i) one red and one white marble;

   (ii) one red and two white marbles?

15. From the box of Problem 14, a sample of size two is drawn at random. What is the probability that it consists of one red and one white marble?

16. (i) Show that the total number of ways of selecting a committee of s from a group of N and naming one as chairman is $\binom{N}{s}\binom{s}{1}$. [Hint: Select first the committee; then from the committee members a chairman.]

   (ii) Show that an alternative answer to (i) is $\binom{N}{1}\binom{N-1}{s-1}$. [Hint: select first the chairman from the whole group; then select the $s-1$ ordinary members.]

   (iii) By comparing the answers to (i) and (ii) show that

(6)
$$\binom{N}{s} = \frac{N}{s}\binom{N-1}{s-1}.$$

17. Using the result of Problem 16, show that the probability that a specified item is included in a sample of size s taken at random from a population of N is

$$\frac{\binom{N-1}{s-1}}{\binom{N}{s}} = \frac{s}{N}.$$

18. By successive application of (6) and the fact that $\binom{k}{1} = k$, find (i) $\binom{5}{2}$; (ii) $\binom{6}{3}$; (iii) $\binom{7}{4}$.

19. Use (6) to show that

   (i) $\binom{N}{2} = \frac{N(N-1)}{2}$;   (ii) $\binom{N}{3} = \frac{N(N-1)(N-2)}{6}$.

20. Show that the probability is $(N-s+1)/\binom{N}{s}$ that a specified $s-1$ items are included in a sample of size s taken at random from a population of N.

21. Suppose that in the drug test of Problem 8, the new drug is given to 10 of the 18 patients, selected at random, and the remaining 8 serve as controls.

   (i) Find the probability that the control group contains at least one of the light cases.

   (ii) Find the probability that the control group contains at least two of the light cases.

[Hint for (ii): Consider the complement of the event in question and use the result of Problem 20.]

2.3 ORDERED SAMPLING

In Section 1, we presented the following method for obtaining a random sample of the desired size from a population of items. Let the N items in the population be numbered from 1 to N, and let a box contain N marbles similarly numbered. After thorough mixing, s marbles are taken from the box, and the sample consists of those s items in the population whose numbers appear on these s marbles.

Unless s is quite small, it is difficult to grab exactly s marbles at once. An obvious method to insure that the desired number of marbles is obtained, is to take them one at a time until just s have been drawn from the box. The result of this procedure is a sample of s marbles *arranged in a particular order*. One of the marbles in the sample was obtained on the first draw, another on the second draw, and so forth.

Definition. A sample arranged in order is called an *ordered sample*.

To distinguish between an ordered sample and the kind discussed previously, which we may now call "unordered," we shall adopt in this section a notational convention. A sample will be indicated by listing its items, in parentheses when the sample is ordered, in braces when it is unordered. Thus an unordered sample consisting of the two items A and B may be denoted indifferently by {A, B} or by {B, A}. On the other hand, (A, B) and (B, A) denote two different ordered samples consisting of the items A and B. In (A, B) item A is first and B second; in (B, A) the order of the two items is reversed.

To illustrate the relation between ordered and unordered samples, consider a population of $N = 4$ marbles labeled A, B, C, D (they could equally well be labeled 1, 2, 3, 4, but the use of letters will provide a clearer notation). An ordered sample of size $s = 2$ is drawn. The 12 possible ordered samples are displayed below, arranged in columns corresponding to the six possible unordered samples of the same size.

(1)

Unordered sample	{A, B}	{A, C}	{A, D}	{B, C}	{B, D}	{C, D}
Corresponding ordered samples	(A, B) (B, A)	(A, C) (C, A)	(A, D) (D, A)	(B, C) (C, B)	(B, D) (D, B)	(C, D) (D, C)

As a second illustration, suppose that from the same set of marbles we draw an ordered sample of size $s = 3$. Each of the $\binom{4}{3} = 4$ unordered samples may be arranged in six different orders, giving the $4 \cdot 6 = 24$ ordered samples shown below.

Unordered sample	{ABC}	{ABD}	{ACD}	{BCD}
Corresponding ordered samples	(ABC)	(ABD)	(ACD)	(BCD)
	(ACB)	(ADB)	(ADC)	(BDC)
	(BAC)	(BAD)	(CAD)	(CBD)
	(BCA)	(BDA)	(CDA)	(CDB)
	(CAB)	(DAB)	(DAC)	(DBC)
	(CBA)	(DBA)	(DCA)	(DCB)

(2)

It will be convenient to introduce the symbol $(N)_s$ to represent the number of different ordered samples of size s that can be drawn from a population of size N. (This quantity is also known as the *number of permutations* of N things taken s at a time.) The illustrations above show that

$$(4)_2 = 12 \quad \text{and} \quad (4)_3 = 24.$$

Table 1 gives the values of $(N)_s$ for populations of size 10 and less. These values are in general much larger than the corresponding values of $\binom{N}{s}$ shown in Table 1.1, as is reasonable since each sample of several items corresponds to many different ordered samples of these items. While the small entries of the present table could be obtained by making systematic lists of all ordered samples for the given values of N and s, this would be too cumbersome for the large entries. All values were actually computed from the simple formula for $(N)_s$ given in the next section.

TABLE 1. THE NUMBER $(N)_s$ OF ORDERED SAMPLES

N \ s	1	2	3	4	5	6	7	8	9	10
1	1									
2	2	2								
3	3	6	6							
4	4	12	24	24						
5	5	20	60	120	120					
6	6	30	120	360	720	720				
7	7	42	210	840	2520	5040	5040			
8	8	56	336	1680	6720	20160	40320	40320		
9	9	72	504	3024	15120	60480	181440	362880	362880	
10	10	90	720	5040	30240	151200	604800	1814400	3628800	3628800

Let us now build a probability model for the random experiment of drawing an ordered sample of size s from a population of size N. It is natural to take as the simple eve. ts all possible different ordered samples. What probabilities should be attacned to them? If the N marbles are of similar size and weight and are thoroughly mixed, and if the drawing is done blindfold, we would expect each of the ordered samples to appear with about equal frequency, and actual experiments with small populations bear out this expectation. Accordingly we shall treat all $(N)_s$ ordered samples as equally likely, and assign to each of them the probability $1/(N)_s$. When an ordered sample is drawn so that this model may reasonably be applied, we shall refer to it as a *random* ordered sample.

The usefulness of ordered sampling for obtaining a random (unordered) sample rests on the following important fact:

(3)   to obtain a random sample of size s from a population of size N, we may first draw an ordered random sample and then disregard the order.

We shall prove this first for the special case of a sample of size $s = 3$ drawn from a population consisting of $N = 4$ items. If the items are numbered A, B, C, D, let us find for example the probability that the sample, without regard to order, is {A, B, D}. Each of the ordered samples shown in the tableau (2) has the same probability $\frac{1}{24}$, and it is therefore only necessary to count the number of ordered samples corresponding to {A, B, D}. These are just the six cases listed in the second column of the tableau, and the desired probability is therefore $\frac{6}{24} = \frac{1}{4}$.

Extending this argument, it is seen that each of the $\binom{4}{3} = 4$ possible unordered samples

$$\{A, B, C\}, \quad \{A, B, D\}, \quad \{A, C, D\}, \quad \{B, C, D\}$$

appears just six times in the tableau (2) of ordered samples. Each of the four possible unordered samples that can be obtained by first drawing an ordered sample of size three and then disregarding the order has probability $\frac{1}{4}$, and the unordered sample obtained in this manner is therefore random.

The above argument can be used to show quite generally that the sample obtained from a random ordered sample by disregarding the order is random. Notice that each unordered sample may be arranged in the same number, in fact $(s)_s$, of different orders. Therefore, in the model for ordered sampling, all (unordered) samples have the same probability $(s)_s/(N)_s$, as was to be proved.

We know, from Section 1, that the probability of each unordered sample is $1/\binom{N}{s}$. Equating this with the expression $(s)_s/(N)_s$ just obtained for the same probability, we see that

(4) $$(N)_s = \binom{N}{s} \cdot (s)_s.$$

That is, the number of ordered samples is the number of unordered samples multiplied by the number of ways in which each sample can be ordered.

To discover another important property of ordered sampling, let us suppose once more that a random ordered sample of size $s = 3$ is drawn from the population of four marbles labeled A, B, C, D and find the probability that marble D appears on the second draw. Inspection of tableau (2) shows the following six ordered samples to have marble D in the second position:

$$(A, D, B) \quad (B, D, A) \quad (A, D, C) \quad (C, D, A) \quad (B, D, C) \quad (C, D, B)$$

The desired probability is therefore $\frac{6}{24} = \frac{1}{4}$.

In the same manner it is easily checked that each of the four marbles has the same chance of appearing on the second draw (Problem 10). How could it be otherwise, since there is nothing to favor any marble over any other as a candidate for the second spot? This result generalizes to the following *equivalence law of ordered sampling:*

(5)
> if a random ordered sample of size s is drawn from a population of size N, then on any particular one of the s draws each of the N items has the same probability $1/N$ of appearing.

As an illustration, suppose that an instructor has announced that he will call at random on three students to present solutions of the three homework problems at the board. One of the ten students is unable to solve Problem 2, and wonders what his chances are of being found out. In effect, the instructor will draw an ordered random sample of size $s = 3$ from a population of size $N = 10$. By the equivalence law, the worried student has probability $1/N = \frac{1}{10}$ of being the person sampled on the second draw, and hence being asked to work Problem 2.

The equivalence law asserts that each item in the population is equally likely to be obtained on any specified draw. It is also true that each *pair* of items in the population is equally likely to be obtained on any *two* specified draws. For example, if we give equal probability to each of the ordered samples in tableau (2), what is the probability that the items {A, B} will be obtained on draws 2 and 4? This event occurs with the following four samples

$$(C, A, D, B), \quad (D, A, C, B), \quad (C, B, D, A), \quad (D, B, C, A),$$

so that the desired probability is $\frac{4}{24} = \frac{1}{6}$. The same probability attaches to each of the six (unordered) pairs. The result also extends beyond two items: each (unordered) triple of items has the same chance $1 \Big/ \binom{N}{3}$ of being obtained on any three specified draws, etc. We shall call this the "generalized equivalence law of ordered sampling."

PROBLEMS

1. Check the values $(5)_2$ and $(5)_3$ of Table 1 by enumeration from tableaus similar to (1) and (2).

2. (i) Check the values $(3)_1$ and $(4)_4$ of Table 1 by enumeration.
  (ii) Explain why it is always true that $(N)_N = (N)_{N-1}$.
  (iii) Is it true that the total number of ways in which N items can be arranged in order is $(N)_N$?

3. Check the values $(3)_1$ and $(4)_1$ by enumeration, and explain why it is always true that $(N)_1 = N$.

4. From Table 1, calculate $(N)_2/(N)_1$ for $N = 2, 3, \ldots, 10$, and conjecture a general formula for $(N)_2$.

5. From Table 1, calculate $(N)_N/(N - 1)_{N-1}$ for $N = 2, 3, \ldots, 10$, and conjecture a general formula for $(N)_N$.

6. In an essay contest, there are three prizes worth \$500, \$200, and \$100 respectively. In how many different ways can these be distributed among (i) nine contestants; (ii) ten contestants?

7. A newly formed club of ten members decides to select its officers by lot. From the ten names, four are drawn out one by one. The first one drawn will serve as President, the second as Vice-President, the third as Treasurer, and the fourth as Secretary.
  (i) How many different possible outcomes are there to this lottery?
  (ii) What is the probability that the only girl in the club will become Vice-President?
  (iii) What is the probability that the four oldest members will be chosen for the four offices (not necessarily in order of importance)?

8. In a food tasting experiment, a subject is asked to rate five brands of coffee which are sold at different prices. The coffee is served to the subject in five cups in random order (i.e., so that all possible orders are equally likely). Find the probability that
  (i) the first three cups served are, in order, the most expensive, the next most expensive, the third most expensive;
  (ii) the first three cups served are the three most expensive brands, but not necessarily in order.

9. Solve the preceding problem if the number of brands being compared is seven rather than five.

10. From tableau (2), check that in a random ordered sample of $s = 3$ drawn from a population of four items labeled A, B, C, D, each of the four items has the same probability of being obtained on the second draw.

11. From the tableau of Problem 1, check that in a random ordered sample of $s = 3$ drawn from a population of five items, each of the five items has the same probability of being obtained on the second draw.

12. A lottery contains one grand prize of $1000; the winning number will be announced after all customers have drawn their tickets. Two customers arrive simultaneously, and customer A is permitted to draw a ticket first. Customer B complains that he is at a disadvantage since if A has drawn the grand prize, B will have no chance to get it. Is B's complaint justified?

13. A box contains four red and five white marbles. To obtain a sample, three marbles are drawn at random, one after another. What is the probability that
  (i) the second marble is red.
  (ii) the second and third marbles are both red;
  (iii) the second marble is white;
  (iv) the second and third marbles are both white?

14. Solve the four parts of the preceding problem if the box contains three red and six white marbles, and if four marbles are drawn from the box.

15. A box contains six red, seven white and ten black marbles. Two marbles are drawn as in Problem 13. Find the probability that
  (i) the second marble is white;
  (ii) the two marbles are of the same color.

16. Four runners in a ski race, wearing numbers 1, 2, 3, 4 on their backs, are started one after another in random order. Make a tableau of all possible starting orders and find the probabilities of the following events.
  (i) Runner No. 1 is started first, then runner No. 2, then No. 3, and lastly No. 4.
  (ii) Runner No. 1 is started first.
  (iii) The starting position of at least two of the runners coincides with the, numbers on their backs.
  (iv) The starting position of exactly one runner coincides with the number on his back.
  (v) The starting positions of exactly two runners agree with the numbers on their backs.

17. A magazine prints the photographs of four movie stars and also (in scrambled order) a baby picture of each. What is the probability that a reader by purely random matching of the pictures gets at least two right?

18. An ordered sample of $s = 5$ is drawn from ten items numbered 1 to 10. What is the probability that the items drawn will appear in order of increasing (not necessarily consecutive) numbers?

19. Consider an ordered sample of size s drawn at random from a population of N items. Let E represent the result that a specified item is included in the sample. Find $P(E)$ by noting that $E = (E_1$ or E_2 or . . . or $E_s)$, where $E_1, E_2, \ldots$ represent the events that the specified item is obtained on the 1st draw, on the 2nd draw, ...

2.4 SOME FORMULAS FOR SAMPLING

In the preceding sections we introduced the methods of ordered and un-ordered random sampling. If the size of the population is N and that of the sample s, we denoted by $(N)_s$ the number of possible ordered samples and by $\binom{N}{s}$ the number of possible samples without regard to order.   We shall first develop general formulas for these quantities and then give some applications of these formulas.

Let us begin by obtaining a formula for $(N)_2$, the number of ordered samples of size two that can be drawn from a population of N items numbered $1, \ldots, N$. Consider the process of drawing such a sample. On the first draw there are N items to choose from. Whichever item is first chosen, there will remain $N - 1$ possibilities for the second draw. Therefore the number of ordered samples of size two is

$$(1) \qquad\qquad (N)_2 = N(N - 1).$$

For example, there are $(5)_2 = 5 \cdot 4 = 20$ ordered samples of size two that can be drawn from a population of five items. These 20 items may be displayed as follows:

———	(1, 2)	(1, 3)	(1, 4)	(1, 5)
(2, 1)	———	(2, 3)	(2, 4)	(2, 5)
(3, 1)	(3, 2)	———	(3, 4)	(3, 5)
(4, 1)	(4, 2)	(4, 3)	———	(4, 5)
(5, 1)	(5, 2)	(5, 3)	(5, 4)	———

where the five rows correspond to the five possible choices on the first draw, and the four entries in each row to the four choices remaining for the second draw after one item has been chosen.

What about $(N)_3$, the number of ordered samples of size three?   Again there will be N possibilities on the first draw, and $N - 1$ possibilities on the second draw regardless of the choice on the first one. Finally, however the first two choices are made, there will remain $N - 2$ possibilities for the third draw. Thus the total number of samples is $(N)_3 = N(N - 1)(N - 2)$. For example, when $N = 10$ there are $10 \cdot 9 \cdot 8 = 720$ different ordered samples of size three.

In a quite similar way it can be argued that the number of ordered samples of size four is $N(N - 1)(N - 2)(N - 3)$, and in general that the number of ordered samples of size s will be the product of s factors, starting with N and decreasing one each time. What is the last factor? By the time the sth factor has been reached, the original N will have been reduced by one $s - 1$ times, so the last factor is $N - (s - 1) = N - s + 1$. The final formula for the number of ordered samples of size s from a population of size N is

$$(2) \qquad\qquad (N)_s = N(N - 1)(N - 2) \cdots (N - s + 1).$$

For example, the number of ordered samples of size five from a population of size ten is $10 \cdot 9 \cdot 8 \cdot 7 \cdot 6 = 30{,}240$.

An important and interesting special case of formula (2) arises when $s = N$, that is, when the entire population is taken into the sample. The expression (2) then becomes

$$(3) \qquad (N)_N = N(N-1)(N-2) \cdots 2 \cdot 1$$

or just the product of the first N positive integers. This quantity is known as N *factorial* and is written $N!$ Thus, $(N)_N = N!$ is the number of ways in which N elements can be arranged in order.

By substituting (2) and (3) into formula (3.4), we obtain

$$(4) \qquad \binom{N}{s} = \frac{N(N-1) \cdots (N-s+1)}{1 \cdot 2 \cdots s}.$$

An easy way to remember this formula is to notice that the numerator is the product of s decreasing factors starting with N, while the denominator contains s increasing factors starting with 1. As an example we compute

$$\binom{10}{4} = \frac{10 \cdot 9 \cdot 8 \cdot 7}{1 \cdot 2 \cdot 3 \cdot 4} = 10 \cdot 3 \cdot 7 = 210,$$

which agrees with the value given in Table 1.1.

Formula (2) can be used to prove the equivalence law of ordered sampling stated in the preceding section. To prove this law, we must show that on any particular draw, each of the N items has the same probability $1/N$ of being drawn. To fix ideas, consider the probability of obtaining a particular item, say item A, on the second draw. There are $(N)_s$ equally likely ordered samples. How many of these will have item A in second place? If item A is specified to occupy second place, there will remain $N-1$ other items for the $s-1$ other places. The remaining places can then be filled in $(N-1)_{s-1}$ different ways. The desired probability is therefore

$$\frac{(N-1)_{s-1}}{(N)_s} = \frac{(N-1)(N-2) \cdots (N-s+1)}{N(N-1)(N-2) \cdots (N-s+1)} = \frac{1}{N}.$$

Clearly, the argument would work equally well for any item and any place, which proves the desired result. A similar argument leads to the generalized equivalence law.

PROBLEMS

1. Use formula (2) to check in Table 3.1 the entries for (i) $(7)_3$; (ii) $(7)_4$; (iii) $(10)_5$.

2. Use formula (4) to check the following entries in Table A: (i) $\binom{11}{4}$; (ii) $\binom{17}{3}$; (iii) $\binom{10}{5}$.

3. Use formula (4) to check your answers to Problem 2.3.

4. Solve the three parts of Problem 3.7 when the club has 13 members instead of 10.

5. Use formula (4) to check that the number of possible poker hands is 2,598,960.

6. Prove the following alternative expression for $\binom{N}{s}$:

(5)
$$\binom{N}{s} = \frac{N!}{s!(N-s)!}.$$

7. Give a meaning to 0! in such a way that (5) will remain valid when $s = 0$ and when $s = N$.

8. Use (5) to prove the relations (2.1), (2.3), (2.5), and (2.6).

9. (i) What is the probability that in three throws with a fair die, different numbers occur on all three throws?

   (ii) What is the probability that at least two of the throws show the same number? [Hint: Assume that all $6^3 = 216$ possible results of the three throws are equally likely.]

10. What is the probability that in six throws with a fair die all six faces occur? [Hint: Assume that all 6^6 possible results of the six throws are equally likely.]

11. What is the probability that in a group of $s = 5$ persons at least two have their birthday on the same day? [Hint: Neglect Feb. 29 and assume that all $(365)^5$ possible sets of five birthdays are equally likely.]
Note: It is a surprising fact that for $s = 23$, the desired probability is greater than $\frac{1}{2}$.

12. The *binomial theorem* states that

(5)
$$(a+b)^N = \binom{N}{0}a^N + \binom{N}{1}a^{N-1}b + \binom{N}{2}a^{N-2}b^2 + \cdots + \binom{N}{N}b^N.$$

In the expansion of $(a+b)^{12}$ find the coefficient of (i) a^3b^9; (ii) a^5b^7; (iii) a^2b^9; (iv) b^{12}.

13. By choosing appropriate values for a and b, prove that

(i) $\binom{N}{0} + \binom{N}{1} + \cdots + \binom{N}{N} = 2^N$;

(ii) $\binom{N}{0} - \binom{N}{1} + \binom{N}{2} - \binom{N}{3} + \cdots = 0.$

14. Use 13(i) and (2.5) to prove that when N is even

$$\binom{N}{0} + \binom{N}{2} + \binom{N}{4} + \cdots + \binom{N}{N} = 2^{N-1}.$$

15. Use (4) to check (2.6).

16. Prove the binomial theorem. [Hint: In multiplying out $(a + b)(a + b) \cdots$ $(a + b)$, the terms $a^s b^{N-s}$ arise if a is selected from s of the parentheses and b from the remaining $N - s$. How many such terms are there?]

CHAPTER 3

PRODUCT MODELS

3.1 PRODUCT MODELS FOR TWO-PART EXPERIMENTS

Random experiments often consist of two or more parts: for example, several throws of a die or throws of several dice; drawing two or more marbles from a box; observing the weather on consecutive days or the sexes of successive children in a family; etc. Frequently it is relatively easy to build a satisfactory model for each of the parts separately. The problem then arises: how should the separate models for the various parts be combined to provide a model for the experiment as a whole? The present chapter is devoted to one type of combined model for the whole experiment, which we shall call the "product model." Very many of the probability models used in practice are of this type.

EXAMPLE 1. *Stratified sampling.* We have discussed in Chapter 2 some aspects of random sampling. Frequently the population from which the sample is drawn is made up of a number of different subpopulations or *strata*. Human populations may for example be stratified according to religion or age, sex, party registration, etc. It may be desirable to avoid samples which are too unrepresentative, such as samples consisting mainly of items coming from a single stratum. This can be achieved by the method of *stratified sampling*, according to which a random sample of specified size is drawn from each of the strata. The separate samples drawn from the different strata then constitute parts of the whole experiment, and each part may be represented by the kind of model considered in Section 2.1. The problem is how to combine these models to produce a model for the whole experiment.

EXAMPLE 2. *Two loaded dice.* In Example 1.3.2 we developed a model for the throw of a loaded die. In many dice games two dice are thrown, both of which may be loaded. If we regard the throw of each die as a part of the whole experiment, the problem arises how to combine two models like that of Example 1.3.2 into a model for the experiment of throwing two loaded dice.

We shall begin by considering product models for experiments consisting of only two parts. Let us first see how to choose an event set for a two-part experiment for both parts of which the event sets have already been selected. Suppose that part A has an event set $\mathcal{E}_A$ consisting of simple events $e_1, e_2, \ldots, e_a$ corresponding to the simple results $r_1, r_2, \ldots, r_a$, while part B has an event set $\mathcal{E}_B$ consisting of simple events $f_1, f_2, \ldots, f_b$ corresponding to the simple results $s_1, s_2, \ldots, s_b$. When the whole experiment is performed, we shall want to record the results of both parts. If, for example, the result of part A is r_2 and the result of part B is s_1, the result of the whole experiment is "r_2 and s_1." Since any of the a simple results $r_1, r_2 \ldots, r_a$, of part A may occur in combination with any of the b simple results $s_1, s_2, \ldots, s_b$ of part B, there will be $a \cdot b$ possibilities for the experiment as a whole. (Thus, in Example 2, die A may show any of its $a = 6$ faces and die B may also show any of its $b = 6$ faces, so that there will be $6 \cdot 6 = 36$ simple results for the experiment of throwing both dice.)

The event set for the whole experiment will need $a \cdot b$ simple events, corresponding to the $a \cdot b$ simple results. These simple events may be conveniently displayed in a tableau of a rows and b columns:

(1)

$$
\begin{array}{cccc}
e_1 \text{ and } f_1 & e_1 \text{ and } f_2 & \ldots & e_1 \text{ and } f_b \\
e_2 \text{ and } f_1 & e_2 \text{ and } f_2 & \ldots & e_2 \text{ and } f_b \\
\ldots & & & \\
e_a \text{ and } f_1 & e_a \text{ and } f_2 & \ldots & e_a \text{ and } f_b.
\end{array}
$$

Here, for example, (e_1 and f_2) is the simple event corresponding to the simple result "r_1 and s_2." We shall refer to (1) as the *product* of the event sets $\mathcal{E}_A$ and $\mathcal{E}_B$, and denote it by $\mathcal{E}_A \times \mathcal{E}_B$.

To complete the model, we must assign probabilities to each of the $a \cdot b$ simple events in $\mathcal{E}_A \times \mathcal{E}_B$. Suppose that for parts A and B of the experiment separately, probability models $\mathcal{A}$ and $\mathcal{B}$ are given with the following probabilities:

(2)

Model $\mathcal{A}$	*Model* $\mathcal{B}$
$e_1, e_2, \ldots, e_a$	$f_1, f_2, \ldots, f_b$
$p_1, p_2, \ldots, p_a$	$q_1, q_2, \ldots, q_b$

where of course

(3) $p_1 + p_2 + \ldots + p_a = 1$ and $q_1 + q_2 + \ldots + q_b = 1$.

The probabilities to be assigned to the simple events (1) of $\mathcal{E}_A \times \mathcal{E}_B$ must be nonnegative and add up to one. This can be done in many different ways. One simple method, which turns out to be realistic in many cases, consists of assigning to each of the simple events of (1) the *product* of the corresponding probabilities of models α and $\mathcal{B}$. Thus, to the simple event (e_1 and f_2) we assign the probability $p_1 q_2$, and so forth. The tableau of probabilities is then as follows:

(4)
$$
\begin{array}{llll}
P(e_1 \text{ and } f_1) = p_1 q_1 & P(e_1 \text{ and } f_2) = p_1 q_2 & \ldots & P(e_1 \text{ and } f_b) = p_1 q_b \\
P(e_2 \text{ and } f_1) = p_2 q_1 & P(e_2 \text{ and } f_2) = p_2 q_2 & \ldots & P(e_2 \text{ and } f_b) = p_2 q_b \\
\ldots & \ldots & & \ldots \\
P(e_a \text{ and } f_1) = p_a q_1 & P(e_a \text{ and } f_2) = p_a q_2 & \ldots & P(e_a \text{ and } f_b) = p_a q_b
\end{array}
$$

It is easy to check that tableau (4) does specify a probability model, as defined in Section 1.4. The probabilities are obviously not negative, and they add up to 1, as the following argument proves. Consider the sum of the probabilities in the first row of (4); using (3) we find that

(5) $p_1 q_1 + p_1 q_2 + \ldots + p_1 q_b = p_1(q_1 + q_2 + \ldots + q_b) = p_1 \cdot 1 = p_1.$

Similarly the probabilities in the second row of (4) add up p_2, and so forth. The sum of the rows is therefore $p_1 + p_2 + \ldots + p_a$, which by (3) equals 1.

The above discussion leads to the following formal definition.

Definition. The model defined by (1) and (4) will be called the *product* of models α and $\mathcal{B}$, and will be denoted by $\alpha \times \mathcal{B}$. The models α and $\mathcal{B}$ will be called the *factors* of model $\alpha \times \mathcal{B}$.

EXAMPLE 3. *Multiple-choice quiz.* A quiz has two multiple choice questions, the first offering three choices a, b, c, and the second five choices A, B, C, D, E. A student guesses at random on each question. Let us consider the guess on the first question as part A, and the guess on the second question as part B, of a two-part experiment. Model α has three simple events a, b, c, and the statement that the student "guesses at random" justifies assigning probabilities $\frac{1}{3}$ to each. Similarly model $\mathcal{B}$ has the five simple events $A, \ldots, E$, to each of which probability $\frac{1}{5}$ is assigned. The product model $\alpha \times \mathcal{B}$ will then have $3 \cdot 5 = 15$ simple events:

(6)
$$
\begin{array}{lllll}
a \text{ and } A & a \text{ and } B & a \text{ and } C & a \text{ and } D & a \text{ and } E \\
b \text{ and } A & b \text{ and } B & b \text{ and } C & b \text{ and } D & b \text{ and } E \\
c \text{ and } A & c \text{ and } B & c \text{ and } C & c \text{ and } D & c \text{ and } E
\end{array}
$$

to each of which is assigned the probability $\frac{1}{3} \cdot \frac{1}{5} = \frac{1}{15}$.

EXAMPLE 2. *Two loaded dice (continued).* An experiment consists of throwing two dice. Suppose that for each throw separately the model (1.3.1) is satisfactory. Let $e_1, \ldots, e_6$ represent the result of getting one, $\ldots$, six points on the first die. Then $p_1 = .21$, $p_2 = \ldots = p_5 = .17$, $p_6 = .11$. Similarly, let $f_1, \ldots, f_6$ represent the results on the second die, with $q_1 = .21$, $q_2 = \ldots = q_5 = .17$, $q_6 = .11$. Then the product model

PROBABILITIES OF THROWS OF TWO LOADED DICE

	f_1	f_2	f_3	f_4	f_5	f_6
e_1	$P(e_1 \text{ and } f_1) =$ $.21 \times .21 = .0441$	$P(e_1 \text{ and } f_2) =$ $.21 \times .17 = .0357$	$P(e_1 \text{ and } f_3) =$ $.21 \times .17 = .0357$	$P(e_1 \text{ and } f_4) =$ $.21 \times .17 = .0357$	$P(e_1 \text{ and } f_5) =$ $.21 \times .17 = .0357$	$P(e_1 \text{ and } f_6) =$ $.21 \times .11 = .0231$
e_2	$P(e_2 \text{ and } f_1) =$ $.17 \times .21 = .0357$	$P(e_2 \text{ and } f_2) =$ $.17 \times .17 = .0289$	$P(e_2 \text{ and } f_3) =$ $.17 \times .17 = .0289$	$P(e_2 \text{ and } f_4) =$ $.17 \times .17 = .0289$	$P(e_2 \text{ and } f_5) =$ $.17 \times .17 = .0289$	$P(e_2 \text{ and } f_6) =$ $.17 \times .11 = .0187$
e_3	$P(e_3 \text{ and } f_1) =$ $.17 \times .21 = .0357$	$P(e_3 \text{ and } f_2) =$ $.17 \times .17 = .0289$	$P(e_3 \text{ and } f_3) =$ $.17 \times .17 = .0289$	$P(e_3 \text{ and } f_4) =$ $.17 \times .17 = .0289$	$P(e_3 \text{ and } f_5) =$ $.17 \times .17 = .0289$	$P(e_3 \text{ and } f_6) =$ $.17 \times .11 = .0187$
e_4	$P(e_4 \text{ and } f_1) =$ $.17 \times .21 = .0357$	$P(e_4 \text{ and } f_2) =$ $.17 \times .17 = .0289$	$P(e_4 \text{ and } f_3) =$ $.17 \times .17 = .0289$	$P(e_4 \text{ and } f_1) =$ $.17 \times .17 = .0289$	$P(e_4 \text{ and } f_5) =$ $.17 \times .17 = .0289$	$P(e_4 \text{ and } f_6) =$ $.17 \times .11 = .0187$
e_5	$P(e_5 \text{ and } f_1) =$ $.17 \times .21 = .0357$	$P(e_5 \text{ and } f_2) =$ $.17 \times .17 = .0289$	$P(e_5 \text{ and } f_3) =$ $.17 \times .17 = .0289$	$P(e_5 \text{ and } f_4) =$ $.17 \times .17 = .0289$	$P(e_5 \text{ and } f_5) =$ $.17 \times .17 = .0289$	$P(e_5 \text{ and } f_6) =$ $.17 \cdot \times .11 = .0187$
e_6	$P(e_6 \text{ and } f_1) =$ $.11 \times .21 = .0231$	$P(e_6 \text{ and } f_2) =$ $.11 \times .17 = .0187$	$P(e_6 \text{ and } f_3) =$ $.11 \times .17 = .0187$	$P(e_6 \text{ and } f_4) =$ $.11 \times .17 = .0187$	$P(e_6 \text{ and } f_5) =$ $.11 \times .17 = .0187$	$P(e_6 \text{ and } f_6) =$ $.11 \times .11 = .0121$

assigns to its 36 simple events the probabilities shown in the tableau on page 65.

As these examples show, it is very easy to construct the product of two given factor models. However, product models provide realistic representations of two-part experiments only if certain conditions are satisfied, and one should not be tempted into indiscriminate use of such models by the simplicity of their construction. We shall investigate in the next section the conditions under which product models may be expected to be appropriate.

For simplicity we have so far restricted the consideration of product models to experiments with only two parts, but the ideas extend to experiments with three or more parts. In analogy with (1) the simple events of the product model will then be of the form

$$(7) \qquad e \text{ and } f \text{ and } g \text{ and } \ldots$$

where e, f, g are simple events in the models for the separate parts. As in (4), the product model assigns to the event (7) the product of the corresponding probabilities of the factor models, that is,

$$(8) \qquad P(e \text{ and } f \text{ and } g \text{ and } \ldots) = P(e) \cdot P(f) \cdot P(g) \ldots .$$

It is again not difficult to check that the probabilities (8) add up to 1 as $e, f, g, \ldots$ range independently over the simple events of the factor models, so that (7) and (8) define a probability model.

We conclude with some remarks on the logical consistency of the notation employed for product models. We have used, for example, $(e_1 \text{ and } f_2)$ to denote the simple event corresponding to the result of getting r_1 on part A and s_2 on part B. Previously (Section 1.6) the word "and" was used to represent the intersection of two events. Can we regard $(e_1 \text{ and } f_2)$ as the intersection of e_1 with f_2? At first glance, it would seem that this is not possible, since e_1 is an event in model α while f_2 is an event in model $\mathcal{B}$. However, these two uses of "and" can be reconciled as follows.

Consider the composite event made up of the simple events in the first row of (1):

$$(9) \qquad \{(e_1 \text{ and } f_1), (e_1 \text{ and } f_2), \ldots, (e_1 \text{ and } f_b)\}.$$

This event corresponds to the occurrence of one of the results $(r_1 \text{ and } s_1)$, $(r_1 \text{ and } s_2)$, $\ldots, (r_1 \text{ and } s_b)$; that is, to the occurrence of the result r_1 on part A (regardless of what happens on part B). In model α, r_1 is represented by e_1, so it would not be unreasonable to use e_1 also to represent r_1 in model $\alpha \times \mathcal{B}$; that is, to use e_1 to denote the event (9). The symbol e_1 would then denote two different events: a simple event in A, and a composite event in $\alpha \times \mathcal{B}$, but both corresponding to the same result r_1. Similarly, f_2 could be used to denote the event in $\alpha \times \mathcal{B}$ consisting of the second column of (1). With this notation, the intersection of e_1 (first row) and f_2 (second column) is just the event denoted in (1) by $(e_1 \text{ and } f_2)$.

With the above convention, two different events are denoted by e_1, and it is then desirable that the same probability be assigned to these two events. This is in fact the case. According to (2) the simple event e_1 of model α has been assigned probability p_1, while (5) shows that the same probability p_1 has been assigned to the composite event e_1 of model $\alpha \times \mathcal{B}$.

PROBLEMS

1. Build a product model for the experiment of throwing two fair dice. Compare this with the model of Example 1.7.1.

2. Build a product model for the sexes of two children in a family, assuming a boy or girl equally likely for both the first and second child. Compare this with the model of Example 1.5.2.

3. Compare the model assumed in Problem 1.5.9 with the product model of Example 3.

4. Suppose model α is a uniform model with simple events $e_1, \ldots, e_a$, and model $\mathcal{B}$ is a uniform model with simple events $f_1, \ldots, f_b$. Show that the product model $\alpha \times \mathcal{B}$ is again a uniform model.

5. Suppose that models α and $\mathcal{B}$ each have three simple events with probabilities

$$P(e_1) = .1 \quad P(e_2) = .2 \quad P(e_3) = .7$$
$$P(f_1) = .2 \quad P(f_2) = .3 \quad P(f_3) = .5;$$

construct the product model $\alpha \times \mathcal{B}$.

6. Suppose that model α has two simple events e_1, e_2 and model $\mathcal{B}$ four simple events f_1, f_2, f_3, f_4 with probabilities

$$P(e_1) = .4 \quad P(e_2) = .6$$
$$P(f_1) = P(f_2) = .1 \quad P(f_3) = .3 \quad P(f_4) = .5.$$

Construct the product model $\alpha \times \mathcal{B}$.

7. Determine which of the following two models is a product model and find its factor models:

(i) $P(e_1 \text{ and } f_1) = 0 \qquad P(e_1 \text{ and } f_2) = .1$
$ \; P(e_2 \text{ and } f_1) = .2 \qquad P(e_2 \text{ and } f_2) = .5$
$ \; P(e_3 \text{ and } f_1) = .2 \qquad P(e_3 \text{ and } f_2) = 0$

(ii) $P(e_1 \text{ and } f_1) = .04 \qquad P(e_1 \text{ and } f_2) = .06$
$ \; P(e_2 \text{ and } f_1) = .28 \qquad P(e_2 \text{ and } f_2) = .42$
$ \; P(e_3 \text{ and } f_1) = .08 \qquad P(e_3 \text{ and } f_2) = .12$

[Hint: Use formulas analogous to (5) to check (4).]

8. Suppose that models α and $\mathcal{B}$ have simple events e_1, e_2, e_3 and f_1, f_2 respectively and that $P(e_2 \text{ and } f_1) = 0$. What can you say about other simple events of the product model having zero probability?

9. Determine which of the following two models is a product model and find its factor models:

(i) $P(e_1 \text{ and } f_1) = .1$ $P(e_1 \text{ and } f_2) = .1$ $P(e_1 \text{ and}(f_3) = .3$
    $P(e_2 \text{ and } f_1) = .2$ $P(e_2 \text{ and } f_2) = .1$ $P(e_2 \text{ and } f_3) = .2$

(ii) $P(e_1 \text{ and } f_1) = .05$ $P(e_1 \text{ and } f_2) = .05$ $P(e_1 \text{ and } f_3) = .1$
    $P(e_2 \text{ and } f_1) = .05$ $P(e_2 \text{ and } f_2) = .05$ $P(e_2 \text{ and } f_3) = .1$
    $P(e_3 \text{ and } f_1) = .15$ $P(e_3 \text{ and } f_2) = .15$ $P(e_3 \text{ and } f_3) = .3$

10. Suppose that models α and $\mathcal{B}$ have simple events $e_1, e_2, \ldots, e_m$ and $f_1, f_2, \ldots, f_n$ respectively and that $P(e_1 \text{ and } f_1) = P(e_1 \text{ and } f_2)$. What can you say about other simple events of the product model having equal probabilities?

11. Suppose that models α and $\mathcal{B}$ have simple events e_1, e_2, e_3 and f_1, f_2, f_3, f_4 respectively. In the product model $\alpha \times \mathcal{B}$, if $P(e_1 \text{ and } f_1) = .01$, $P(e_1 \text{ and } f_2) = .02$, $P(e_2 \text{ and } f_1) = .04$, determine $P(e_2 \text{ and } f_2)$.

12. In the preceding problem, determine $P(e_2 \text{ and } f_2)$ if

(i) $P(e_1 \text{ and } f_1) = .01$ $P(e_1 \text{ and } f_2) = .02$ $P(e_2 \text{ and } f_1) = .05$
(ii) $P(e_1 \text{ and } f_1) = .01$ $P(e_1 \text{ and } f_2) = .03$ $P(e_2 \text{ and } f_1) = .04$
(iii) $P(e_1 \text{ and } f_1) = .02$ $P(e_1 \text{ and } f_2) = .01$ $P(e_2 \text{ and } f_1) = .04$.

13. Suppose that models α and $\mathcal{B}$ have simple events e_1, e_2 and f_1, f_2, f_3. In the product model $\alpha \times \mathcal{B}$, if

$$P(e_1 \text{ and } f_1) = .03 \quad P(e_1 \text{ and } f_2) = .1 \quad P(e_1 \text{ and } f_3) = .12$$

determine the remaining probabilities.

14. Solve the preceding problem if

$$P(e_1 \text{ and } f_1) = .03 \quad P(e_1 \text{ and } f_2) = .07 \quad P(e_1 \text{ and } f_3) = .1.$$

15. Using the model of Example 2, compute the probability that the total number of points showing on the two dice is four.

16. Let models α, $\mathcal{B}$, $\mathcal{C}$ each have two simple events with probabilities $\frac{1}{2}$. Construct the product model $\alpha \times \mathcal{B} \times \mathcal{C}$.

17. If models α, $\mathcal{B}$, $\mathcal{C}$ have a, b, c simple events respectively, how many simple events are there in the product model $\alpha \times \mathcal{B} \times \mathcal{C}$?

18. In Problem 5, let $E = \{e_1, e_2\}$, $F = \{f_2, f_3\}$. In the product model $\alpha \times \mathcal{B}$, find the probability of the events (i) E and F; (ii) E or F.

19. In Problem 6, let $E = \{e_2\}$, $F = \{f_1, f_3\}$. In the product model $\alpha \times \mathcal{B}$, find the probability of the events (i) E and F; (ii) E or F.

3.2 REALISM OF PRODUCT MODELS

Suppose that α and $\mathcal{B}$ are realistic models for parts A and B of a two-part experiment. It is now necessary to investigate under what conditions the product model $\alpha \times \mathcal{B}$ will then be realistic for the experiment as a whole.

Let us consider to this end a simple event e of α corresponding to a simple result r of part A of the experiment, and suppose that probability p has been assigned to e. Since α is assumed to be a realistic model for part A, it follows that the probability p is at least approximately equal to the long-run frequency with which r occurs; that is, if $f(r)$ is the frequency with which r occurs in a large number n of trials, then

$$(1) \qquad p \sim f(r),$$

where the symbol $\sim$ means that the two sides are approximately equal. Similarly, if f is one of the simple events of $\mathfrak{B}$, corresponding to the simple result s of part B, with probability q, the assumed realism of model $\mathfrak{B}$ implies that

$$(2) \qquad q \sim f(s).$$

Consider now the simple event $(e \text{ and } f)$ in the product model. By (1.4), the product model $\mathfrak{A} \times \mathfrak{B}$ assigns to the event $(e \text{ and } f)$ the probability

$$(3) \qquad P(e \text{ and } f) = pq.$$

This assignment will be realistic if, and only if, the frequency of the corresponding result "r and s" is approximately equal to pq; that is, if

$$(4) \qquad f(r \text{ and } s) \sim pq.$$

Substituting (1) and (2) into (4) gives

$$(5) \qquad f(r \text{ and } s) \sim f(r)f(s).$$

Recalling that the frequency of a result is the number of occurrences of the result, divided by the number n of trials, requirement (5) may be written as

$$(6) \qquad \frac{\#(r \text{ and } s)}{n} \sim \frac{\#(r)}{n} \cdot f(s).$$

If both sides are multiplied by $n/\#(r)$, equation (6) becomes

$$(7) \qquad \frac{\#(r \text{ and } s)}{\#(r)} \sim f(s).$$

This is the form of the requirement which is often most convenient to apply.

What is the meaning of the ratio on the left-hand side of (7)? In a long sequence of n trials of the whole experiment, certain trials will produce the result r on part A of the experiment, and $\#(r)$ is the number of trials that do so. Restrict attention to these $\#(r)$ trials. Some of them may also happen to produce the result s on part B; these trials are just the ones which produce result r on part A and result s on part B. Their number is $\#(r \text{ and } s)$. The ratio on the left-hand side of (7) is therefore the fraction of the trials producing r which also produce s. In other words, it is the frequency of s among those trials which produce r. We shall refer to this ratio as the *conditional frequency* of s, given that r has occurred, and denote it by

$$(8) \qquad f(s|r) = \frac{\#(r \text{ and } s)}{\#(r)}.$$

EXAMPLE 1. *Penny and dime.* Suppose that an experiment consists of tossing a penny and a dime and recording for each coin whether it falls heads or tails. If we regard the toss of the penny as part A of the experiment, and the toss of the dime as part B, then $a = b = 2$. In $n = 1000$ trials of the whole experiment, suppose that the number of occurrences of the $2 \cdot 2 = 4$ simple results are as follows:

#(penny heads and dime heads) = 238,

#(penny heads and dime tails) = 247

#(penny tails and dime heads) = 267,

#(penny tails and dime tails) = 248.

The penny fell heads in $238 + 247 = 485$ of the trials. On 238 of these 485 trials, the dime also fell heads. Therefore the conditional frequency of "dime heads" given "penny heads" is f(dime heads|penny heads) = $\frac{238}{485} = .491$.

In the notation of conditional frequency, the requirement (7) may be written as

$$(9) \qquad\qquad f(s|r) \sim f(s).$$

In words, the product model is realistic only if the frequency $f(s)$ of s in the whole long sequence of trials is approximately equal to the conditional frequency $f(s|r)$ of s among those trials which produce r. In our illustration, f(dime heads) $= (238 + 267)/1000 = .505$, which is approximately equal to f(dime heads|penny heads) $= .491$. We would of course not expect exact equality because of chance fluctuations.

In a product model the two factors play symmetrical roles, and it is easy to show (Problem 4) that in fact relation (9) is equivalent to

$$(10) \qquad\qquad f(r|s) \sim f(r).$$

The equivalent relations (9) and (10) assert that the frequency of a simple result in one part of the experiment is not much altered by the fact that a specified simple result has occurred in the other part.

Definition. The parts of a two-part experiment are said to have *unrelated frequencies* or, for short, to be *unrelated*, if the long-run frequency of any result in one part of the experiment is approximately equal to the long-run conditional frequency of that result, given that any specified result has occurred in the other part of the experiment.

In terms of this definition we can answer the question raised at the beginning of the section. Assuming always that the factor models are realistic, the argument leading to (9) and (10) shows that the product model will be realistic if, and only if, formulas (9) and (10) hold for all simple results r and s, that is, if the two parts of the experiment are un-related in the sense of our definition.

EXAMPLE 2. *Two draws from a box.* A box contains six marbles, alike except for color: one is red, two are white, and three are green. An ordered sample of size two is drawn from the box, and the colors of the marbles are noted. The two drawings may be considered as two parts of the experiment.

Let us consider the first drawing separately, as part A of the whole experiment. The marble drawn must be one of three colors, which in model α may be represented by the three simple events r_A (for red), w_A (for white), g_A (for green). If the marbles are thoroughly mixed and the drawing is done blindfold, it seems reasonable to suppose that the six marbles would occur with about equal frequency, and to complete model α by assigning to these events the probabilities $\frac{1}{6}$, $\frac{2}{6}$, $\frac{3}{6}$ respectively.

Now consider part B of the experiment, which consists of the second drawing. It is represented by model $\mathcal{B}$, for which we shall also need three simple events, say r_B, w_B, g_B. Because of the equivalence law of ordered sampling (Section 2.3) the six marbles should also show up with about equal frequency on the second draw, so that in model $\mathcal{B}$ we should wish to assign probabilities $\frac{1}{6}$, $\frac{2}{6}$, $\frac{3}{6}$ to the events r_B, w_B, g_B respectively.

The product of the two models gives us model $\alpha \times \mathcal{B}$ in which there are $3 \cdot 3 = 9$ simple events, with probabilities as follows:

$$P(r_A \text{ and } r_B) = \frac{1}{6} \cdot \frac{1}{6} = \frac{1}{36} \qquad P(r_A \text{ and } w_B) = \frac{1}{6} \cdot \frac{2}{6} = \frac{2}{36}$$

(11) $\qquad P(w_A \text{ and } r_B) = \frac{2}{6} \cdot \frac{1}{6} = \frac{2}{36} \qquad P(w_A \text{ and } w_B) = \frac{2}{6} \cdot \frac{2}{6} = \frac{4}{36}$

$$P(g_A \text{ and } r_B) = \frac{3}{6} \cdot \frac{1}{6} = \frac{3}{36} \qquad P(g_A \text{ and } w_B) = \frac{3}{6} \cdot \frac{2}{6} = \frac{6}{36}$$

$$P(r_A \text{ and } g_B) = \frac{1}{6} \cdot \frac{3}{6} = \frac{3}{36}$$

$$P(w_A \text{ and } g_B) = \frac{2}{6} \cdot \frac{3}{6} = \frac{6}{36}$$

$$P(g_A \text{ and } g_B) = \frac{3}{6} \cdot \frac{3}{6} = \frac{9}{36}.$$

Would the product model be realistic for the experiment as a whole? Obviously not, for consider the event (r_A and r_B). Since the box contains only one red marble, we cannot get red on both draws, and the realistic probability for (r_A and r_B) is not $\frac{1}{36}$ but 0. In this experiment, the long-run frequency of "red on second draw" among all trials (which is $\frac{1}{6}$) is not the same as the conditional frequency of "red on second draw" given "red on first draw" (which is zero), so that the two parts of the experiment are not unrelated. In fact, the result of the first draw changes the conditions under which the second marble is drawn. In this case, the result of part A exerts a direct influence on part B.

EXAMPLE 3. *Sex of twins.* Consider the "experiment" of observing the sex of the first-born and second-born of two twins. We may regard the sex of the first-born twin as part A, and that of the second-born as part B, of a two-part experiment. Part A has two possible results, which in model α may be represented by events M_A and F_A, corresponding to "first-born is male" and "first-born is female," respectively. Similarly, model β has the two simple events M_B and F_B, and model $\alpha \times \beta$ for the whole experiment has $2 \cdot 2 = 4$ simple events: M_A and M_B, M_A and F_B, F_A and M_B, F_A and F_B.

Would a product model be realistic here? That is, would the frequency of, say, "second-born is male" in a long sequence of observations be about the same as its frequency among those births on which "first-born is male"? One might say "the first-born cannot exert any influence on the second-born, whose sex is already determined, and therefore the answer is Yes." But experience would not bear this out: actual observations show that "second-born is male" occurs about 52% of the time among all twin births, while it occurs about 67% of the time among those twin births for which the first-born is male. It is true that the sex of the first-born does not exert any direct influence on the sex of the second-born, but in some cases the sex of both twins is subject to a common influence. In about one third of all twin births, the two twins originate from a single egg (identical twins), and in these cases both are always of the same sex. If we were to change the experiment by restricting the observations to two-egg (or fraternal) twins, then a product model would be found to work reasonably well.

This example illustrates the possibility that there may be a relation between the frequencies of the two parts of an experiment, arising not from the direct influence of the result of the first part on the chances of the second part (as in Example 2), but indirectly from the fact that both parts are subject to influence from a common "outside" factor. The outside factor in many cases is not known to the experimenter, who may then be tempted to assume a product model in cases where it is quite unrealistic.

Let us now give an example for which a product model is realistic.

EXAMPLE 4. *Sampling with replacement.* Consider once more two draws from the box of Example 2, but carried out in a different way. After the first marble has been drawn and its color noted, it is returned to the box, and the marbles are stirred again before the second draw. When a sample is drawn in this way, with the box restored to its original condition after each draw, we say that the sampling is *with replacement.*

In sampling with replacement, it seems reasonable to suppose that the result of the first draw will not in any way alter the chances on the second draw, since the box is restored to its original condition before the second marble is drawn.   One may expect that the conditional frequency of "red on second draw" given "red on first draw" would be about the same as the frequency of "red on second draw" among all trials.   Consequently, the product model (11) should be realistic for the present experiment.   Of course, such speculations can never be definitive, and the final test of the realism of any model must be experimental.

Since product models are so simple, it is fortunate that in many two-part experiments experience shows the frequencies to be unrelated, so that the product of realistic factor models is realistic for the experiment as a whole. These cases are characterized by the fact that there is no connection, direct or indirect, between the parts.   In addition to the present Example 4, Examples 1, 2 and 3 of the preceding section illustrate situations in which product models are appropriate.

Product models are characterized by the fact that

$$(12) \qquad\qquad P(e \text{ and } f) = P(e)P(f)$$

for any simple events e pertaining to the first factor and f pertaining to the second factor.   It follows from (12), (see Problem 11) that

$$(13) \qquad\qquad P(E \text{ and } F) = P(E)P(F)$$

where E and F are any events (simple or composite) pertaining respectively to the first and second factors of a product model.   Equation (13) has an interpretation in terms of unrelated frequencies exactly analogous to that given earlier in the section for equation (3).

It is customary to refer to events E and F for which (13) holds as "independent."

Definition.   In any probability model, two events E and F satisfying (13) are said to be *independent.*

The independent events that we shall have occasion to discuss will nearly always pertain to separate factors of a product model, but the terminology is used also in other cases (see Section 4.2).

The results of the present section extend in a natural way to experiments with more than two parts.   Suppose that $\mathfrak{a}$, $\mathfrak{B}$, $\mathfrak{C}$, . . . are realistic models for parts A, B, C, . . . of an experiment.   Then the product model $\mathfrak{a} \times \mathfrak{B} \times \mathfrak{C} \times \cdots$ defined by (1.7) and (1.8) will be realistic provided the parts are unrelated according to the following definition.

Definition.   The different parts of an experiment with several parts are said to have *unrelated frequencies* or, for short, to be *unrelated* if the long-run frequency of occurrence of any result on one part is approximately equal to the long-run conditional frequency of that result, given that any specified results have occurred on the other parts.

Analogously to the case of two factors it follows from (1.8) (see Problem 12) that

(14) $P(E \text{ and } F \text{ and } G \cdots) = P(E)P(F)P(G) \cdots$

for any events (simple or composite) E, F, G, ... pertaining respectively to the first, second, third, ⋯ factor of a product model.

Throughout this section we have assumed that the factor models α and $\mathcal{B}$ are realistic for parts A and B of the experiment. To complete the discussion, we shall prove below that model $\alpha \times \mathcal{B}$ cannot be realistic unless the factor models α and $\mathcal{B}$ are realistic. Combining this fact with the earlier result of this section, we can then state that the product model $\alpha \times \mathcal{B}$ is realistic if, and only if, (i) the factor models α and $\mathcal{B}$ are realistic, and (ii) the two parts of the experiment have unrelated frequencies.

To prove that model $\alpha \times \mathcal{B}$ can be realistic only if both α and $\mathcal{B}$ are realistic, consider once more a simple event e of model α corresponding to the simple result r of part A, and suppose that model α assigns probability p to e. We have then seen (at the end of Section 1 in small print) that model $\alpha \times \mathcal{B}$ assigns probability p to the composite event e which represents the result r in model $\alpha \times \mathcal{B}$. Realism of model $\alpha \times \mathcal{B}$ then requires $p \sim f(r)$ and hence that the probability assigned to e in model α be realistic. This shows that realism of $\alpha \times \mathcal{B}$ implies that α is realistic and the same argument applies to $\mathcal{B}$.

PROBLEMS

1. The following table shows the number of students in a class of 50 getting 0–10 points, 11–20 points, 21–30 points respectively on the first and second problems of a test.

Prob. 1 \ Prob. 2	0–10	11–20	21–30
0–10	7	4	2
11–20	4	15	5
21–30	1	3	9

(i) Find the conditional frequency with which students get 21–30 points on the second problem, given that they got 21–30 points on the first problem.

(ii) Find the frequency with which students get 21–30 points on the second problem among the totality of students.

(iii) Describe in words analogous to (i) the frequency given by $2/(7 + 4 + 2)$.

2. The following table gives the distribution of 100 couples being married, according to their marital status.

Bride \ Groom	single	widowed	divorced
single	70	2	5
widowed	1	3	2
divorced	6	2	9

  (i) Find the conditional frequency with which the bride is a divorcee given that the groom is a widower.

  (ii) What is the frequency with which the groom is a widower among all cases?

  (iii) Describe in words analogous to (i) the frequency given by $6/(6 + 2 + 9)$.

3. In an election involving four candidates, a polling organization decides to interview 200 persons. These are classified in the following table according to whether they were reached on the first, second, or a later call of the interviewer and according to their preference for candidate A, B, C or D.

Interviewed on \ Prefers Candidate	A	B	C	D
1st call	24	38	58	20
2nd call	7	18	20	5
3rd or later call	2	3	4	1

  (i) Find the conditional frequency with which candidate C was preferred among those reached on the first call.

  (ii) Find the frequency with which candidate B was preferred among all those interviewed.

  (iii) Find the conditional frequency with which a voter was reached on the first call among those preferring candidate A.

  (iv) Find the frequency with which three or more calls were required among all those interviewed.

4. Show that relation (9) implies relation (10).

5. For each of the experiments whose results are described
  (i) in Problem 1
  (ii) in Problem 2,
check whether the two parts appear to be unrelated. Do your findings agree with what you would expect?

6. In a table giving the age distribution of all married couples in a city, would you expect the frequencies of the results "husband over 60" and "wife over 50" to be related or unrelated?

7. In a table giving the distribution of grades for all students taking Statistics 1B during the last two years, would you expect the frequency of the result "getting an A in the course" to be related or unrelated to the following results:

    (i) getting an A in Statistics 1A;

    (ii) getting an A in English 1A;

    (iii) being male;

    (iv) being a freshman?

8. (i) In a table giving the frequency distribution for pairs of consecutive letters in works of English prose, would you expect the frequencies of the results "first letter is a vowel" and "second letter is a consonant" to be related or unrelated?

    (ii) Make a frequency count of the first 100 pairs of letters of this section to illustrate the point.

9. (i) In a table giving the frequency distribution for pairs of consecutive words in works of English prose, would you expect the frequencies of the results "first word has at most three letters" and "second word has more than three letters" to be related or unrelated?

    (ii) Make a frequency count of the first 100 pairs of words of the Preface of this book to check your conjecture.

10. Would you expect a product model to be appropriate for an experiment whose two parts consist of

    (i) one throw each with two dice,

    (ii) two successive throws with the same die?

11. State whether you would expect a product model to fit well an experiment whose two parts consist in observing the weather (rain or shine)

    (i) on February 1 and 2 of the same year,

    (ii) on February 1 of two successive years.

12. Build a realistic model for the sex of the first- and that of the second-born of identical twins. Is this a product model?

13. Use the results of Section 2.3 to build a realistic model for Example 2. Is this a product model?

14. In the model of Example 1.2 (continued), let E and F denote respectively the events, "the number of points showing on the first die is even" and "the number of points showing on the second die is even." Check that (13) holds by finding $P(E)$, $P(F)$ and $P(E \text{ and } F)$.

15. Let E consist of the simple events (e_1, e_3) of model α, and F of the simple events (f_2, f_3, f_5) of model $\mathcal{B}$.

    (i) List the simple events in model $\alpha \times \mathcal{B}$ which make up the event $(E \text{ and } F)$.

    (ii) If $P(e_1) = p_1$, $P(e_3) = p_3$, $P(f_2) = q_2$, $P(f_3) = q_3$, $P(f_5) = q_5$, find $P(E \text{ and } F)$.

    (iii) Show that $P(E \text{ and } F) = P(E)P(F)$.

[Hint: see small print at end of Section 2. For (iii), collect together the terms of $P(E \text{ and } F)$ involving p_1 and those involving p_3, and factor out p_1 from the first and p_3 from the second of these sums. Note that the resulting factors of p_1 and p_3 are the same.]

16. Solve the preceding problem if

(i) E consists of the simple events (e_1, e_3, e_6) and F of the simple events (f_2, f_3, f_5);

(ii) E consists of the simple events (e_1, e_3) and F of the simple events (f_2, f_3, f_5, f_7).

17. Let E consist of the simple events (e_1, e_3) of model α, F of the simple events (f_2, f_5) of model $\mathbb{B}$, and G of the simple events (g_1, g_4) of model $\mathbb{C}$. If $P(e_1) = p_1$, $P(e_3) = p_3$, $P(f_2) = q_2$, $P(f_5) = q_5$, $P(g_1) = r_1$, $P(g_4) = r_4$, find $P(E$ and F and $G)$ and show that it is equal to $P(E)P(F)P(G)$. [Hint: Use the method of Problem 15.]

18. In Example 2, compare P(red on first draw) with P(red on first draw|red on second draw). Would you say that "the result of the second draw changes the conditions under which the first marble is drawn"?

3.3 BINOMIAL TRIALS

We shall now consider a simple but very important application of product models. Suppose that a sequence of trials is performed, on each of which a certain result may or may not happen. The occurrence of the result is called a *success*, its nonoccurrence a *failure*. In a sequence of tosses of a penny, for example, "heads" might be designated a success in which case "tails" would constitute a failure. This terminology is purely conventional, and the result called success need not be desirable.

Each trial may be considered as one part of the whole experiment. If the trials are unrelated, in the sense of the preceding section, a product model is appropriate. Each factor model corresponding to a particular trial then has two simple events corresponding to success and failure. In the simplest case, in which success has the same probability, say p, and hence failure has the same probability $q = 1 - p$, in each factor model, the product model is called the *binomial trials model*.

EXAMPLE 1. *Three throws with a die.* Consider a dice game in which the player wins each time the die shows either 1 or 2 points (success) and loses each time it shows 3, 4, 5, or 6 points (failure). If the experiment consists of three throws, the results may be represented in the model by the events

SSS   SSF   SFS   SFF   FSS   FSF   FFS   FFF

with SFS, for example, representing the result that the player wins on the first and third throws but loses on the second. The binomial trials model assigns to these events the probabilities

$$p \cdot p \cdot p \quad p \cdot p \cdot q \quad p \cdot q \cdot p \quad p \cdot q \cdot q \quad q \cdot p \cdot p \quad q \cdot p \cdot q \quad q \cdot q \cdot p \quad q \cdot q \cdot q.$$

The probability of the player winning, for example, in exactly two of the three throws is then

$$P(\text{SSF}) + P(\text{SFS}) + P(\text{FSS}) = 3p^2q.$$

If the die is assumed to be fair, then $p = \frac{1}{3}$, $q = \frac{2}{3}$, and the desired probability is $\frac{2}{9}$.

Under what conditions is the binomial trials model realistic for a sequence of trials? Since it is a product model, the various trials (parts) should have unrelated frequencies in the sense of the preceding section. Thus if for example the occurrence of success on one trial tends to increase or decrease the chance of success on the next trial to a material degree, a product model would not be realistic. In addition, the binomial trials model assumes that the probability of success is the same on each trial.

When considering whether to use the model for a sequence of trials, the following two questions must therefore be asked.

(a) Is the chance of success the same on each trial?

(b) Will the chance of success on any trial be affected by success or failure on the others?

We shall now consider several examples, and discuss for each whether the binomial trials model might be expected to work.

EXAMPLE 2. *Sex of children in three-child families.* Consider a family with three children. Each birth may be considered a trial (especially by the mother) that produces either a male (M) or female (F) child. When the records of thousands of such families are assembled, it is found (a) that the frequency of male on each of the three births is very nearly the same, being about .514, and (b) that there is little or no relation between the frequency of males on the different births. For example, among the families in which the first child is of a specified sex (say female), the frequency of males on the second birth is still about .514. Thus, a binomial trials model with $p = .514$ is quite realistic. Using this model, we find that the probability that all three children are female is about $(.486)(.486)(.486) = .115$; while the probability that all three are male is about $(.514)^3 = .136$.

When very large numbers of records are assembled, slight departures from this model become apparent. For example, the frequency of male births differs slightly in different populations. In general, no mathematical model corresponds exactly to reality, and any probability model will show deficiencies when a sufficiently great body of data is collected.

EXAMPLE 3. *Defectives in lot sampling.* Suppose an ordered sample of ten items is drawn from a lot of 50 items, and each sampled item is inspected to see if it is defective. Would it be realistic to use a binomial model? It follows from the equivalence law for ordered sampling (2.3.5) that, if the sampling is random, the probability of a defective is the same on each draw, so that condition (a) is met. But condition (b) fails, since the results on the earlier draws will influence the chances on the later ones. To make this point concrete, suppose the lot contains exactly one defective item. Then each of the ten draws has the same probability $\frac{1}{50}$ of producing a defective, but if the first-drawn item is defective, there is no possibility of getting a defective on any subsequent draw.

If the sampling were done with replacement, however, the ten trials would be unrelated, and the binomial trials model would serve very well. Furthermore, if the size of the lot instead of being 50 were very large compared with the size of the sample, sampling with or without replacement would nearly always lead to the same result (since the chance of drawing the same item more than once would then be very small even when sampling with replacement). In this case, the binomial model would therefore be quite satisfactory.

EXAMPLE 4.   *Winning at tennis.*   The games played by two tennis players may be thought of as a sequence of trials, on which player A either wins or loses. Typically, the server has a considerable advantage, so that the frequency of wins by A will be materially higher when he serves than when he receives. Condition (a) fails, and the binomial trials model does not work.

EXAMPLE 5.   *Rain on successive days.*   We may regard successive days as trials, on each of which it either does or does not rain in a certain city. To be specific, consider the Saturday, Sunday, and Monday of the first weekend in July. If we regard the successive years as repetitions of this three-part experiment, examination of the record may show that the frequency of rain is about the same on the Saturday, Sunday, and Monday; thus condition (a) is met. However, the frequencies may be found to be heavily related. A possible explanation is that in some areas rain tends to come in storms that last for several days. If it rains on a Saturday, this means that a storm is occurring, and the risk of rain on the next day is considerably greater than if the Saturday were fair.

EXAMPLE 6.   *Red at roulette.*   A roulette wheel has 37 slots, of which 18 are red. Can a binomial trials model be used for the occurrence or non-occurrence of red on, say, five successive spins of the wheel? Records kept over long periods of the play at Monte Carlo and elsewhere can be broken up into successive sequences of five spins to provide data of repetitions of the experiment. Such data justify the use of the binomial model, since the five spins turn out to be unrelated in the frequency sense, and the frequency of red is the same (about $\frac{18}{37}$) on each of the five trials.

This finding contradicts the intuitive belief of many gamblers, who play according to systems that imply a relationship among the frequencies. For example, a gambler may wait until the wheel has shown non-red four times and then bet on red. He feels that "the law of averages" will force the wheel to tend to show red after a run of non-red; or he will argue that a run of five non-reds is so unlikely that after four non-reds the next spin should give red. But the wheel has no memory, and (except where the wheel is dishonestly controlled by a magnet or other device) the record supports the assumption of unrelated frequencies.

EXAMPLE 7. *Defective items in mass production.* Suppose an automatic machine produces certain items one after the other. Occasionally the machine turns out a defective item. Frequency counts appear to justify the use of a binomial trials model for many such experiments. Over a very long period of time, however, an essential part of the machine may gradually wear out. In this case the frequency of defectives rises and condition (a) fails if a long sequence of trials is considered.

The simple events in a sequence of n binomial trials are sequences of successes (S) and failures (F). If $n = 5$, the outcome of the experiment might for example be SSFSF or SSSSF, etc. Since the binomial trials model is a product model, the probabilities of these events are obtained by multiplying together the probabilities of the indicated events of the individual trials. Thus the probability of SSFSF, for example, is

$$P(\text{SSFSF}) = p \cdot p \cdot q \cdot p \cdot q = p^3 q^2.$$

Similarly

$$P(\text{FFSSS}) = q \cdot q \cdot p \cdot p \cdot p = p^3 q^2.$$

In this manner we see that the probability of three successes and two failures in any specified order will be equal to $p^3 q^2$.

More generally, in a sequence of n binomial trials the probability of any specified sequence of b successes and $n - b$ failures is a product of b factors equal to p and $n - b$ factors equal to $q = 1 - p$, and hence is equal to

(1) $$p^b q^{n-b}.$$

EXAMPLE 8. *Comparing two drugs.* To compare two drugs A and B a doctor decides to give some of his patients drug A and some drug B. In order to rule out the possibility of his assigning the drugs to the patients in a manner which might favor one or the other, he selects for each patient one of the drugs at random, that is with probabilities $\frac{1}{2}$. If the experiment involves ten patients, let us find the probability that all will get the same drug so that no comparison is possible. The ten patients constitute ten binomial trials with probability $\frac{1}{2}$ of success (being assigned drug A). The probability of getting either ten successes or ten failures is

$$P(10 \text{ successes}) + P(10 \text{ failures})$$

since the two events are exclusive. The first of these terms is $p^{10} = (\frac{1}{2})^{10}$ since $p = \frac{1}{2}$; the second one is $q^{10} = (\frac{1}{2})^{10}$ since $q = \frac{1}{2}$. The desired probability is therefore $(\frac{1}{2})^{10} + (\frac{1}{2})^{10} = 2(\frac{1}{2})^{10} = (\frac{1}{2})^9 = \frac{1}{512}$.

EXAMPLE 9. *The shooting gallery.* At a fair, a prize is offered to any person who in a sequence of four shots hits the target three times in a row. What is the probability of a person getting the prize if his probability of hitting on any given shot is $\frac{3}{4}$? If the shots can be considered to be unrelated, they constitute four binomial trials with probability $p = \frac{3}{4}$ of success (hitting the target). The prize will be won in the cases SSSS, SSSF, FSSS, and in no others. The probabilities of these cases are

$$P(\text{SSSS}) = (\tfrac{3}{4})^4 = \tfrac{81}{256}; \quad P(\text{SSSF}) = P(\text{FSSS}) = (\tfrac{3}{4})^3\tfrac{1}{4} = \tfrac{27}{256}.$$

Since the three cases are exclusive, the desired probability is

$$\tfrac{81}{256} + \tfrac{27}{256} + \tfrac{27}{256} = \tfrac{135}{256} = .527.$$

The binomial trials model extends in a natural way to situations like those of Examples 4 and 7, where the trials are unrelated but have unequal probabilities. To see what becomes of formula (1) in this case, consider three unrelated trials with success probabilities p_1, p_2, p_3 and failure probabilities $q_1 = 1 - p_1$, $q_2 = 1 - p_2$, $q_3 = 1 - p_3$. There are then eight possible patterns of successes and failures, with the following probabilities:

$$P(\text{SSS}) = p_1p_2p_3, \ P(\text{SSF}) = p_1p_2q_3, \ P(\text{SFS}) = p_1q_2p_3, \ P(\text{SFF}) = p_1q_2q_3$$

$$P(\text{FSS}) = q_1p_2p_3, \ P(\text{FSF}) = q_1p_2q_3, \ P(\text{FFS}) = q_1q_2p_3, \ P(\text{FFF}) = q_1q_2q_3.$$

This model easily generalizes to more than three trials.

Another extension of the idea of binomial trials is to unrelated trials with more than two possible outcomes, the probabilities of which do not change from trial to trial. In the next section we shall consider an example of such *multinomial* trials. More general aspects will be taken up in Section 7.3.

PROBLEMS

1. In Example 1, find the probability that the player who uses a fair die will win (i) at least two of the three throws; (ii) at most two of the three throws; (iii) exactly two of the three throws.

2. In Example 1, suppose that the player wins when the die shows six points and loses otherwise. Find the probabilities of the three events of Problem 1.

3. For the following sequences of trials discuss whether you would expect the binomial model to work and give your reasons.

(i) A carton of eggs is taken at random from a stack at the grocery without opening it. At home, the first three eggs are taken out, and for each it is noted whether or not the egg is cracked.

(ii) A five-letter word is selected at random from a work of English prose, and for each of the five consecutive letters it is noted whether it is a vowel or a consonant.

(iii) From a conveyor belt one item is taken every hour and it is classified as defective or nondefective.

(iv) At a given time and place, you observe for the next ten passing cars whether each car has two or four doors.

(v) At a given time and place, you observe for the first minute of each hour whether or not a car is passing.

(vi) On a number of successive weeks it is observed whether or not there occurs a fatal traffic accident in the town.

(vii) You repeat a number of times the experiment of bisecting a line segment by eye. After each trial, before you attempt the next one, you are informed whether your point is to the left or to the right of the center, which are the two results of interest.

4. Give an example of a sequence of trials (different from those in the text) such that you would expect the binomial model

(i) to work well;

(ii) to fail for reason (a) but not (b);

(iii) to fail for reason (b) but not (a).

5. In a sequence of five tosses with a fair penny, find the probability of the following events: (i) HHHHT (four Heads then one Tail); (ii) HHHHH; (iii) HTHTH; (iv) HTTTH.

6. In four binomial trials with success probability p, find the probability of (i) four successes; (ii) four failures; (iii) exactly one success; (iv) exactly two successes; (v) exactly three successes; (vi) exactly one failure; (vii) at most two successes; (viii) at least one success.

7. In Example 8, what is the probability that as a result of the random assignment the two drugs will be used alternately?

8. In a sequence of five tosses with a fair penny,

(i) what is the probability of observing a run of four successive heads;

(ii) what is the probability of never having a head after a tail?

9. A farmer wishes to determine the value of a fertilizer. He divides his field into 15 plots, five rows of three plots each. In each row he selects one of the three plots at random and applies the fertilizer to it, while giving no fertilizer to the other two in the row. What is the probability

(i) that all five fertilized plots are in the first column;

(ii) that all fertilized plots are in the same column?

10. Solve the preceding problem under the assumption that the field is divided into 16 plots, four rows of four plots each.

11. (i) In n binomial trials with success probability p find the probability that all n trials will be successful.

(ii) If $p = \frac{1}{2}$ how large must n be before the probability exceeds .9 that among the n trials there is at least one success and one failure?

12. (i) For n binomial trials, find the probability of exactly $n - 1$ successes and one failure for $n = 2, 3, 4$.

(ii) Conjecture a formula for this probability for several n and check it for $n = 5$.

13. In a tennis match suppose that the individual games are unrelated, and that player A has probability $\frac{2}{3}$ of winning a game in which he serves and probability $\frac{1}{2}$ of winning a game in which he receives. The players alternate in serving. What is the probability that player A will win the first six games?

14. A set in tennis is won by the first player who wins six games provided by that time his opponent has won at most four games. Under the assumptions of Problem 13, find the probability that player A will win the set with a score of 6:1 if

(i) he serves in the first game;

(ii) he receives in the first game.

15. Suppose that in training new workers to perform a delicate mechanical operation, the probabilities that the worker will be successful on his first, second, and third attempt are $p_1 = .03$, $p_2 = .06$, $p_3 = .11$ respectively, and that the three attempts may be considered to be unrelated. Find to three decimals the probabilities that the worker will be successful (i) all three times; (ii) exactly twice; (iii) exactly once; (iv) on none of his three attempts.

16. In the preceding problem, let there be four attempts with success probabilities $p_1 = .01$, $p_2 = .02$, $p_3 = .03$, $p_4 = .05$. Find the probabilities that the worker will succeed (i) all four times; (ii) exactly three times; (iii) exactly once; (iv) on none of the four attempts; (v) exactly twice.

3.4 THE USE OF RANDOM NUMBERS TO DRAW SAMPLES

In Section 2.1 we emphasized the desirability of drawing samples at random and mentioned that this is most commonly done with the aid of a table of random numbers. We shall now explain how such tables can be made, and how they may be used to obtain random samples.

Suppose a box contains ten marbles, identical except that each is labeled with a different one of the ten digits $0, 1, 2, \ldots, 9$. The marbles are stirred, one of them is drawn from the box, and the digit on it is written down—suppose it happens to be "1." The marble is returned to the box, and the process repeated—perhaps "0" is obtained on the second trial, and then "0," "9," "7," and so forth. The results of 1000 drawings might look like the following table.

TABLE 1. RANDOM DIGITS

10097	32533	76520	13586	34673	54876	80959	09117	39292	74945
37542	04805	64894	74296	24805	24037	20636	10402	00822	91665
08422	68953	19645	09303	23209	02560	15953	34764	35080	33606
99019	02529	09376	70715	38311	31165	88676	74397	04436	27659
12807	99970	80157	36147	64032	36653	98951	16877	12171	76833
66065	74717	34072	76850	36697	36170	65813	39885	11199	29170
31060	10805	45571	82406	35303	42614	86799	07439	23403	09732
85269	77602	02051	65692	68665	74818	73053	85247	18623	88579
63573	32135	05325	47048	90553	57548	28468	28709	83491	25624
73796	45753	03529	64778	35808	34282	60935	20344	35273	88435
98520	17767	14905	68607	22109	40558	60970	93433	50500	73998
11805	05431	39808	27732	50725	68248	29405	24201	52775	67851
83452	99634	06288	98083	13746	70078	18475	40610	68711	77817
88685	40200	86507	58401	36766	67951	90364	76493	29609	11062
99594	67348	87517	64969	91826	08928	93785	61368	23478	34113
65481	17674	17468	50950	58047	76974	73039	57186	40218	16544
80124	35635	17727	08015	45318	22374	21115	78253	14385	53763
74350	99817	77402	77214	43236	00210	45521	64237	96286	02655
69916	26803	66252	29148	36936	87203	76621	13990	94400	56418
09893	20505	14225	68514	46427	56788	96297	78822	54382	14598

This is a *table of random digits.* Because the ten marbles are alike, it is reasonable to assume that, in any given place in the table, each of the ten digits has the same probability $\frac{1}{10}$ of appearing. Furthermore, because the drawn marble is replaced and the marbles are restirred after each drawing, it may be assumed that the frequencies of the digits appearing in the various places are unrelated. The digits therefore constitute the results of a sequence of multinomial trials, where each trial has ten possible outcomes. Large random digit tables are produced by electronic machines rather than by hand drawing. The table* from which our Table 1 is taken contains one million digits, which have been carefully examined by various frequency counts to check empirically the assumptions of unrelatedness and equal probability.

We shall explain how random numbers are used in drawing random samples by taking up a sequence of examples, starting with the simplest case of a sample of size one.

* "A Million Random Digits with 100,000 Normal Deviates," by The RAND Corporation. The Free Press (1955). (Table 1 reproduced by permission.)

EXAMPLE 1. *Choosing one of ten items.* A poll taker wishes to choose at random one of the ten houses on a certain street. After making a sketch map on which the houses are labeled 0, 1, 2, . . . , 9, he opens his random number table and reads off the first digit, "1." The house to be sampled is then the one labeled "1" on the map. This house may be assumed to have been drawn at random since each house had the same chance $\frac{1}{10}$ to be selected.

EXAMPLE 2. *Choosing one of one hundred items.* The digits appearing in two consecutive places in the table form a two-digit number—for example "10" occupies the first two places in our table. There are 100 two-digit numbers that might occupy the two places, beginning with 00 and going up to 99. The assumptions of randomness and unrelatedness of the digits imply that each of these 100 numbers has the same probability $\frac{1}{100}$ of appearing.

Suppose the Census Bureau wishes to choose at random one of the 100 counties of the State of North Carolina. A list of the counties is obtained, and the counties labeled successively 00, 01, 02, . . . , 99. The random number table is consulted—let us use the second line this time—and the number "37" read off. From the county list, the county bearing the number 37 is selected.

EXAMPLE 3. *Choosing one of an arbitrary number of items.* Suppose we wish to choose at random one of the 876 students attending a certain college. The registrar provides a list of the students, and they are numbered consecutively from 000 to 875. By an easy extension of the argument in Example 2, it appears that a three-place entry in the table has probability $\frac{1}{1000}$ of being any of the digits from 000 to 999. Suppose we use the fourth line, getting the number "990." Unfortunately the highest number on the list is 875, so we continue along the line, taking the next number, "190" which is usable. The student whose name is opposite 190 on the list is therefore selected.

The above procedure will always eventually yield one of the 876 usable numbers. Since all 1000 three-digit numbers are equally likely, it is plausible that each of the 876 usable numbers has the same probability $\frac{1}{876}$ of being selected in this way. However, a proof of this fact is beyond the scope of this book.

The procedure illustrated in this example easily generalizes to a method for selecting one of an arbitrary number of items. One continues drawing from the table numbers of the appropriate length until one is found that appears on the list. With this method, each item on the list has the same chance of being drawn.

EXAMPLE *4*. *Choosing two of ten items.* Let us modify Example 1 by requiring the poll taker to choose two of the ten houses. He takes two digits from the table—say "1" and "0" if the first line is used—and then proceeds to the houses labeled "0" and "1" on his map.

A difficulty would arise if the second digit had also been "1." In that case it would be necessary to continue until a digit appears that differs from the first digit obtained. It is intuitively clear that the first digit found to be different from "1" has the same chance, $\frac{1}{9}$, of being each of the 9 digits other than "1." However, a formal argument that justifies regarding the two houses thus drawn as forming a random sample will again have to be omitted.

Let us now summarize in a general rule the procedure for drawing a random sample of s items from a population of N items. A list of the N items is made and the items are numbered consecutively. Random numbers of the required length are read from the table, until s distinct numbers of the list are obtained. The items having these s numbers form the sample.

EXAMPLE *5*. *Choosing ten of the fifty American states.* Number the 50 states of the United States in an alphabetical list beginning with Alabama 01 and ending with Wyoming 50. It is desired to select ten of them at random. For illustration, let us begin with the ninth row of the table, getting the two-digit numbers

$$63, 57, 33, 21, 35, 05, 32, 54, 70, 48, 90,$$
$$55, 35, 75, 48, 28, 46, 82, 87, 09, 83, 49.$$

Ignoring 00 and numbers above 50, as well as repeats, we obtain the following ten distinct usable numbers:

$$33, 21, 35, 05, 32, 48, 28, 46, 09, 49.$$

From the alphabetical list we now read off the ten state names; these are a random sample of size 10:

California (05), Florida (09), Massachusetts (21), Nevada (28),
New York (32), North Carolina (33), Ohio (35), Virginia (46),
West Virginia (48), Wisconsin (49).

As in the above examples, it is customary in practice to use a different line or page of the random number table for each application. Otherwise, the user may remember that the first digit in the table happens to be a "1," and may use this information to destroy the randomness of the sample. The poll taker of Example 1, for example, may number the houses on his sketch map so as to avoid attaching "1" to a house he does not wish to visit, perhaps because he noticed a large dog in the yard. This obviously leads to bias—dog owners will tend to be underrepresented in the sample!

PROBLEMS

1. Use Table 1 to obtain a random sample of 10 pages from the text.

2. Use Table 1 to obtain a random sample of six lines from page (i) 12; (ii) 43; (iii) 56; (iv) 69; (v) 71; (vi) 76.

3. (i) For each part of the preceding problem, count the number of words on each line and use this to estimate the total number of words on the given page.

(ii) Count the total number of words on the page and compare the count with your estimate.

4. Obtain a sample of ten last digits in Table E (the last digit of the entry .5398 in the first column, for instance, is 8)

(i) as a random sample of the 350 last digits given in the table;

(ii) by selecting one last digit at random from each of the ten columns.

5. (i) Obtain a stratified sample of pages from Part I of the text by selecting two pages at random from each chapter.

(ii) For each page of the sample count the number of paragraphs starting on that page.

(iii) Use the counts of part (ii) to estimate the total number of paragraphs in Part I of the text.

6. (i) Select one item at random from 200 items numbered 0, 1, ..., 199 by using the following device: To obtain the first digit (0 or 1), select a digit at random from Table 1 and call it 0 if it is even and 1 if it is odd.

(ii) Use this idea to make the sampling of Example 5 more efficient.

7. A chessboard has eight rows and eight columns. One of the 64 squares could be selected (a) by selecting at random one of the digits 1, ..., 64; (b) by selecting at random one of the rows, and independently one of the columns. In what sense are these two methods equivalent?

8. (i) To select a sample of five telephone subscribers, carry out the following process five times: Select at random a page of the directory; on the selected page, select at random one of the columns; from the selected column, select at random one of the names.

(ii) Discuss whether the resulting sample is a random sample of telephone subscribers in the sense of Section 2.1.

9. (i) By throwing a die and tossing a penny, one may get 12 equally likely events. How may this experiment be used to produce a random digit?

(ii) Devise an efficient scheme for using the throw of two dice to obtain a random digit.

(iii) Produce three random digits by repeated tosses of a coin.

CHAPTER 4

CONDITIONAL PROBABILITY

4.1 THE CONCEPT OF CONDITIONAL PROBABILITY

It sometimes happens that we acquire partial knowledge of the way a random experiment is turning out before the complete result becomes known. During the course of the experiment we may see that certain results are precluded by what has happened so far, or the experiment may be completed but its result revealed to us only in part. Examples will make it clear that when such partial information is obtained, the original probability model should be modified.

EXAMPLE 1. *Poker hands.* Suppose a poker player happens, by accident, to catch a glimpse of the hand dealt to an opponent. The glimpse is too fleeting for individual cards to be distinguishable, but the player does perceive that all the cards are red. It is then certain that the opponent cannot have "four of a kind," as that would require him to have at least two black cards. In our earlier model for poker hands (Example 2.2.1), every hand is assigned a positive probability; this model will no longer be realistic, since it assigns positive probability to an event we know to be impossible. One also feels that the chance of the opponent holding a "heart flush" is now higher than it was before the new information was obtained.

EXAMPLE 2. *Life table.* Suppose we are observing a life span for which Table 1.4.1 is appropriate, and that the individual in question has just celebrated his 20th birthday. The experiment is still in progress, but we know that neither of the results $A = 1$ or $A = 2$ (death in the first or second decade) can happen. These results are assigned probabilities .064 and .013 in Table 1.4.1, as was appropriate at birth, but we now want to give them probability zero since they are impossible. On the other hand, we should want to assign a higher probability to $A > 6$ than at birth, since the individual has now survived the dangers of childhood and adolescence.

In each of these examples, the new information serves to rule out certain of the simple results that were originally possible. Let us denote by R the set of simple results that are still possible on the basis of what is now known: in the first example, R consists of all poker hands formed of red cards only while in the second example R consists of death in any decade from the third to the tenth. We shall denote by E the set of simple events that correspond to the simple results in R. It is clear that the original probability model, according to which events in $\bar{E}$ have positive probability, should be replaced by a new model.

We have emphasized in Section 1.2 that the frequency of a result depends on the conditions of the experiment. Our basic idea is that the probability attributed to a result on a particular trial of an experiment should correspond to the frequency of the result in a long sequence of trials of the experiment *carried out under essentially the same conditions* as the particular experiment under consideration. We may think of the new information as serving to change the experimental conditions: in addition to the requirements previously given, we now require that the result must be one of those in R. Any trial that gives a result in $\bar{R}$ is not considered a proper trial under the new conditions and is thrown out. The original sequence of trials is reduced to the sequence consisting of those trials that gave a result in R, and it is the frequency of results in this sequence that we want to represent in the new model. In the terminology of Section 3.2, what is now of interest is the conditional frequency, given that R has occurred. The relation between the probabilities in the two models will depend on the relation between the frequencies in the two sequences. This can best be explained by a simple example.

EXAMPLE 3. *The fair die.* A fair die has been thrown, and the top face shows an even number of points. Under this condition, what is the probability that the die shows two points?

To motivate a model for this problem, let us imagine that the fair die is thrown 6,000 times. We would then expect each face to appear about 1,000 times. The given condition (top face shows an even number of points) rules out of consideration the 3,000 or so appearances of one, three or five points—these trials do not conform to the new conditions of our experiment. There remain about 3,000 trials that do conform; of these, about 1,000 each yield the results: two points, four points, six points. Each of these three outcomes therefore has relative frequency about $\frac{1}{3}$ among the trials satisfying the given condition. This reasoning motivates the construction of a new model with the probabilities shown in the last line below.

Face	1	2	3	4	5	6
Probability without condition	$\frac{1}{6}$	$\frac{1}{6}$	$\frac{1}{6}$	$\frac{1}{6}$	$\frac{1}{6}$	$\frac{1}{6}$
Probability with even-points condition	0	$\frac{1}{3}$	0	$\frac{1}{3}$	0	$\frac{1}{3}$

For simplicity we took as our first example a case where the original model is uniform (Section 1.5), but the basic idea works quite generally.

EXAMPLE 4. *Loaded die.* Suppose that the die of the preceding example is loaded, and that the experiment of throwing it is adequately represented by the following model:

Face	1	2	3	4	5	6
Probability	.2	.1	.1	.2	.3	.1

Again the die is thrown and the top face shows an even number of points. Under this condition, what is the probability that the die shows two points?

In the original model, $P(\text{even}) = P(2) + P(4) + P(6) = .1 + .2 + .1 = .4$. This means that in a long sequence of throws an even face will occur about 40% of the time. Since the face 2 will occur in about 10% of all throws, it follows that it occurs in about $(.1) \div (.4) = \frac{1}{4}$ of those throws that result in an even face. Thus, we would want to say that the chance of the face 2 is .25, once it is given that the face is even.

Similarly, given that the face is even there is a 50% chance that it shows four points and a 25% chance that it shows six points. Faces 1, 3, and 5 are impossible on this throw. We are led to the following model for the experiment when it is known that the result is an even face:

Face	1	2	3	4	5	6
Probability	0	.25	0	.50	0	.25

We are now prepared to give the general definition of conditional probability. Suppose we have built a model for the original experiment, consisting of an event set $\mathcal{E}$ that corresponds to the list $\mathcal{R}$ of simple results of the experiment, and that a probability $P(e)$ has been assigned to each simple event e in $\mathcal{E}$. Then new information becomes available: the experiment has in fact produced one of the results on a sub-list R of the original list $\mathcal{R}$. Let E be the set of simple events that correspond to the results of R. We must construct a new model, using the same event set as before, but assigning to the simple events new probabilities that will be reasonable in the light of the new information. We shall denote by $P(e|E)$ the probability assigned by the new model to the simple event e. If e does not belong to E, then of course we shall want to put $P(e|E) = 0$.

To see what value would be reasonable if e does belong to E, imagine a sequence of n trials of the original experiment. Suppose that in this sequence the composite result R occurred $\#(R)$ times, while the simple result r corresponding to e occurred on $\#(r)$ trials. Those trials on which R did not occur are irrelevant to our present problem (since in our particular case R did occur), and by eliminating them from consideration we obtain a new and more relevant sequence consisting of $\#(R)$ trials. The frequency with which r occurred in this sequence, that is the *conditional frequency of r given R*, is

$$ f(r|R) = \frac{\#(r)}{\#(R)} = \frac{\#(r)}{n} \div \frac{\#(R)}{n} = \frac{f(r)}{f(R)} $$

where $f(r)$ and $f(R)$ are frequencies in the original sequence of n trials. If n is large, we may reasonably expect $f(r)$ to be close to $P(e)$ and $f(R)$ to be close to $P(E)$, so that $f(r|R)$ should be close to $P(e)/P(E)$. Since probability is supposed to correspond to frequency, this motivates the following definition.

Definition. The *conditional probability* of a simple event e given an event E is

(1)
$$ \begin{array}{ll} P(e|E) = P(e)/P(E) & \text{if } e \text{ is in } E \\ P(e|E) = 0 & \text{if } e \text{ is not in } E. \end{array} $$

With this definition, conditional probability given E corresponds in the model to the long-term frequency of occurrence of r in a sequence of experiments in which the result R has occurred; that is, to the conditional frequency of r given R.

Equation (1) defines a probability model which is known as the *conditional probability model given E*. (We shall check at the end of the section that (1) does define a probability model.)

Let us illustrate the definition of conditional probability model on two further examples.

EXAMPLE 5. *Sampling with and without replacement.* A box contains four marbles labeled A, B, C, D. Two marbles are drawn with replacement. Given that the same marble does not appear on both draws, what is the probability that neither of the two marbles is marble B?

Before imposing the condition, it would be reasonable (in accordance with the discussion of Example 3.2.4) to attribute equal probabilities of $\frac{1}{16}$ to each of the 16 possible results shown below:

AA	AB	AC	AD
BA	BB	BC	BD
CA	CB	CC	CD
DA	DB	DC	DD

Here, for example, CB represents the result "marble C on first draw, marble B on second draw." But now we impose the condition that the same marble does not appear twice. This reduces to 0 the probabilities of the diagonal entries of our array: AA, BB, CC, DD. The twelve remaining simple events are still to have equal probability, so that each is assigned probability $\frac{1}{12}$ by the conditional model. A simple count now shows that in the conditional model the event "B does not occur" has probability P(B does not occur|same marble does not appear twice) = $\frac{6}{12} = \frac{1}{2}$.

It is interesting to note that the above conditional model has the same probabilities as the model for ordered sampling (without replacement) discussed in Section 2.3. The same argument shows that, quite generally, if we impose on a model for sampling with replacement the condition that no item is drawn more than once, the resulting conditional model is equivalent to the model for ordered sampling without replacement.

EXAMPLE 2. *Life table (continued)*. Let us illustrate on the life table example the procedure of computing a conditional probability model. Since the values 1 and 2 of A have been ruled out by the fact that the person has reached his 20th birthday, E consists of the simple events $A = 3, 4, \ldots, 10$. From Table 1.4.1 we compute $P(E) = P(A > 2) = .022 + .032 + \ldots + .028 = .923$. (This value could also be obtained by subtracting from 1 the probabilities $P(A = 1) = .064$ and $P(A = 2) = .013$.) Therefore $P(A = 3|A > 2) = P(A = 3)/P(A > 2) = .022/.923 = .024$, and similarly for the other values. The table below compares the probability distribution of decade of death at birth with the conditional distribution at age 20.

TABLE 1. LIFE TABLE AT BIRTH AND FOR 20 YEAR OLDS

a	$P(A = a)$	$P(A = a\|A > 2)$	a	$P(A = a)$	$P(A = a\|A > 2)$
1	.064	0	6	.127	.138
2	.013	0	7	.211	.229
3	.022	.024	8	.269	.291
4	.032	.035	9	.170	.184
5	.064	.069	10	.028	.030

We still must check that (1) really defines a probability model. The quantities (1) are certainly nonnegative; it is therefore only necessary to show that the sum of the probabilities $P(e|E)$ extended over all simple events of the event set ε is equal to 1. Since the zero probabilities of the simple events outside E do not contribute anything, it is enough to add the probabilities

$$(2) \qquad P(e|E) = \frac{P(e)}{P(E)}$$

over all simple events e in E.   In doing so, we can place the fraction over the common denominator $P(E)$ and observe that the numerator, which is the sum of the probabilities $P(e)$ for all e in E, is just $P(E)$.   We thus find that the sum of the terms (2) for all e in E is

$$\frac{P(E)}{P(E)} = 1,$$

as was to be proved.

Conditional models are appropriate only under more restrictive circumstances than may at first be apparent. The requirements are that (i) there is an original unconditional model that is realistic for a sequence of repetitions of a random experiment, and (ii) partial information eliminates from this sequence all trials yielding results not in R (the outcome on which we are conditioning) without eliminating any of the remaining trials.   In practice, the partial information will often not only eliminate all trials of the non-R type, but also selectively some of the others.   If this selection occurs at the same rate for each trial of the R-type, so that each has the same chance of being eliminated, the conditional model will continue to be realistic.   It is, however, hard to be sure that "differential selection" is not at work.

To illustrate the difficulty, let us reconsider Example 2.   The original life table is assumed to be realistic for a certain well-specified population, that is, for a certain sequence of observable life spans.   The conditional table, given that $A \geq 2$, is then appropriate for the subsequence consisting of *all* those life spans in the original sequence that do not terminate in the first two decades.   In most practical applications, however, one will be concerned with a smaller subsequence consisting of only some of the members of the population who reach age 20.   The realism of the conditional model then depends on whether the sub-sequence in question is representative of the subsequence of all members of the population who reach age 20.   This is usually not an easy question to answer.

Suppose that an individual is admitted to a hospital on his twentieth birthday to undergo tests for leukemia.   Everyone would agree that a sequence of such cases is not well-represented by our conditional model.   Next, suppose that someone appears on his twentieth birthday at an insurance office to purchase a large policy on his life.   Some insurance agents suspect that the survival chances of such persons are below those of the totality of persons reaching age twenty (while purchasers of annuities tend to live longer than the average person).   If this suspicion is correct, the conditional model would again not be realistic for a sequence of such cases.   Finally, suppose that we are attending the twentieth-birthday party of a friend.   Are our friends representative of the whole population covered by the table, or do they perhaps belong to a class with above-normal life expectancy?

EXAMPLE 6.   *Two-child families.*   As a second example of the problem of differential selection, suppose that a two-child family is moving into the neighborhood. We learn that not both children are girls.   Under this condition, what is the probability that both are boys?   Let us distinguish four family types, say BB, BG, GB, GG, where for example BG represents "older child boy, younger child girl."

If we ignore the slightly higher frequency of boys, we might reasonably use the following as the original model.

Family type	BB	BG	GB	GG
Probability	$\frac{1}{4}$	$\frac{1}{4}$	$\frac{1}{4}$	$\frac{1}{4}$

(We are here tacitly assuming that the sequence of two-child families moving into our neighborhood is representative of all such families with respect to sex of children. This in itself may be questionable, but it is not the point at issue here.)

Consider now the partial information: the family is not GG. The conditional model is then, by (1):

Family type	BB	BG	GB	GG
Probability	$\frac{1}{3}$	$\frac{1}{3}$	$\frac{1}{3}$	0

Whether this is realistic depends on how the partial information comes to our attention. If it comes to us in a way that would eliminate any two-girl family but would preserve all other cases, then the conditional model is reasonable. But if the partial information implies a selection among the non-GG families, the conditional model would be realistic only if the selection affects the other three types equally.

For example, suppose the partial information consists in our hearing that the father of the family paid a call on the Director of Admissions of a private school for boys, which is interpreted (correctly) to mean that he has at least one son. This information does eliminate the GG possibility—but beyond that, it restricts attention to families which at least consider enrolling a son in this school. In the subsequence of such families, do BB cases occur with the same frequency as for example BG cases? Not if enrollment in the boys' school is more likely when there are two boys in the family than when there is only one, as may well be the case (if only because it is then more likely that at least one boy will be of school age).

To change the circumstances of the example, suppose the partial information consists in our happening to see one of the children, and he is a boy. Logically, "There is at least one boy" is the same as "Not both are girls," and it is therefore tempting to apply our conditional model. But in fact, if we select from the sequence of all two-child families moving into the neighborhood the subsequence of those in which the first child seen is a boy, will the BB and BG families be equally represented in this subsequence? Clearly not, since the selection will eliminate those BG families in which the first child seen is the younger one without eliminating any of the BB families. (For an alternative model in this case, see Problem 3.17.)

PROBLEMS

1. What is the probability that the loaded die of Example 4 will show two points if we know only that the number of points showing is greater than one?

2. Let the sample space consist of five points with probabilities $P(e_1) = .1$, $P(e_2) = .15$, $P(e_3) = .2$, $P(e_4) = .2$, $P(e_5) = .35$. Let $E = \{e_1, e_3, e_4\}$. Find the conditional probability model given E, and check that it is a model.

3. Recall Problem 3.2.1. If the instructor picks up one of the tests at random and finds that the number of points on the first problem is between 21 and 30, what is the conditional probability that the number of points on the second problem will also be between 21 and 30?

4. Suppose we are observing a life span for which Table 1.4.1 is appropriate and that the individual in question has survived to his 60th birthday.
  (i) Construct the conditional probability model.
  (ii) What is the conditional probability that the individual will survive to the age of 80?

5. Two fair dice are thrown and we are told that the sum of points is 4.
  (i) Construct the conditional probability model given this information.
  (ii) Find the conditional probability P(number of points showing on first die $\leq 2|$ sum of the number of points $= 4$).

6. Solve the two parts of the preceding problem if we are told that the sum of points is ≤ 4.

7. Suppose that of two unrelated tosses with a coin whose probability of heads is p, one falls heads and the other tails. Show that the conditional probability that the first one is heads is $\frac{1}{2}$. [Hint: First construct the appropriate product model for the original experiment.]

8. Suppose a poker hand is drawn and it is observed that all five cards are red. In the conditional model, what is the probability of
  (i) a hand consisting of any five specified red cards;
  (ii) a heart flush?

9. Show that, if the original model is uniform, the nonzero probabilities in any conditional model will all be equal.

10. (i) Suppose that $P(e_2)/P(e_1) = \frac{1}{2}$ and that e_1, e_2 both belong to E. What can you say about $P(e_2|E)/P(e_1|E)$?
  (ii) Formulate a generalization of the result found in part (i).
  (iii) Use part (ii) to give an alternative solution of Problem 9.

11. (i) In Example 5, construct the conditional probability model given that the first draw produces marble B.
  (ii) Find P(2nd draw $=$ A|1st draw $=$ B).

12. (i) In Example 5, construct the conditional probability model given that the 2nd draw produces marble B.
  (ii) Find P(1st draw $=$ A|2nd draw $=$ B).

13. (i) In Example 5, construct the conditional model given that the first draw produces either marble A or B.
  (ii) Find P(2nd draw $=$ B|1st draw $=$ A or B).
  (iii) Find P(2nd draw $=$ C|1st draw $=$ A or B).

14. (i) In Example 5, construct the conditional model given that marble A occurred on at least one of the two draws.
  (ii) Find P(both draws produced marble A|A occurred on at least one of the two draws).

4.2 INDEPENDENCE

In the preceding section we have seen how to modify a probability model to obtain the conditional model corresponding to a given event E. We shall now derive an important formula which expresses the probability of a composite event in the conditional model in terms of probabilities in the original model.

Let F be any composite event in $\mathcal{E}$. The probability attached to F in the conditional probability model (1.1) will be called *the conditional probability of F given E* and will be denoted by $P(F|E)$. By the general definition of the probability of a composite event (Section 1.4), $P(F|E)$ is just the sum of the conditional probabilities $P(e|E)$ for all the simple events e in F. Of course, the simple events that are in F but outside E will contribute nothing to the sum. It is therefore enough to add up the probabilities $P(e|E)$ for the simple events that are in both E and F, that is, for the simple events in (E and F).

When adding up the numbers $P(e|E) = P(e)/P(E)$ for all e in (E and F), we can place the fractions over the common denominator $P(E)$ and observe that the numerators add up to $P(E$ and $F)$. This proves the important formula

$$(1) \qquad\qquad P(F|E) = \frac{P(E \text{ and } F)}{P(E)}$$

which permits one to calculate conditional probabilities without first constructing the conditional model.

The reader may find it helpful to trace through the argument leading to (1) in the example of Figure 1.6.2a. There F consists of e_2, e_4, e_6, and e_7. In the conditional distribution given E, we have $P(e_2|E) = P(e_7|E) = 0$, while $P(e_4|E) = P(e_4)/P(E)$ and $P(e_6|E) = P(e_6)/P(E)$. Therefore $P(F|E) = 0 + 0 + P(e_4)/P(E) + P(e_6)/P(E) = [P(e_4) + P(e_6)]/P(E) = P(E \text{ and } F)/P(E)$.

EXAMPLE 1. *Poker hands.* What is the conditional probability that a poker hand is a heart flush, given that it consists only of red cards? We shall assume that the hand constitutes a sample of 5 cards drawn at random from a deck of 52, so that all $\binom{52}{5}$ possible hands are equally likely. The event E represents the result "all five cards in the hand are red" and F the result "the hand is a heart flush," so that the event (E and F) stands for the result "the hand is a heart flush." Clearly the number of simple events in E is $\binom{26}{5}$ so that

$$P(E) = \binom{26}{5} \bigg/ \binom{52}{5}.$$

The number of heart flushes is $\binom{13}{5}$, and hence

$$P(E \text{ and } F) = \binom{13}{5} \Big/ \binom{52}{5}.$$

The desired probability is therefore

$$P(F|E) = P(E \text{ and } F)/P(E) = \binom{13}{5} \Big/ \binom{26}{5}.$$

Using Table A, we find this to be

$$1287/65{,}780 = .0196.$$

While not a large number, this is relatively much greater than the probability of a heart flush when nothing is known about the hand, which in Example 2.2.1 was found to be .000495. These calculations support the feeling expressed at the end of Example 1.1.

The conditional probability $P(F|E)$ has a frequency interpretation exactly analogous to that of $P(e|E)$ in Section 1. Let R and S be the results of the experiment represented in the event set $\mathcal{E}$ by E and F. Then in a long sequence of trials of the experiment, one has (in the notation of Section 3.2)

$$P(E) \sim f(R) \quad \text{and} \quad P(E \text{ and } F) \sim f(R \text{ and } S),$$

and hence by (1)

$$P(F|E) \sim \frac{f(R \text{ and } S)}{f(R)} = \frac{\#(R \text{ and } S)}{\#(R)}.$$

Here the right-hand side is equal to the frequency with which the result S occurs among those trials in which the result R occurred; that is, the *conditional frequency* $f(S|R)$ *of S given R*. The last displayed relation may therefore be rewritten as

$$(2) \qquad\qquad P(F|E) \sim f(S|R).$$

With a realistic model, a conditional probability is approximately equal to the corresponding conditional frequency.

Equation (1), on multiplication of both sides by $P(E)$, becomes

$$(3) \qquad\qquad P(E \text{ and } F) = P(E)P(F|E).$$

In this form, the equation is known as the *multiplication law*. Of course, the two forms are equivalent (except that the present form is valid even when $P(E) = 0$), but their uses are entirely different. Equation (1) is used to compute conditional probabilities from the probabilities of the original model, while the multiplication law (3) is used to compute the (unconditional) probability $P(E \text{ and } F)$ when we know $P(E)$ and the conditional probability $P(F|E)$. We shall give several illustrations of the use of (3) in the next section.

Both $P(F)$ and $P(F|E)$ are probabilities of the event F, but they are the probabilities assigned to this event by two different models, and consequently they will in general have different values. However, as we shall see below, there is an important class of cases in which

$$(4) \qquad\qquad P(F|E) = P(F).$$

When relation (4) holds, the imposition of the condition E does not change the probability of F, and we then say that F is *independent* of E.

EXAMPLE 2. *Heart face card.* Consider the experiment of drawing a card from a shuffled bridge deck, where each of the 52 cards is assigned probability $\frac{1}{52}$ of being drawn. Let E be the event corresponding to the result that the card drawn is a heart while F is the event corresponding to the draw of a face card (King, Queen, or Knave). We see that $P(E) = \frac{13}{52}$, $P(F) = \frac{12}{52} = \frac{3}{13}$, and $P(E \text{ and } F) = \frac{3}{52}$. Thus, by (1), $P(F|E) = (\frac{3}{52}) \div (\frac{13}{52}) = \frac{3}{13} = P(F)$, so that F is independent of E. This means that the frequency with which a face card is drawn among all trials is approximately the same as the conditional frequency with which a face card is drawn on those trials resulting in a heart.

It is interesting to note that, in the above example, $P(E|F) = (\frac{3}{52}) \div (\frac{3}{13}) = \frac{13}{52} = P(E)$, so that it is also true that E is independent of F. This illustrates the general phenomenon that independence is symmetric: if F is independent of E, then* E is also independent of F. In fact, it is easy to see (Problem 12) that each of the conditions $P(F|E) = P(F)$ and $P(E|F) = P(E)$ is equivalent to the symmetric condition

$$(5) \qquad\qquad P(E \text{ and } F) = P(E)P(F).$$

Definition. Two events E and F are said to be *independent* (of each other) if they satisfy condition (5).

The concept of independence finds its most important use in connection with experiments having several parts (Section 3.1). In fact, as was pointed out in Section 3.2, if such an experiment is represented by a product model, and if E and F are events relating to different factors of the model, then E and F are independent.

When applying a product model to a specific experiment, the usual approach (Section 3.1) is first to build the factor models and then to combine them by means of formula (3.1.4) into a product model. The probabilities such as $P(E)$ and $P(F)$ may then be thought of either as probabilities in the factor models α and $\mathcal{B}$ or as the equivalent probabilities in model $\alpha \times \mathcal{B}$. It is usually convenient to consider $P(E)$ and $P(F)$ as probabilities in the factor models, and then use (5) to obtain the probability $P(E \text{ and } F)$ in the product model, without the necessity of explicitly constructing the latter.

---

* Strictly speaking, these statements are correct only if we exclude the trivial cases $P(E) = 0$ and $P(F) = 0$, in which $P(F|E)$ and $P(E|F)$ are not defined.

EXAMPLE *3*. *Stratified sampling.* Consider a residential block having six houses on the north side and seven houses on the south side (Figure 1).

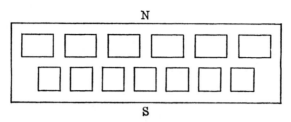

N

S

FIGURE 1. SKETCH MAP OF BLOCK

A sample of four houses is to be drawn in a survey to check the accuracy of tax assessments. The survey leader notices that the houses on the north side are more substantial than those on the south side. If a sample of four is drawn randomly from all 13 houses, it could happen by chance that all four in the sample come from one side or the other. The sample could thus lead to a quite mistaken idea of the block as a whole, as discussed in Example 3.1.1.

To forestall this difficulty, it may instead be decided to draw a sample of two houses from each side. The drawing of two north-side houses may be considered as part A, and the drawing of two south-side houses as part B, of a two-part experiment. For part A, a model α can be built as in Section 2.1, in which the $\binom{6}{2} = 15$ possible samples of two north-side houses are regarded as equally likely. Similarly, in model $\mathcal{B}$ the $\binom{7}{2} = 21$ possible samples of two south-side houses are treated as equally likely. Finally, if the samples are drawn by separate applications of some random mechanism, it seems reasonable to suppose that the frequencies of the two parts are unrelated, and to represent the experiment as a whole by the product of the two models.

As an illustration of this product model, let us find the probability that none of the corner houses appear in the stratified sample. Let E represent the result "no corner house appears in the sample from the north side." Then $P(E) = \binom{4}{2} / \binom{6}{2} = \frac{6}{15} = \frac{2}{5}$. Similarly, if F represents "no corner house appears in the sample from the south side," $P(F) = \binom{5}{2} / \binom{7}{2} = \frac{10}{21}$. According to (5), the probability that no corner house appears in either sample is then given by

$$P(E \text{ and } F) = P(E) \cdot P(F) = \tfrac{2}{5} \cdot \tfrac{10}{21} = \tfrac{4}{21} = .190.$$

EXAMPLE 4. *Two throws of a die.* What is the probability that in two unrelated throws with a fair die we will get at least five points on each throw? Let E represent "at least five points on the first throw," and F represent "at least five points on the second throw." In the factor models, we have $P(E) = P(F) = \frac{2}{6} = \frac{1}{3}$ and $P(E$ and $F) = P(E) \cdot P(F) = (\frac{1}{3})(\frac{1}{3}) = \frac{1}{9}$. This result can also be obtained directly from the uniform model of Example 1.5.3 by counting cases.

When the two sides $P(F|E)$ and $P(F)$ of (4) are not equal, the event F is said to *depend* on the event E. When this occurs, if the model is realistic the corresponding frequencies $f(S|R)$ and $f(S)$ will also be unequal. Suppose for example that $f(S|R)$ is much larger than $f(S)$; i.e., that S occurs much more frequently in those cases when R occurs, than it does in general. It is then tempting to conclude that the event R "causes" the event S to tend to occur. It has, for example, been observed that the conditional frequency of lung cancer is much higher among persons who are heavy cigarette smokers than in the population at large. Again, the conditional frequency of cavities is lower among children in communities with a high fluorine content in the water supply. It seems natural to conclude that cigarette smoking tends to cause lung cancer and that fluorine helps to prevent tooth decay. Many useful causal relations have been discovered as a result of noticing such empirical frequency relations, but it is important to realize the pitfalls of the argument. When results R and S tend to go together, it may not mean that R causes S, but rather that S causes R, or that both are causally connected with other factors while not exerting any direct influence on each other. Some examples may make this clear.

EXAMPLE 5. *Sampling without replacement.* Two marbles are drawn without replacement from a box of three marbles, one of which is red. The result "red on first draw" will occur in about one third of the trials. But its conditional frequency, given that "red on second draw" has occurred, will be zero. Yet we would not want to say that "red on second draw" exerts a causal influence on "red on first draw," since cause operates only forward in time.

EXAMPLE 6. *Insulin and diabetes.* It may be observed that the conditional frequency of diabetes is much higher among persons taking insulin than in the general population. Should one conclude that insulin causes diabetes?

EXAMPLE 7. *Income and politics.* When a college faculty was voting on a controversial issue, it was found that among the faculty members in the upper half of the income bracket a much higher proportion took the conservative side than among those in the lower half. This does not prove that higher income tends to make a person more conservative. The observations may reflect mainly the fact that as persons get older they tend toward both more conservative outlooks and higher income.

EXAMPLE 8. *Smoking and lung cancer.* As mentioned above, the incidence of lung cancer is much higher among persons who are heavy cigarette smokers than in the population at large. This fact, taken by itself, would equally well support the conclusions either that lung cancer causes smoking or that both are caused by some common constitutional or environmental factors. The mechanism causing the dependence may in part be like that of Example 7. For example, it is known that people in cities smoke more heavily than people on farms, and that the air in cities contains more carcinogens than rural air. The rural-urban difference is thus a factor that could tend to increase the dependence between smoking and cancer, and there may be other factors of a similar nature. We do not wish to imply that these considerations exonerate smoking as a cause of lung cancer, but merely want to emphasize once again that dependence does not imply a causal connection. Causal relationships can sometimes be demonstrated by means of experiments of a kind to be discussed in Chapters 12 and 13. The suspected cause must be applied in some cases and not in others, this assignment being made with the aid of a random mechanism.

PROBLEMS

1. In the product model $\alpha \times \mathfrak{B}$ specified by (3.1.1) and (3.1.4), list the simple events of which the event (E and F) is composed, and check the validity of equation (5) for the following cases.

  (i) $E = \{e_2, e_3\}$ and $F = \{f_1, f_2, f_3\}$;

  (ii) $E = \{e_2, e_3, e_4\}$ and $F = \{f_1, f_3, f_5\}$.

2. Suppose you attend two classes consisting of eight and ten students respectively. In the first class, two students are selected at random for recitation, and in the second class three students are selected at random, without reference to the selection in the first class.

  (i) Construct a probability model for this experiment.

  (ii) Use this model to find the probability of your not being called upon for recitation in either class.

3. Suppose that in the situation of the preceding problem, you have the choice between the following two schemes:

  (i) three students are selected from each class;

  (ii) two students are selected from the first and four from the second class.

If you are anxious not to be called, which scheme would you prefer?

4. In Example 3, suppose that four houses were drawn from the 13 houses on the block without stratification. What are the probabilities that

     (i) all four come from the same side of the block;
     (ii) just two come from each side of the block;
     (iii) no corner house is chosen?

5. In Example 3, suppose that there are seven houses on the north side and nine on the south side. Find the probability that a sample of five houses does not contain any of the corner houses if
     (i) the sample is a stratified sample, consisting of two houses drawn at random from those on the north side and three from those on the south side;
     (ii) the sample is drawn at random without stratification from all 16 houses.

6. A classroom consists of two rows: in the first row, there are three freshmen and six sophomores; in the second row, there are seven freshmen and two sophomores. Find the probability that a sample of four consists of freshmen only if
     (i) the sample is a stratified sample consisting of two students drawn at random from each row;
     (ii) the sample is drawn at random without stratification from all 18 students.

7. In the preceding problem, find the probability that the four students in the sample are either all freshmen or all sophomores, for the two sampling schemes.

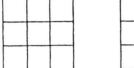

FIGURE 2.

8. Two fields contain nine and twelve plots respectively, as shown in Fig. 2. From each field, one plot is selected at random for a soil analysis.
     (i) Construct a probability model for this experiment.
     (ii) What is the probability that both selected plots are corner plots?
     (iii) What is the probability that at least one of the selected plots is a corner plot?

9. Twenty percent of the patients on which a certain heart operation is performed die during the operation. Of those that survive, ten percent die from the after-effects. What is the over-all proportion of cases dying from one or the other of these causes?

10. Let E and F be exclusive. Show that $P(F|\overline{E}) = P(F)/[1 - P(E)]$. Is this relation necessarily correct if E and F are not exclusive?

11. For two unrelated tosses with a coin, let E, F stand for

$$E: \text{heads on first toss}$$
$$F: \text{heads on one toss and tails on the other.}$$

Determine whether E and F are independent
     (i) if the coin is fair;
     (ii) if the coin is not fair.
[Hint: See Problem 1.7.]

12. A card is drawn at random from a bridge deck. Determine whether the events "red" and "has an even number of spots" are independent.

13. A box contains three red and four white marbles. Two marbles are drawn with replacement. Assuming the $7^2 = 49$ choices for the two marbles to be equally likely, determine whether the events "red on first draw" and "white on second draw" are independent.

14. At a lecture there are four freshman boys, six freshman girls and six sophomore boys. How many sophomore girls must be present if sex and class are to be independent when a student is drawn at random?

15. Three square fields each contain nine plots as in the left-hand side of Figure 2. From each, a plot is selected at random for a soil analysis, the selections being performed without reference to each other.

(i) Construct a probability model for this experiment.

(ii) What is the probability that all three selected plots are corner plots?

16. A classroom consists of three rows, with a distribution of freshmen, sophomores and juniors as indicated

	freshmen	sophomores	juniors
1st row	5	2	1
2nd row	2	5	2
3rd row	2	2	5

Find the probability that a sample of five consists either only of freshmen or only of sophomores or only of juniors if

(i) the sample is a stratified sample consisting of one student drawn at random from the first row and two students drawn at random each from the second and from the third row;

(ii) the sample is drawn at random without stratification.

17. Solve the preceding problem if the distribution of the students is as follows

	freshmen	sophomores	juniors
1st row	5	2	2
2nd row	2	4	3
3rd row	2	2	5

and if

(i) the sample is a stratified sample consisting of two students drawn at random from each row;

(ii) a sample of six students drawn at random without stratification.

[Hint: Use relation (2.5.5) to extend Table A as far as is necessary for part (ii).]

18. Solve the problem of Example 4 for two unrelated throws with a loaded die, if for each throw separately the model (1.3.1) is appropriate.

19. Find the probability of at least one six in three unrelated tosses with a fair die. [Hint: Find the probability of the complementary outcome.]

20. Five judges each are asked to compare three brands of cigarettes A, B, C. Suppose that the three brands are indistinguishable so that each of the judges without any relation to the decisions of the other judges is equally likely to rank the brands in any of the six possible orders ABC, ACB, BAC, BCA, CAB, CBA. What is the probability that all judges (i) assign the ranking ABC? (ii) rank A highest, (iii) prefer A to B?

21. (i) Use (1) to prove that the relation $P(F|E) = P(F)$ is equivalent to (5) provided $P(E)$ is not zero.

    (ii) Prove that the relation $P(E|F) = P(E)$ is equivalent to (5) provided $P(F)$ is not zero.

22. Show that if E, F are independent, then E, $\overline{F}$ are independent. [Hint: $P(E \text{ and } \overline{F}) = P(E) - P(E \text{ and } F).$]

23. Show that if E, F are independent, then $\overline{E}$, $\overline{F}$ are independent. [Hint: Use Problem 22 twice.]

24. Show that if E and F are exclusive, they cannot be independent unless either $P(E) = 0$ or $P(F) = 0$.

4.3 TWO-STAGE EXPERIMENTS

Many interesting random experiments are performed in successive stages, with the chance results of each stage determining the conditions under which the next stage will be carried out. We shall in this section present the main ideas involved in building probability models for the simplest such experiments, those carried out in only two stages. A few illustrations will indicate the nature of the applications.

(a) When drawing a sample of the labor force of an industry, it is frequently convenient first to draw a sample of factories, and then to visit the selected factories to draw samples of workers. The first stage is the sampling of factories, the second stage the sampling of workers within the selected factories. This is an example of the method of *two-stage sampling*.

(b) In genetic experiments, animals may be bred to produce a first generation of offspring. These in turn are bred to produce a second generation. The chance inheritance of genes by the first generation determines the genetic constitutions involved in breeding the second generation (see Section 5).

(c) When designing an experiment it often happens that not enough is known about the situation to permit an efficient design, or even to decide whether a full-scale experiment is worthwhile. In such cases it may pay to conduct a preliminary or pilot experiment whose results may be used to make this decision and to plan the main experiment if one proves necessary. This is known as a *two-stage sequential design*, the first stage consisting of the pilot study, and the second stage of the main experiment.

(d) When checking on the quality of a lot of mass-produced articles, it is frequently possible to decrease the average sample size by carrying out the. inspection in two stages. One may for example first take a small sample and accept the lot if all articles in the sample are satisfactory; otherwise a larger second sample is inspected.

All of these examples may be thought of as experiments having two parts, but it would be quite unrealistic to represent them by product models, as we did for the two-part experiments considered in Chapter 3. In fact, the use of a product model is equivalent to the assumption that the parts are unrelated, that is, that the probabilities for the second part are not influenced by the outcome of the first part. This assumption clearly is not appropriate in the present examples, where the results on the first stage will alter the conditions under which the second stage is carried out.

Before embarking on a general discussion of models for two-stage experiments, we shall consider a simple artificial example.

EXAMPLE 1. *The two boxes.* Suppose that we have two boxes, for example two drawers of a desk, of which one contains one red and one white marble, and the other three red and one green marble. A box is chosen at random, and from it a marble is drawn at random. What is the probability of getting a red marble?

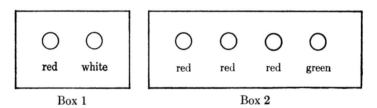

Box 1   Box 2

FIGURE 1. THE TWO BOXES

An enthusiastic devotee of "equally likely cases" might argue as follows. There are six marbles, any one of which may be drawn; since four of them are red, the probability of getting a red marble is $\frac{4}{6} = \frac{2}{3}$. To this analysis it could be objected that one has no right to treat the six marbles as equally likely. The fact that there are fewer marbles in the first than in the second box may give those in the first box a greater chance to be chosen. We shall now present a method for obtaining an alternative model.

The approach is similar to that used in the construction of product models (Sections 3.1 and 3.2). A product model $\mathcal{A} \times \mathcal{B}$ was built up by combining two simpler models $\mathcal{A}$ and $\mathcal{B}$ for parts A and B of the experiment. Analogously, let us begin here by considering a model $\mathcal{A}$ for the first *stage* of the two-stage experiment. Suppose that the possible results of the first stage are $r_1, r_2, \ldots, r_a$. These will be represented in model $\mathcal{A}$ by the events $e_1, e_2, \ldots, e_a$ with probabilities $p_1, p_2, \ldots, p_a$ satisfying

(1) $$p_1 + p_2 + \ldots + p_a = 1.$$

This model will be realistic only if its probabilities are approximately equal to the frequencies they represent, that is, if

$$(2) \qquad p_1 \sim f(r_1), \quad p_2 \sim f(r_2), \quad \ldots, \quad p_a \sim f(r_a).$$

The construction of model $\mathcal{C}$ may be illustrated by Example 1. The first stage consists of drawing a box and it has $a = 2$ results, "first box chosen" and "second box chosen." The statement that the box is chosen at random means that the boxes would be chosen with about equal frequency, which justifies assigning the values $p_1 = \frac{1}{2}$ and $p_2 = \frac{1}{2}$.

The construction of a model for the first stage has been completely parallel to that of a model for part A of a two-part experiment, with even the same notation being used. However, the analogy between the second stage and part B is less close, since a product model was based on the assumption that the two parts are unrelated while in general the second stage will depend on the outcome of the first stage. Thus, in Example 1, if box 1 has been drawn, the marble obtained on the second stage must be either red or white, while if box 2 has been drawn, it must be either red or green. The list of results of the second stage must therefore consist of the three possibilities "red," "white," "green" if it is to cover all cases that could arise. In general, for the second stage the list of results, say s_1, $s_2, \ldots, s_b$, is a comprehensive list which covers the various second-stage experiments that may arise. These results will be represented by the events $f_1, f_2, \ldots, f_b$. (In Example 1, let f_1 represent "red," f_2 "white," and f_3 "green.")

Because of the dependence of the experimental conditions of the second stage on the result of the first stage, we have to build a separate second-stage model corresponding to each possible result of the first stage. Let us see for example how to build a model $\mathcal{B}^{(1)}$ for the second stage when the result of the first stage is r_1. The frequencies with which the results $s_1, s_2, \ldots, s_b$ occur, among the cases in which r_1 has occurred on the first stage, are the conditional frequencies $f(s_1|r_1), f(s_2|r_1), \ldots, f(s_b|r_1)$. Let us denote the corresponding probabilities by $q_1^{(1)}, q_2^{(1)}, \ldots, q_b^{(1)}$, where of course we must have

$$(3) \qquad q_1^{(1)} + q_2^{(1)} + \ldots + q_b^{(1)} = 1.$$

The model

$$\mathcal{B}^{(1)}: \begin{array}{c} f_1, \quad f_2, \quad \ldots, f_b \\ q_1^{(1)}, q_2^{(1)}, \ldots, q_b^{(1)} \end{array}$$

will be a realistic model for the second stage given that the result of the first stage is r_1, provided its probabilities are approximately equal to the corresponding conditional frequencies:

$$(4) \qquad q_1^{(1)} \sim f(s_1|r_1), \quad q_2^{(1)} \sim f(s_2|r_1), \quad \ldots, \quad q_b^{(1)} \sim f(s_b|r_1).$$

Similarly, when the result of the first stage is r_2, a model $\mathcal{B}^{(2)}$ can be built, which assigns to $f_1, f_2, \ldots, f_b$ the probabilities $q_1^{(2)}, q_2^{(2)}, \ldots, q_b^{(2)}$; and so forth up to model $\mathcal{B}^{(a)}$ when the result of the first stage is r_a.

These models may again be illustrated by Example 1, with model $\mathcal{B}^{(1)}$ corresponding to the cases in which box 1 is selected on the first stage, and $\mathcal{B}^{(2)}$ to the cases in which box 2 is selected. In those trials of the experiment in which box 1 is selected, a marble is drawn at random from a box containing one red and one white marble. It is natural to expect that these marbles will occur with about equal (conditional) frequency, which justifies the assignment in model $\mathcal{B}^{(1)}$ of

$$q_1^{(1)} = \tfrac{1}{2}, \quad q_2^{(1)} = \tfrac{1}{2}, \quad q_3^{(1)} = 0.$$

(The zero probability reflects the fact that there are no green marbles in box 1.) Similarly, appropriate probabilities for model $\mathcal{B}^{(2)}$ are

$$q_1^{(2)} = \tfrac{3}{4}, \quad q_2^{(2)} = 0, \quad q_3^{(2)} = \tfrac{1}{4}.$$

Since the record of a two-stage experiment as a whole consists of noting the results of both stages, the simple results may conveniently be represented by the events

(5)
$$
\begin{array}{llcl}
e_1 \text{ and } f_1 & e_1 \text{ and } f_2 & \ldots & e_1 \text{ and } f_b \\
e_2 \text{ and } f_1 & e_2 \text{ and } f_2 & \ldots & e_2 \text{ and } f_b \\
\cdots & \cdots & & \cdots \\
e_a \text{ and } f_1 & e_a \text{ and } f_2 & \ldots & e_a \text{ and } f_b
\end{array}
$$

where (e_1 and f_1) represents the result "r_1 and s_1," (e_1 and f_2) represents the result "r_1 and s_2," etc. (This is the same event set that was used with the product model.) It may happen in the present case that some of the events correspond to results which are impossible (e.g. "box 1 and green"), but since our model will assign probability zero to such events, no trouble will result.

What probabilities should be assigned to the simple events of (5), for example to (e_1 and f_2)? If the model is to be realistic, $P(e_1$ and $f_2)$ must be close to $f(r_1$ and $s_2)$. This frequency, in a sequence of n trials, may be written as

$$f(r_1 \text{ and } s_2) = \frac{\#(r_1 \text{ and } s_2)}{n} = \frac{\#(r_1)}{n} \cdot \frac{\#(r_1 \text{ and } s_2)}{\#(r_1)},$$

so that

(6) $$f(r_1 \text{ and } s_2) = f(r_1)f(s_2|r_1).$$

If models $\mathcal{C}$ and $\mathcal{B}^{(1)}$ are realistic, we have by (2) and (4)

$$f(r_1) \sim p_1 \quad \text{and} \quad f(s_2|r_1) \sim q_2^{(1)}$$

and hence

$$f(r_1 \text{ and } s_2) \sim p_1 q_2^{(1)}.$$

This shows that $p_1 q_2^{(1)}$ will, under the assumptions made, be a realistic value for $P(e_1 \text{ and } f_2)$. Application of similar considerations to the other cases leads to the assignment

(7)
$$
\begin{aligned}
P(e_1 \text{ and } f_1) &= p_1 q_1^{(1)} & P(e_1 \text{ and } f_2) &= p_1 q_2^{(1)} & \ldots & & P(e_1 \text{ and } f_b) &= p_1 q_b^{(1)} \\
P(e_2 \text{ and } f_1) &= p_2 q_1^{(2)} & P(e_2 \text{ and } f_2) &= p_2 q_2^{(2)} & \ldots & & P(e_2 \text{ and } f_b) &= p_2 q_b^{(2)} \\
&\ldots & &\ldots & & & &\ldots \\
P(e_a \text{ and } f_1) &= p_a q_1^{(a)} & P(e_a \text{ and } f_2) &= p_a q_2^{(a)} & \ldots & & P(e_a \text{ and } f_b) &= p_a q_b^{(a)}
\end{aligned}
$$

The model for two-stage experiments defined by (5) and (7) will be denoted by model 3.

EXAMPLE 1. *The two boxes* (*continued*). Let us illustrate (7) with Example 1, where e_1 and e_2 represent "box 1" and "box 2," while f_1, f_2, f_3 represent "red," "white," and "green." The probabilities are

(8)
$$
\begin{aligned}
P(e_1 \text{ and } f_1) &= \tfrac{1}{2} \cdot \tfrac{1}{2} = \tfrac{1}{4}; & P(e_1 \text{ and } f_2) &= \tfrac{1}{2} \cdot \tfrac{1}{2} = \tfrac{1}{4}; \\
& & P(e_1 \text{ and } f_3) &= \tfrac{1}{2} \cdot 0 = 0 \\
P(e_2 \text{ and } f_1) &= \tfrac{1}{2} \cdot \tfrac{3}{4} = \tfrac{3}{8}; & P(e_2 \text{ and } f_2) &= \tfrac{1}{2} \cdot 0 = 0; \\
& & P(e_2 \text{ and } f_3) &= \tfrac{1}{2} \cdot \tfrac{1}{4} = \tfrac{1}{8}.
\end{aligned}
$$

Here for example the assignment of probability $\tfrac{3}{8}$ to $(e_2 \text{ and } f_1)$ reflects our assumptions that box 2 will be chosen in about $\tfrac{1}{2}$ of the trials, and that in those cases when box 2 is chosen, a red marble will be drawn about $\tfrac{3}{4}$ of the time.

Having built the model, we can now answer the original question and find the probability that the marble drawn is red. The event corresponding to the result of getting a red marble consists of the first column of (8) and its probability is therefore $\tfrac{1}{4} + \tfrac{3}{8} = \tfrac{5}{8}$.

Two different models (the uniform model discussed at the beginning of the section and that defined by (8)) have given to the probability of getting a red marble the values $\tfrac{2}{3}$ and $\tfrac{5}{8}$. Which is right? As usual, the only sure test is experimental. The frequency with which a red marble is obtained depends on just how the boxes and marbles are chosen. Many persons approaching a desk with two drawers side by side might tend to open the one on the right. Again, if the red marbles are larger than the others, this might tend to favor their choice. But if the box, and then the marble from the box, is chosen with equal probabilities, for example by means of a table of random numbers, we would expect to get a red marble with frequency close to $\tfrac{5}{8}$.

EXAMPLE 2. *Sex of twins.* Consider the "experiment" of observing the sex of the first-born and second-born of a pair of human twins. We have pointed out in Example 3.2.3 that a product model is not suitable for analyzing this experiment, and shall now see how it may be represented by a two-stage model.

Twins are of two types: *identical* twins originate from a single fertilized egg and are consequently always of the same sex; *fraternal* twins originate from two separately fertilized eggs, and their sexes are unrelated. Approximately one third of all human twin pairs are identical. If we treat the type of twin as the result of the first stage of the experiment, and represent identical twins by e_1 and fraternal twins by e_2, then a reasonable model α is obtained by setting $p_1 = \frac{1}{3}$, $p_2 = \frac{2}{3}$.

We shall view the observation of the sex of first- and second-born twins as the result of the second stage. Just as in Example 1.5.2 and Problem 3.1.2, we shall represent the four possibilities by f_1: MM, f_2: MF, f_3: FM, f_4: FF, where for instance MF means "first born is male and second born is female." Observations show that the following probabilities are roughly correct:

	f_1: MM	f_2: MF	f_3: FM	f_4: FF
$\mathcal{B}^{(1)}$:	$q_1^{(1)} = \frac{1}{2}$	$q_2^{(1)} = 0$	$q_3^{(1)} = 0$	$q_4^{(1)} = \frac{1}{2}$
$\mathcal{B}^{(2)}$:	$q_1^{(2)} = \frac{1}{4}$	$q_2^{(2)} = \frac{1}{4}$	$q_3^{(2)} = \frac{1}{4}$	$q_4^{(2)} = \frac{1}{4}$

The zero values assigned to $q_2^{(1)}$ and $q_3^{(1)}$ reflect the fact mentioned above that identical twins, arising from a single fertilized egg, cannot be of mixed sex. At the time of fertilization this egg, like any other, is about equally likely to receive a male-determining or a female-determining sperm, which explains the assignments $q_1^{(1)} = \frac{1}{2}$, $q_4^{(1)} = \frac{1}{2}$. With fraternal (two-egg) twins, the two fertilizations are unrelated so that frequencies represented in the second row behave like those in ordinary two-child families.

Now let us combine models α, $\mathcal{B}^{(1)}$, and $\mathcal{B}^{(2)}$. The resulting probabilities are, in the usual rectangular array (7)

	f_1: MM	f_2: MF	f_3: FM	f_4: FF
e_1: identical	$\frac{1}{3} \cdot \frac{1}{2} = \frac{1}{6}$	$\frac{1}{3} \cdot 0 = 0$	$\frac{1}{3} \cdot 0 = 0$	$\frac{1}{3} \cdot \frac{1}{2} = \frac{1}{6}$
e_2: fraternal	$\frac{2}{3} \cdot \frac{1}{4} = \frac{1}{6}$	$\frac{2}{3} \cdot \frac{1}{4} = \frac{1}{6}$	$\frac{2}{3} \cdot \frac{1}{4} = \frac{1}{6}$	$\frac{2}{3} \cdot \frac{1}{4} = \frac{1}{6}$

From this model we see at once that the probability that both twins are male is $\frac{1}{6} + \frac{1}{6} = \frac{1}{3}$. For comparison, with ordinary two-child families the probability that both children are male is about $\frac{1}{4}$ (Problem 3.1.2).

Several other applications of two-stage models are given in the next section.

PROBLEMS

1. In Example 1, find the probability that
   (i) the only white marble is selected,
   (ii) the only green marble is selected.
Give an intuitive explanation of why these probabilities differ.

2. In Example 1, what value of p_1 would make the six marbles equally likely to be drawn?

3. Solve parts (i) and (ii) of Problem 1 for each of the following three situations:
  (i) box 1 contains two red and one white marble and box 2 is as in Example 1;
  (ii) box 2 contains four red and one green marble and box 1 is as in Example 1;
  (iii) box 1 contains three red and one white marble and box 2 is as in Example 1.

4. Suppose that we have two boxes, of which one contains three red and two white marbles, and the other four red and two green marbles. A box is chosen at random and from it two marbles are drawn at random without replacement. What is the probability that both are (i) white; (ii) green; (iii) red?

5. In the preceding problem, find the probability that both marbles are red if one marble is chosen at random from each of the boxes, the two drawings being unrelated.

6. A classroom consists of four rows, with the following distribution of boys and girls

row	1	2	3	4
boys	3	9	9	2
girls	8	4	5	10

To obtain a sample of three, the instructor selects three of the rows at random and one student at random from each of the selected rows. Find the probability that all three are (i) boys; (ii) of the same sex.

7. In the model of Example 2, what is the probability that the twins will be of the same sex?

8. In a game of Russian Roulette, a person selects (at random) one of three guns, each containing five chambers. The number of empty chambers is four, three, and two respectively.
  (i) Construct a model for this two-stage experiment, with the result of interest at the second stage being whether the selected chamber is empty or not.
  (ii) Find the probability that the person survives this "experiment."

9. Suppose an ordered sample of size two is drawn at random without replacement from a collection of N items, say marbles numbered $1, 2, \ldots, N$.
  (i) If $e_1, e_2, \ldots, e_N$ represent the results that the first marble drawn is marble number 1, number 2, ..., number N, what would be reasonable values for $P(e_1), P(e_2), \ldots, P(e_N)$?
  (ii) If $f_1, f_2, \ldots, f_N$ represent the results that the second marble drawn is marble number 1, number 2, ..., number N, what would be reasonable values for $P(f_1|e_1), P(f_2|e_1), \ldots, P(f_N|e_1)$? For $P(f_1|e_2), P(f_2|e_2), \ldots, P(f_N|e_2)$? Etc.
  (iii) Using (i) and (ii), build a model for this two-stage experiment.
  (iv) Compare the model of (iii) with the uniform model built for this experiment in Sections 2.3 and 2.4.

10. A firm which has eight factories in widely different locations wishes to interview a sample of its workers. For administrative convenience it is decided to interview workers from only three of the factories. A two-stage sample is therefore taken: first three of the eight factories are selected at random; then at each of the chosen factories, a random sample of 10% of the workers of that factory is obtained. In the model for this experiment given by (5) and (7):

    (i) What results are represented by $e_1, e_2, \ldots$?

    (ii) What are the possible second stages of this experiment? Is it correct to say that each possible second stage is a stratified sample?

    (iii) What results are represented by $f_1, f_2, \ldots$?

11. In Problem 2.1.18, find the probability that the selected lawyer is a specified lawyer from one of the firms consisting of (i) only one lawyer; (ii) two lawyers; (iii) three lawyers.

12. Suppose that a page in a telephone directory has four columns and that the same number (say, n) of subscribers is listed in each column. In order to select a subscriber from the page, you select at random first one of the four columns and then one of these subscribers from that column. Has the subscriber been selected at random from the page, i.e., does each subscriber have the same probability of being selected?

13. Suppose that on each play of slot machine A there is probability $\frac{1}{2}$ of winning \$1 and probability $\frac{1}{2}$ of not winning anything, while on slot machine B there is probability $\frac{1}{10}$ of winning \$1 and probability $\frac{9}{10}$ of winning nothing. Suppose further that the probability of winning on any play is not influenced by the outcome of the preceding play. Build two-stage models for the following two experiments:

    (i) A player begins with machine A; if he wins, he plays machine A a second time, otherwise he switches to machine B for his second play.

    (ii) A player begins with machine B; if he wins, he plays machine B a second time, otherwise he switches to machine A for his second play.

[Hint: As simple events of the two-stage model use win-win, win-lose, lose-win, lose-lose.]

14. (i) Build a two-stage model for the experiment of selecting at random (i.e. with probability $\frac{1}{2}$ each) one of the two machines A and B of Problem 13, playing it twice if the first play wins and otherwise switching to the other machine for the second play. [Hint: Use models (i) and (ii) of the preceding problem for the two possible second stages.]

    (ii) Find the probability of winning at least once in the two plays.

15. (i) Build a two-stage model for the experiment of selecting at random one of the slot machines of Problem 13 and playing it twice.

    (ii) Find the probability of winning at least once in the two plays.

    (iii) Which of the two methods of playing, described in Problems 14(i) and 15(i), do you prefer, and why?

[Hint: Note that a binomial trials model is appropriate for the second stage of 15(i).]

16. (i) Build a two-stage model if the first stage consists of a two-child family moving into a neighborhood, with the equally likely outcomes BB, BG, GB, GG (see Example 1.6), and the second stage of our seeing one of the children of this family, it being equally likely that this is either one of the two children.

    (ii) In this model, find the probability that the family is of type BB, given that the child we saw was a boy.

4.4 PROPERTIES OF TWO-STAGE MODELS; BAYES' LAW

We shall now derive certain properties of the model $\mathfrak{I}$ defined by (5) and (7) of the preceding section. As was the case with the product model $\mathfrak{A} \times \mathfrak{B}$, the symbols $e_1, e_2, \ldots, f_1, f_2, \ldots$ are being used in more than one sense. In model $\mathfrak{A}$, e_1 denotes the simple event that corresponds to the result r_1 of the first stage; in model $\mathfrak{I}$, e_1 denotes the composite event consisting of the first row of (3.5), but still represents r_1. Since e_1 represents the same experimental result in the two models, it should be assigned the same probability in both models, and this has in fact been done; by (3.7), the sum of the probabilities in the top row of (3.5) is

$$(1) \qquad P(e_1) = p_1 q_1^{(1)} + p_1 q_2^{(1)} + \ldots + p_1 q_b^{(1)} = p_1(q_1^{(1)} + q_2^{(1)} + \ldots + q_b^{(1)})$$
$$= p_1 \cdot 1 = p_1$$

which is just the probability of e_1 in model $\mathfrak{A}$.

The same argument applies to $e_2, \ldots, e_a$. As an immediate consequence of this fact we see that (3.7) is a legitimate model. The sum of all of the probabilities of (3.7) is the sum of the row-sums, which by the above result and (3.1) is $p_1 + p_2 + \ldots + p_a = 1$.

In building model $\mathfrak{I}$ we introduced the q's to represent certain conditional frequencies, for example $q_2^{(1)}$ to represent $f(s_2 | r_1)$. Since in model $\mathfrak{I}$ the result r_1 is represented by e_1 and the result s_2 by f_2, one might hope that $q_2^{(1)}$ would equal $P(f_2 | e_1)$. This is in fact the case, as may be seen by combining (2.1), (3.7), and (1):

$$P(f_2 | e_1) = P(e_1 \text{ and } f_2)/P(e_1) = (p_1 q_2^{(1)})/p_1 = q_2^{(1)}.$$

Just as was the case with the e's, the f's also denote more than one event. Thus f_2 represents the result s_2 not only in model $\mathfrak{I}$, where it is the composite event consisting of the second column of (3.5), but also in models $\mathfrak{B}^{(1)}, \ldots,$ $\mathfrak{B}^{(a)}$, in each of which it is a simple event. However, while e_1 has the same probability in both models $\mathfrak{A}$ and $\mathfrak{I}$, the probability of f_2 will be different in the different models, reflecting the dependence of the second stage on the outcome of the first. Thus, in model $\mathfrak{B}^{(1)}$ the probability assigned to f_2 was $q_2^{(1)}$, in model $\mathfrak{B}^{(2)}$ it was $q_2^{(2)}$, etc. What is the probability of f_2 in model $\mathfrak{I}$? Since in model $\mathfrak{I}$ the event f_2 consists of the second column of (3.5), we have

$$(2) \qquad \begin{aligned} P(f_2) &= p_1 q_2^{(1)} + p_2 q_2^{(2)} + \ldots + p_a q_2^{(a)} \\ &= P(e_1)P(f_2 | e_1) + P(e_2)P(f_2 | e_2) + \ldots + P(e_a)P(f_2 | e_a). \end{aligned}$$

The probability assigned to f_2 in model $\mathfrak{I}$ is thus a "weighted average" of the values assigned in model $\mathfrak{B}^{(1)}, \ldots, \mathfrak{B}^{(a)}$, the weights being $p_1, \ldots, p_a$. Similar formulas hold for $f_1, f_3, \ldots, f_b$. To illustrate this formula, recall Example 3.1. The conditional probability of a white marble is $q_2^{(1)} = \frac{1}{2}$ if box 1 has been drawn, and is $q_2^{(2)} = 0$ if box 2 has been drawn. The overall probability of a white marble is therefore by (2)

$$P(f_2) = \tfrac{1}{2}\cdot\tfrac{1}{2} + \tfrac{1}{2}\cdot 0 = \tfrac{1}{4}.$$

There are many interesting and important two-stage experiments in which one cannot observe the result of the first stage but only that of the second stage. In such cases, the second-stage result may provide indirect information about the unobserved first stage, and in fact it is frequently the purpose of the second stage to provide this indirect information.

EXAMPLE 1. *Diagnostic tests.* For many diseases, diagnostic tests have been developed that are helpful in detecting the presence of the disease. There is, for example, a skin test for tuberculosis, a blood test for syphilis, a cytological test for cancer, etc. Unfortunately such tests are never perfect. A test may on occasion give a positive reaction, i.e., indicate that the disease is present, even when it is not; such a reaction is called a *false positive.* Similarly the test may result in a *false negative,* by showing a negative reaction when applied to an individual who is in fact suffering from the disease.

The application of a diagnostic test may be viewed as a two-stage experiment. The selection of the individual who receives the test constitutes the first stage; this individual may be either "sick" or "healthy," i.e., he may either have or not have the disease in question. The application of the test to the individual constitutes the second stage, the result of which is either a positive or a negative reaction.

In the routine application of a diagnostic test, the result of the first stage ordinarily is not observed. To determine whether or not the disease is present may require an extensive and expensive period of careful clinical observation; it may in some cases not become definitely known until much later when an autopsy is performed. However, the result of the second stage, a positive or negative reaction on the test, is directly observed. The purpose of the second stage is in fact to provide (indirect) information about the individual's health.

EXAMPLE 2. *College admission.* Not all of the applicants for admission to a college would, if admitted, be able to do successful work. We may accordingly classify the applicants as "able" or "unable" on this basis. We shall regard this classification as the first stage of an experiment, although, of course, when an applicant presents himself, the registrar cannot observe into which category he falls. In order to decide whether to admit the applicant, the registrar gives him an entrance examination which he may either pass or fail. This result constitutes the observable second stage. The purpose of the examination is in fact to provide indirect evidence about the ability of the student to do successful work.

Suppose that (3.5) and (3.7) provide a satisfactory model for a certain two-stage experiment, for which the first-stage results, represented by $e_1, e_2, \ldots$, are of particular interest but cannot be directly observed.

Prior to the experiment these events have the probabilities $P(e_1) = p_1$, $P(e_2) = p_2, \ldots$. We shall call these the *prior* probabilities of $e_1, e_2, \ldots$. Now the experiment is performed, and it is seen that the result represented by f_2 has occurred. Once this partial information is in hand, the principles of Section 1 show that the relevant probabilities for $e_1, e_2, \ldots$ are now the *conditional* probabilities given f_2: $P(e_1|f_2)$, $P(e_2|f_2), \ldots$. We shall call these the *posterior* probabilities of $e_1, e_2, \ldots$ since they are relevant *after* the experiment. The posterior probabilities can be computed from formula (2.1) for conditional probabilities. Thus we see, for example, from (3.7) that

$$P(e_1|f_2) = P(e_1 \text{ and } f_2)/P(f_2) = P(e_1)P(f_2|e_1)/P(f_2)$$

and hence from (2) that

$$(3) \quad P(e_1|f_2) = \frac{P(e_1)P(f_2|e_1)}{P(e_1)P(f_2|e_1) + P(e_2)P(f_2|e_2) + \cdots + P(e_a)P(f_2|e_a)}$$

Similar formulas hold in the other cases. This formula, which expresses a posterior probability in terms of prior and conditional probabilities, is known as *Bayes' law*.

EXAMPLE 3. *The two boxes.* Recall Example 3.1, but suppose now that we are permitted to observe not which box was chosen but only the color of the selected marble. Prior to this observation, each box has probability $\frac{1}{2}$, or $P(e_1) = P(e_2) = \frac{1}{2}$. Suppose that the experiment is performed and that the selected marble is red. This result is represented in the model by f_1. Since $P(f_1|e_1) = \frac{1}{2}$, and $P(f_1|e_2) = \frac{3}{4}$, application of Bayes' law gives

$$P(e_1|f_1) = \frac{P(e_1)P(f_1|e_1)}{P(e_1)P(f_1|e_1) + P(e_2)P(f_1|e_2)} = \frac{\frac{1}{2} \cdot \frac{1}{2}}{\frac{1}{2} \cdot \frac{1}{2} + \frac{1}{2} \cdot \frac{3}{4}} = \frac{2}{5}$$

$$P(e_2|f_1) = \frac{P(e_2)P(f_1|e_2)}{P(e_1)P(f_1|e_1) + P(e_2)P(f_1|e_2)} = \frac{\frac{1}{2} \cdot \frac{3}{4}}{\frac{1}{2} \cdot \frac{1}{2} + \frac{1}{2} \cdot \frac{3}{4}} = \frac{3}{5}.$$

These are the posterior probabilities of the two boxes, after it is known that a red marble was obtained. Before the experiment the two boxes are equally likely, but after observing a red marble the second box is more likely than the first. The reason for this is that the probability of a red marble is higher when drawing from the second box than from the first.

EXAMPLE 1. *Diagnostic tests (continued).* The authorities of a college are considering giving a diagnostic test to the entire student body in order to identify those students who have a certain infectious disease. It is known that the test gives some false positive reactions, and the plan calls for subjecting all students with positive reactions to an expensive clinical examination to determine whether they do in fact have the disease. The infirmary wants to know what fraction of the students will have to be examined clinically, i.e. what fraction will give a positive reaction. By (2) we have

(4) $P(\text{positive})$
$$= P(\text{sick})P(\text{positive}|\text{sick}) + P(\text{healthy})P(\text{positive}|\text{healthy}).$$

This formula will be useful only if numerical values can be supplied for the terms on the right. Suppose that previous studies of the test indicate that it gives a positive reaction to about 80% of the persons having the disease, and to about 10% of the persons who do not, so that it is reasonable to assume $P(\text{positive}|\text{sick}) = .8$ and $P(\text{positive}|\text{healthy}) = .1$. Suppose further that general experience with the disease suggests that its incidence among college students is about 1%, so that one may put $P(\text{sick}) = .01$. With these values, (4) gives $P(\text{positive}) = .01 \times .8 + .99 \times .1 = .008 + .099 = .107$. Thus, it may be anticipated that about 11% of the students will have to be examined, and that as a result of the test about 80% of the students having the disease will be identified.

From the point of view of a student who takes the diagnostic test and gets a positive reaction, the interesting question (before he undergoes the clinical examination) is: "How likely am I to have the disease?" The relevant probability is provided by Bayes' law:

$$P(\text{sick}|\text{positive}) = P(\text{sick})P(\text{positive}|\text{sick})/P(\text{positive})$$
$$= (.01 \times .8)/.107 = .075.$$

Consider now the frequency interpretation of this probability. It shows that about 92% of the students who have a positive reaction and must therefore undergo the clinical examination, are in fact healthy. The reason for this somewhat startling conclusion is that the disease is quite rare (incidence of 1%), so that even with a rather low rate of false positives (10%), the bulk of the positive reactions come from the healthy group. This illustrates the general fact that, when diagnostic tests for rare diseases are applied in mass screening surveys, the majority of the positive reactors are healthy. Nevertheless such surveys may be useful, since they permit the expensive examination to be restricted to a small fraction of the whole population.

PROBLEMS

1. Of two boxes, one contains one red and one white marble, the other two red and three white marbles. A box is selected at random and a marble is drawn at random from the selected box. Use (2) to find the probability that the marble is red.

2. Using the model for ordered sampling of Problem 3.9 and (2)
  (i) find the probabilities $P(f_1), P(f_2), \ldots, P(f_N)$;
  (ii) compare $P(f_2)$ with $P(f_2|e_1)$ by giving the frequency interpretation of each.

3. In any model (not necessarily of a two-stage experiment) suppose that the event set ε is broken down into the exclusive event sets $E_1, E_2, \ldots$, so that $\varepsilon = (E_1 \text{ or } E_2 \text{ or } \ldots)$. Prove that for any event F, the probability of F can be computed by the following "breakdown law,"

(5) $P(F) = P(E_1)P(F|E_1) + P(E_2)P(F|E_2) + \ldots .$

[Hint: Apply first the addition law and then the multiplication law to the right side of (1.6.1).]

4. A box contains three red and seven white marbles. A marble is drawn at random and in its place a marble of the other color is put in the box. The marbles are stirred and another one is drawn at random. Find the probability that it is red. [Hint: Apply (5), with E_1, E_2 representing the two possible colors of the first marble drawn.]

5. A box contains three pennies, of which two are fair and the third is two-headed.
   (i) A penny is selected at random and tossed. What is the probability that it will fall heads?
   (ii) If the selected penny is tossed twice, what is the probability that it gives heads both times?
[Hint: Use (5) with E_1, E_2 representing the selection of a fair penny or the two-headed penny respectively. In (ii), assume that, if a fair penny has been selected, the two tosses with this penny are unrelated.]

6. A box contains six pennies, which may be fair, two-headed, or two-tailed. A penny is selected at random and is tossed twice. How many pennies should there be of each of the three types so that the experiment simulates that of Example 3.2?

7. Suppose that the probability is p that the weather (sunshine or rain) is the same on any given day as it was on the preceding day. It is raining today. What is the probability that it will rain the day after tomorrow? [Hint: Use (5), with E_1 and E_2 representing the possible states of the weather tomorrow.]

8. (i) Under the assumptions of Example 1 (continued), find the probability of an incorrect diagnosis.
   (ii) Compare the probability obtained in (i) with the probability of an incorrect diagnosis for the "diagnostic procedure" of declaring every student healthy without performing a test.

9. A lot of 25 items is inspected by the following two-stage sampling plan. A first sample of five items is drawn. If one or more is bad, the lot is rejected; if all are good, a second sample of ten items is drawn (from the 20 items remaining). The lot is rejected if any of the items in the second sample is bad, and is accepted if all are good. Find the probability of accepting a lot containing two bad items. [Hint: Use (5), with E_1, E_2 representing the possibilities that the first sample does or does not contain at least one bad item.]

10. Under the assumptions of Problem 3, prove the following generalization of Bayes' law:

(6) $P(E_1|F) = \dfrac{P(E_1)P(F|E_1)}{P(F)} = \dfrac{P(E_1)P(F|E_1)}{P(E_1)P(F|E_1) + P(E_2)P(F|E_2) + \ldots} .$

11. Use (6) to find
   (i) in Problem 1 the probability $P(\text{box } 1|\text{red})$;
   (ii) in Problem 5(i) the probability that the two-headed penny was selected given that the penny fell heads;
   (iii) in Problem 5(ii) the probability that the two-headed penny was selected given that the penny fell heads both times.

12. In Example 1 (continued), find the conditional probability $P(\text{sick}|\text{negative})$.

4.5 APPLICATIONS OF PROBABILITY TO GENETICS

That the inheritance of certain traits could be regarded as a random experiment was first pointed out by Mendel* in 1866. on the basis of his studies of the culinary pea.   To take a somewhat more explicative example than any of Mendel's, suppose that a certain plant may have red (R), pink (P), or white (W) flowers.   It is found that the seeds obtained from crossing two red-flowered plants will always produce red-flowered plants: let us represent this fact schematically by the formula R × R → R. Similarly, we find W × W → W and R × W → P.   But the other crosses will give variable results.   From R × P we may obtain either R or P, and the record of the successive results will look like the record of a sequence of penny tosses, the red- and pink-flowering plants occurring in about equal frequency in a large experiment.   Similarly W × P produces W and P with about equal frequency.   Finally the cross P × P will give all three types, P about half the time and R and W each about one quarter of the time.

To explain such observations Mendel postulated that there were certain entities, now called *genes*, responsible for flower color and passed on from the plants to their progeny.   The present example involves two types of genes, which we may denote by A and a.   Each plant has two genes so that there are three kinds of plants, or genotypes: AA, Aa, and aa.   A plant (AA), both of whose genes are of type A, will have red flowers.   A plant (aa), both of whose genes are of type a, will have white flowers.   Finally, a plant (Aa) with one gene of each type will have pink flowers.

Each offspring obtains one gene from each parent to make up its own pair.   This explains why R × R → R, since when we cross an AA plant with another AA plant, the offspring must also be AA.   Similarly (aa) × (aa) → (aa), and (AA) × (aa) → (Aa).   The varying results are now also explained: in the (AA) × (Aa) cross, the offspring is sure to get A from the first parent, but may receive either A or a from the other, and thus may be either AA or Aa.   Similarly the (Aa) × (Aa) cross may produce each of the three types.

To explain the stable frequencies, we now introduce probability assumptions into the model.   We suppose
   (i)   from an (Aa) parent, an offspring has probability $\frac{1}{2}$ of receiving each of the genes A and a;
   (ii)   the genes obtained by an individual from his two parents are unrelated.
Breeding experiments also justify another assumption of unrelatedness:
   (iii) the genes passed on by a parent to different offspring are unrelated.
   With these simple assumptions, all the Mendelian frequencies are explained.   Thus in an (Aa) × (Aa) cross, the probability is $\frac{1}{4}$ that the

* Gregor Mendel, 1822–1884.

offspring will be (AA), since there is probability $\frac{1}{2}$ of obtaining the A gene from each parent, and by assumption (ii) these probabilities are to be multiplied. The table below shows the nine possible crosses, and the probabilities of each of the three possible offspring in each case.

TABLE 1.   PROBABILITIES OF GENOTYPES

Father \ Mother	AA		Aa		aa	
AA	AA	1	AA	$\frac{1}{2}$	AA	0
	Aa	0	Aa	$\frac{1}{2}$	Aa	1
	aa	0	aa	0	aa	0
Aa	AA	$\frac{1}{2}$	AA	$\frac{1}{4}$	AA	0
	Aa	$\frac{1}{2}$	Aa	$\frac{1}{2}$	Aa	$\frac{1}{2}$
	aa	0	aa	$\frac{1}{4}$	aa	$\frac{1}{2}$
aa	AA	0	AA	0	AA	0
	Aa	1	Aa	$\frac{1}{2}$	Aa	0
	aa	0	aa	$\frac{1}{2}$	aa	1

The Mendelian theory is one of the most satisfying in all of science. It is simple, elegant, and far-reaching. It is now known that mechanisms of this sort occur throughout the plant and animal kingdoms. Things are not always quite as simple as we have here suggested. Many traits depend on more than two types of genes—for example, the A-B-O blood groups in man seem to involve three types, disregarding subtypes. Many traits are determined by more than one pair of genes (hair color, for example). And the development of an individual depends not only on his genetic inheritance, but also on his environment. But there are many traits in man that can be understood in terms of the simple model we have presented above.

Among the properties inherited according to the simple mechanism described above are a number of diseases or crippling defects which appear in childhood and lead to the death of an affected individual before maturity. We shall assume, as is frequently the case, that all AA- and Aa-individuals are healthy, showing no signs of the defect, but that all aa-individuals are affected. Because of its death-bringing qualities, the gene a in such cases is called *lethal*. The lethal gene is harmless in a single dose (Aa) becoming effective only in a double dose (aa). An individual of the (Aa) type may be called a *carrier* of the lethal gene, which is harmless to him but which he may pass on to his descendants. It follows from the assumptions made that affected individuals can result only from marriages of two Aa individuals. Thus there is strong reason for a carrier to marry only an AA individual, as only this will insure that all his children will escape the lethal combination aa.

Unfortunately it is usually the case that AA- and Aa-individuals are not distinguishable: when A is present in even a single dose, it determines the physical appearance of the individual. The gene A is then said to be dominant over the gene a. In this case, the Aa individual will not know that he is a carrier of the lethal gene since a healthy individual may be either AA or Aa. Can we attach probabilities to these two possibilities?

Let us suppose that the fraction of healthy adults in the population who carry the lethal gene is the same for males and females, and let us denote this fraction by λ. (This supposition would not be realistic if the lethal gene were "sex-linked.") We shall indicate in the next section how the value of λ may be estimated, and will now only say that typically λ is small. If an adult male or female individual I is then selected either at random from the population, or in such a way that the factors leading to his selection are unrelated to his genetic status, an appropriate model for the genotype of I is

(1)

genotype of I	AA	Aa
probability	$1 - \lambda$	λ

Suppose now however that the person in question comes to our attention because he is calling on a genetic counselor for advice. The reason for his seeking advice is frequently that something is known about his genetic status, and such information may change the probability dramatically. For example, if the individual has an affected child, it becomes certain that both he and his spouse are Aa—the probability that he is a carrier rises from λ to 1. If a person has an affected relative, such as a brother, uncle, or cousin, it follows that the lethal gene runs in the family, and the probability of his being a carrier also rises. We shall now compute this probability for two such cases.

EXAMPLE 1. *Affected sib.* Suppose that an adult individual I had an affected brother or sister. This implies that both of I's parents are carriers, and hence that the appropriate model for the genotype of I *at birth* is given by the central block of Table 1:

(2)

genotype of I	AA	Aa	aa
probability at birth	$\frac{1}{4}$	$\frac{1}{2}$	$\frac{1}{4}$

However, since I has survived to adulthood, he cannot be aa, and the appropriate conditional model, given this information, is by (1.1)

(3)

genotype of I	AA	Aa
probability in adulthood	$\frac{1}{3}$	$\frac{2}{3}$

where in writing (3) we have for simplicity omitted the genotype aa whose probability is zero.

EXAMPLE 2. *Affected uncle.* As a second example, suppose that an adult individual I had an uncle or aunt who was affected, say his father's brother. This is a considerably more complicated problem than that treated in Example 1. Its analysis requires a two-stage model: the first stage pertains to the genotypes of I's parents, and the second stage to those of I himself. Since I's father F had an affected sib, the appropriate model for F's genotype is, by Example 1,

(4)

genotype of F	AA	Aa
probab lity	$\frac{1}{3}$	$\frac{2}{3}$

Since I's mother M is not known to have any affected relatives, we may use model (1) for her genotype

(5)

genotype of M	AA	Aa
probability	$1 - \lambda$	λ

If the genotypes of F and M are assumed to be unrelated, the product of models (4) and (5) may serve as model α for the first stage:

(6)

Event	e_1 (F is AA, M is AA)	e_2 (F is Aa, M is AA)	e_3 (F is AA, M is Aa)	e_4 (F is Aa, M is Aa)
Probability	$\frac{1}{3}(1 - \lambda)$	$\frac{2}{3}(1 - \lambda)$	$\frac{1}{3}\lambda$	$\frac{2}{3}\lambda$

The second stage consists of I's "choice" of genes from his two parents. Corresponding to each possible result of the first stage, the conditional model for the second stage is the appropriate block of Table 1. The probabilities of the four conditional models $\mathcal{B}^{(1)}, \ldots, \mathcal{B}^{(4)}$ corresponding to $e_1, \ldots, e_4$ are shown in the following array.

(7)

	f_1 (I is AA)	f_2 (I is Aa)	f_3 (I is aa)
$\mathcal{B}^{(1)}$	1	0	0
$\mathcal{B}^{(2)}$	$\frac{1}{2}$	$\frac{1}{2}$	0
$\mathcal{B}^{(3)}$	$\frac{1}{2}$	$\frac{1}{2}$	0
$\mathcal{B}^{(4)}$	$\frac{1}{4}$	$\frac{1}{2}$	$\frac{1}{4}$

Combining (6) and (7) with the aid of (3.7) gives for the 12 events of the two-stage model the following probabilities, where for example $P(e_2 \text{ and } f_1)$ $= P(e_2)P(f_1|e_2) = \frac{2}{3}(1 - \lambda) \cdot \frac{1}{2} = \frac{1}{3}(1 - \lambda)$.

	f_1	f_2	f_3
e_1	$\frac{1}{3}(1-\lambda)$	0	0
e_2	$\frac{1}{3}(1-\lambda)$	$\frac{1}{3}(1-\lambda)$	0
e_3	$\frac{1}{6}\lambda$	$\frac{1}{6}\lambda$	0
e_4	$\frac{1}{6}\lambda$	$\frac{1}{3}\lambda$	$\frac{1}{6}\lambda$

(8)

Summing the columns of (8) we obtain by (4.2) the probabilities for I's genotype at birth:

$$P(f_1) = \tfrac{2}{3} - \tfrac{1}{3}\lambda, \qquad P(f_2) = \tfrac{1}{3} + \tfrac{1}{6}\lambda, \qquad P(f_3) = \tfrac{1}{6}\lambda.$$

But since I has reached adulthood, we know that he cannot be aa, and therefore by (2.1) his probability of being a carrier is the conditional probability

$$P(I \text{ is Aa}|I \text{ is not aa}) = \frac{\tfrac{1}{3} + \tfrac{1}{6}\lambda}{1 - \tfrac{1}{6}\lambda}.$$

As remarked above, λ is usually small, so that to a first approximation one may neglect $\tfrac{1}{6}\lambda$ and say that this probability is approximately $\tfrac{1}{3}$. Roughly speaking, it is half as dangerous to have an affected uncle or aunt as to have an affected brother or sister.

A simple intuitive argument leads to this same approximate result. There is probability $\tfrac{2}{3}$ that F carries the gene a, and if he does, I has half a chance of getting it. Therefore I has probability $\tfrac{1}{3}$ of inheriting the gene a from his father. Compared with this, the chance of getting gene a from his mother is negligible.

Similar calculations can be made for other affected relatives. We summarize in Table 2 the approximate probabilities (assuming λ to be small) that a healthy adult is a carrier given that certain relatives are affected.

TABLE 2.   RISK OF BEING A CARRIER

Affected relative	Approximate probability
Child	1
Brother or sister	$\tfrac{2}{3}$
Uncle or aunt	$\tfrac{1}{3}$
First cousin	$\tfrac{1}{4}$
Great uncle or aunt	$\tfrac{1}{6}$

The probabilities in this table become very important when two individuals, both of whom have a family history of the same genetic disease, are contemplating marriage. For example, if both individuals have an affected brother or sister, each has two chances in three of being a carrier. It is reasonable to treat these results as unrelated; and the probability that both are carriers is therefore $\frac{2}{3} \cdot \frac{2}{3} = \frac{4}{9}$. Any child they might have would thus have probability $\frac{4}{9} \cdot \frac{1}{4} = \frac{1}{9}$ of being affected. This probability might be large enough to discourage the marriage of such persons.

PROBLEMS

1. Suppose a gene can occur in any of three types, say A, a or α.
    (i) What are the possible genotypes of an individual?
    (ii) If the father is Aa and the mother is aα, what are the probabilities of the genotypes of the child?

2. Two carriers marry and have two children. What is the probability that just one of these children is affected?

3. A genetics counsellor advises a client, whose aunt was affected, that the client has a 38 percent chance of carrying the gene. How frequent does the counsellor believe carriers to be in the population at large?

4. (i) What is the probability that a healthy person is a carrier, given that his half-brother was affected?
    (ii) What is the approximate value of the probability of part (i) if λ is small?

5. An individual F is mated with an individual M known to be a carrier. They have n children all of which are healthy. How unlikely is this event if F is a carrier?

6. An individual F is randomly selected from the adult population (whose proportion of carriers is λ) and mated with an individual M known to be a carrier. What is the probability that F is a carrier if it is given that
    (i) they have a child C who is healthy;
    (ii) they have two children, both of whom are healthy;
    (iii) they have n children, all of whom are healthy?
[Hint: Use Bayes' law.]

7. Verify the entries of (8).

8. A man F is known to be the carrier of the lethal gene a. His wife M is randomly chosen from a population of which a proportion λ are carriers. How does the probability that their child C is a carrier depend on λ? Explain.

9. An adult woman (S) whose brother (B) has an affected child (and is therefore a carrier) wishes to know the chance that she is a carrier.

(i) Find $P(B = \text{Aa and } S = \text{Aa})$.

(ii) Find $P(B = \text{Aa and } S = \text{AA})$.

(iii) Find $P(S = \text{Aa}|B = \text{Aa})$.

[Hints: (i) Use the break-down law (4.5) with

$$F: (B = \text{Aa and } S = \text{Aa}), \qquad F_1: (F = \text{AA and } M = \text{Aa}),$$
$$F_2: (F = \text{Aa and } M = \text{AA}), \qquad F_3: (F = \text{Aa and } M = \text{Aa}).$$

(iii) Use (i).]

10. (i) Show that if λ is sufficiently small that terms in λ^2 may be neglected, the probability in Problem 6(iii) is approximately equal to $\frac{1}{2}$.

   (ii) Use part (i) to explain the entry $\frac{1}{4}$ in Table 2.

11. A man F had an affected uncle and his wife M an affected first cousin. What approximately is the probability of being diseased for each of their children?

4.6 MARRIAGE OF RELATIVES

In most human societies there is a prohibition against marriages of close relatives. In this connection it is interesting to observe that the probability of affected children (in the sense of the preceding section) is typically much larger in such marriages than in marriages of unrelated persons. For simplicity, consider the case of brother-sister marriages, which were actually favored in the royal families of ancient Egypt. What is the probability that a brother B and sister S are both carriers of the lethal gene a, so that their marriage would be genetically dangerous, assuming that their parents are unrelated healthy persons? Again a two-stage model is required. For the first stage, we assume that the father F and mother M each has genotype given by model (5.1), and since they are unrelated (in both senses!), the product model is appropriate for the genotypes of the parents:

(1)

event	e_1 (F is AA, M is AA)	e_2 (F is Aa, M is AA)	e_3 (F is AA, M is Aa)	e_4 (F is Aa, M is Aa)
probability	$(1 - \lambda)^2$	$(1 - \lambda)\lambda$	$\lambda(1 - \lambda)$	λ^2

Consider now the second stage, pertaining to the genotypes of B and S. For each result of the first stage (i.e., for each combination of parental genotypes), the genotypes at birth of both children are governed by the appropriate block of Table 5.1. By assumption (iii) of the preceding section, the model for the genotypes of B and S is given by the product of this block with itself. It will be enough to distinguish the following three events on the second stage.

f_1: neither B nor S is affected, and at most one is a carrier (in which case there is no genetic counterindication to their marriage);

f_2: both B and S are carriers;

f_3: at least one of B and S is affected.

To compute for example $P(f_2|e_2)$, we note that when F is Aa and M is AA, then B and S each has probability $\frac{1}{2}$ of being a carrier, so that the conditional probability of both being carriers is $\frac{1}{2} \cdot \frac{1}{2} = \frac{1}{4}$. In this way the following four conditional models are obtained.

(2)

	f_1	f_2	f_3
$\mathcal{B}^{(1)}$	1	0	0
$\mathcal{B}^{(2)}$	$\frac{3}{4}$	$\frac{1}{4}$	0
$\mathcal{B}^{(3)}$	$\frac{3}{4}$	$\frac{1}{4}$	0
$\mathcal{B}^{(4)}$	$\frac{5}{16}$	$\frac{4}{16}$	$\frac{7}{16}$

By summing the columns we find that at birth the genotypes of B and S have probabilities

$$P(f_1) = (1 - \lambda)^2 + \tfrac{3}{2}\lambda(1 - \lambda) + \tfrac{5}{16}\lambda^2$$
$$P(f_2) = \qquad\qquad \tfrac{1}{2}\lambda(1 - \lambda) + \tfrac{4}{16}\lambda^2$$
$$P(f_3) = \qquad\qquad\qquad\qquad \tfrac{7}{16}\lambda^2.$$

However, once the children are grown, f_3 is excluded and the relevant probability of both being carriers is the conditional probability

(3) $$P(f_2|\text{not } f_3) = \frac{\tfrac{1}{2}\lambda - \tfrac{1}{4}\lambda^2}{1 - \tfrac{7}{16}\lambda^2}.$$

Since λ is small, λ^2 will be very small and negligible in comparison with λ, so that an approximate value is

(4) $$P(f_2|\text{not } f_3) \sim \tfrac{1}{2}\lambda.$$

For example, if $\lambda = .01$, the exact value (3) is .00498 while the approximate value (4) is .005.

For comparison with this value, let us see what is the probability that a healthy man I and a healthy woman J, who are not related to each other, are both carriers. Since for both males and females the fraction of healthy adults who are carriers is λ, we may use model (5.1) for both I and J. If the genotypes of I and J are assumed to be unrelated, these models may be multiplied together. In the resulting product model,

(5) $$P(I \text{ and } J \text{ are both carriers}) = \lambda^2.$$

When λ is small, both of the probabilities (4) and (5) will be small, but (4) will be *relatively* very much greater than (5). For example, if $\lambda = .01$, then $\tfrac{1}{2}\lambda = .005$ and $\lambda^2 = .0001$, so that (4) is 50 times as large as (5).

This result may seem surprising at first since we have not supposed that the family of B and S has any history of the disease. The mere fact that they are sibs increases by a factor of 50 the risk that both are carriers, as compared with the risk that both of the unrelated individuals I and J are carriers. The explanation rests on the fact that B and S draw their genes from the same source (F and M), while I and J draw from different sources. F and M are no more likely to be carriers than are I and J, but if one of them is a carrier (the probability of which is about 2λ), there is a substantial chance ($\frac{1}{4}$) that *both* B and S will draw the lethal gene.

Calculations similar to those leading to (4) can be made for marriages among other close relatives. Some results of this kind are shown in Table 1. The approximate probability of two relatives both being carriers is seen to be proportional to λ, with the factor of proportionality decreasing as the relationship becomes more distant. The value of this factor can in fact be used to define the closeness of the relationship.

TABLE 1

If I and J are	the approximate probability that both I and J are carriers is
Unrelated	λ^2
Brother and sister	$\frac{1}{2}\lambda$
Father and daughter	$\frac{1}{2}\lambda$
Half brother and half sister	$\frac{1}{4}\lambda$
Uncle and niece	$\frac{1}{4}\lambda$
First cousins	$\frac{1}{8}\lambda$

The fact that brother-sister marriages are such bad genetic risks is thought by some anthropologists to explain the taboo against such marriages that has existed in most human societies. If in a primitive society brother-sister marriages were common, it would soon be observed that diseased children were primarily to be found among their offspring, and the conclusion might be drawn that such marriages were not favored by the gods.

The conclusion that brother-sister marriages are relatively bad genetic risks rests on the assumption that λ is small. For sufficiently large values of λ, the probability (3) actually will be smaller than the probability (5). For example, in some regions of West Africa, the gene responsible for sickle-cell anemia is carried by a considerable proportion of the adult population. If this proportion were for example 50%, a brother-sister marriage would be a slightly better genetic risk than the marriage of unrelated persons (Problem 2).

Let us consider next how one may determine the value of λ for a particular lethal gene. Since carriers Aa do not in general differ in appearance from AA individuals, one cannot conduct a census to determine the frequency λ of carriers in the adult population. However, by (5) we may expect the proportion of couples Aa $\times$ Aa to be λ^2. From Table 5.1, about $\frac{1}{4}$ of the children of such couples will be affected, and no other children can be. This suggests that the frequency of affected children in the population will be about $\frac{1}{4}\lambda^2$. If for example it is observed that one child in 40,000 is affected, we may estimate that

(6) $$\frac{\lambda^2}{4} = \frac{1}{40,000} \quad \text{or} \quad \lambda = .01.$$

This approach is based on several assumptions that may not be satisfied in practice. When a child is found to be affected, and the parents thus learn that they are both carriers, they may well decide to have no further children. This could result in smaller families among such couples than in the population at large, and hence to a fraction of affected children lower than $\lambda^2/4$, so that the estimate (6) for λ would be too small. This difficulty could be overcome by restricting the frequency count of affected children to first-born children.

Another assumption which may not be satisfied is that leading to (5). The use of a product is valid only if there is no relation between the genotypes of I and J. At first this seems reasonable, since it was assumed that the possession of a single a gene does not affect appearance. To see that the assumption of (statistical) unrelatedness may nevertheless not be at all realistic, consider once more the case of sickle-cell anemia. In the population of the United States, the gene a for this disease is carried much more frequently by Negroes than by persons of other races. Since people tend to marry others of the same race, equation (5) is grossly wrong in this case. This is an example of indirect dependence (Section 2); the genotypes of husband and wife become dependent through the indirect effect of the factor of race.

PROBLEMS

1. Verify the entries of (2).

2. Verify that (5) is bigger than (3) when $\lambda = \frac{1}{2}$.

3. (i) What is the probability that a child is affected if his parents are brother and sister?

   (ii) What is the approximate value of this probability if λ is sufficiently small so that terms in λ^2 can be neglected?

4. In breeding experiments with mice, 17 affected offspring are observed out of 1835 births from brother-sister matings. Use the result of Problem 3(ii) to find the approximate value of λ.

5. A man F is known to be a carrier; the genotype of his wife M is given by model (5.1).

   (i) What are the probabilities of the genotypes of their daughter D at birth?

   (ii) What is the conditional probability of D being a carrier given that she is healthy?

6. (i) Suppose that the genotypes of a man F and his wife M are unrelated and both given by model (5.1). Show that the probability of both F and his healthy daughter D being carriers is $(\lambda/2) \div (1 - \lambda/4)$.

(ii) Use (i) to verify an entry in Table 1.

(iii) Use (i) to find for what values of λ father-daughter marriages are genetically safer than the marriages of unrelated persons.

[Hint: (i) Use Problem 5(ii).]

4.7 PERSONAL PROBABILITY

So far we have been regarding the probability of a result in a random experiment as the mathematical abstraction of the frequency with which the result would occur in a long sequence of similar and unrelated experiments. In addition to this interpretation of probability as frequency, there is a quite different interpretation according to which the probability of a statement represents a certain numerical measure of a person's degree of belief in the statement. We shall in this section discuss this alternative "personal" or "subjective" view of probability and illustrate its use in several problems.

The use of the word probability in everyday speech often has a personal aspect. If an astronomer says "there is probably vegetable life on Venus," or if a sports enthusiast says "the chances are two out of five that the home team will win tomorrow's game," or if a juror thinks "it is unlikely that the witness told the truth," they are not thinking of a frequency interpretation, but are giving a personal assessment of the situation based on a subjective evaluation of the evidence.

How can one arrive at the numerical value of one's personal probability for a given statement? In principle this can always be done by asking whether or not one would accept a bet offered at various odds. Suppose that two horses H and L are to run a race, and you wish to discover your personal probability that horse H will win. If you would be willing to take either side of an even money bet that H will win, this can be interpreted as meaning that you think horse H has half a chance to win, so that for you, P(H will win) $= \frac{1}{2}$. Now consider another spectator who thinks that H is more likely to win than L. Suppose he would be indifferent between the alternatives

(i) getting \$2 if H wins,

(ii) getting \$3 if L wins.

It is intuitively plausible that the numbers of dollars he might expect to win in the two cases are

(i') $2 \cdot P$(H will win),

(ii') $3 \cdot P$(L will win).

His indifference may be interpreted as meaning that, for him, (i') and (ii') are equal. Since

$$P(\text{H will win}) + P(\text{L will win}) = 1,$$

it follows that *for him*

$$P(\text{H will win}) = \tfrac{3}{5}, \quad P(\text{L will win}) = \tfrac{2}{5}.$$

By such considerations it is theoretically possible for any person to work out his personal probabilities for any set of possible results. In practice, many people find it very difficult to make the required assessment, especially when the set of results is large.

From the mathematical point of view, the model just developed is indistinguishable from the probability models we have been using: as before, there is an event set, and to each event is assigned a nonnegative probability in such a way that the probabilities add up to 1. Although the conceptual interpretations of personal and frequency probability differ, the formal calculus is the same. The addition law, the multiplication law, Bayes' law, etc. are used in the same way regardless of the meaning attached to the probabilities.

One important aspect of personal probability is that it is a function of the person involved: for two different people, the probabilities of a given statement may be markedly different, reflecting the different information, experience, attitudes, or even prejudices, on which they base their beliefs. To the frequentist, on the other hand, the probability of a result depends on the conditions of the experiments, but at least ideally ought not to differ according to personal knowledge and attitudes.

In contrast with the subjective approach, the frequency interpretation is sometimes called "objective." This term is somewhat misleading, since in reality both approaches involve subjective elements. The frequentist exercises judgement in deciding what assumptions to make in building his probability model, so the choice of model will inevitably involve subjective evaluations. The subjectivist, on the other hand, will presumably consider any relevant frequency experience in formulating his personal probability.

One of the most important differences is that the subjectivist is willing to apply probability ideas to a wider class of problems than is the frequentist. Suppose you meet a stranger in a bar, who offers to toss a coin to decide who pays for the drinks. He will pay if the coin falls tails, you will pay if it falls heads. The thought crosses your mind that the coin may be two-headed, but you are too delicate to ask to examine it. What is the probability that the coin is two-headed? The frequentist may hesitate to answer this question, he may even say that the question is meaningless, while the subjectivist need in principle only ask himself which way he would bet at various odds to arrive at his personal probability for the coin to be two-headed.

An interesting class of problems arises when a subjective probability has to be modified in the light of fresh evidence to which a frequency interpretation can be given. Suppose for example you accept the stranger's offer, he tosses, and the coin falls heads. This experimental result will reinforce to some extent your prior suspicions that the coin is two-headed. (If the coin had fallen tails, your suspicions would by now have disappeared.) Let us see by how much the suspicion should be increased by the result "heads," using a subjective approach.

Suppose that, before the coin was tossed, your personal probability for the hypothesis "coin is two-headed" was equal to π, so that your personal probability for "coin is fair" was $1 - \pi$. (We assume for simplicity that there are only these two possibilities.) If the coin is fair, everyone would agree that the probability of "heads" is $\frac{1}{2}$. By Bayes' law,

$$P(\text{Coin is two-headed}|\text{Heads}) = \frac{\pi \cdot 1}{\pi \cdot 1 + (1 - \pi) \cdot \frac{1}{2}} = \frac{2\pi}{1 + \pi}.$$

For example, if originally you thought there was a ten percent chance that the coin is two-headed, after the coin fell heads you should have increased the chance to $2(.1)/(1 + .1) = .182$. The following table shows how the probability of two heads after the toss (the *posterior* probability) depends on the probability of two heads before the toss (the *prior* probability):

Prior	0	.01	.1	.5	.9	.99	1
Posterior	0	.0198	.182	.667	.947	.9950	1

Notice that if you are certain before the experiment that the coin is two-headed ($\pi = 1$) or that the coin is fair ($\pi = 0$), then the experiment will not change your probability. It is generally true that if the prior probability for any statement is 1, the same is true of the posterior probability (Problem 3). If a person's mind is made up, he will not be confused by the facts!

We conclude with two examples of greater applicational interest.

EXAMPLE 1. *Disputed authorship.* There are a number of literary works whose authorship is in dispute; for example, certain of the Federalist Papers are assumed to have been written either by Madison or by Hamilton. Efforts have been made to resolve the doubt by frequency counts of stylistic elements in the disputed papers and in known works of the two men.

To simplify the example drastically, suppose that a work whose author is known to be either M or H contains a certain peculiar stylistic construction. Author M is fonder of this construction than is author H. In fact, this construction may be found in 60% of the similar works known to be by M, but in only 10% of those by H. Intuitively, its occurrence in the disputed work tends to support the claims of M, but how strongly?

Suppose a historian, before the stylistic analysis, attaches personal probability .3 to the statement "Author is M" and .7 to "Author is H." After the stylistic analysis, his posterior probability for "Author is M" will according to Bayes' law be given by

$$P(\text{Author is M}|\text{Stylistic analysis}) = \frac{.3 \times .6}{.3 \times .6 + .7 \times .1} = \frac{.18}{.25} = .72.$$

This calculation could not of course be defended if the prior assessment that $P(\text{Author is M}) = .3$ had been based on a consideration of style, for in that case the stylistic analysis would not constitute entirely new evidence. In practice it might be difficult for the historian to know the extent to which his prior ideas were influenced by the style of the work.

EXAMPLE 2. *Probability of guilt.* Another interesting use of personal probability is in criminology, where one wishes to know how likely it is that a man accused of committing a crime is in fact guilty. Suppose it is known that the criminal has a certain property A (he may, for example, be known to be left handed and to belong to a certain blood group). If the accused is guilty, he is certain to possess property A. However, even if he is innocent he may possess property A. If the fraction of persons with property A in the population is p, one might assign to $P(\text{A}|\text{innocent})$ the value p. Suppose that the accused is found to possess property A. If for a police official the personal prior probability before this observation was π that the accused is guilty, his posterior probability is by Bayes' law

$$\frac{\pi \cdot 1}{\pi \cdot 1 + (1 - \pi) \cdot p}.$$

The following table shows how this quantity depends on p and π.

(1)

p \ π	0	.1	.3	.5	.7	.9	1
.1	0	.53	.81	.91	.96	.99	1
.3	0	.27	.59	.77	.89	.97	1
.5	0	.18	.46	.67	.81	.94	1
.7	0	.14	.38	.59	.77	.93	1
.9	0	.11	.32	.53	.72	.91	1

We wish to emphasize that the calculation above could be defended only if the information about property A was not involved in bringing the original accusation. In building the model, we assigned to $P(\text{A}|\text{innocent})$ the frequency p with which property A occurs in the population. This assignment would certainly not be reasonable if for example the police had merely looked around for the nearest person with property A and then accused him of the crime.

One must also be careful to see that the value of p is drawn from relevant experience. A blood group or type of hair may for example be rare in the population as a whole, but perhaps be common in the village in which the

crime was committed.   As with all cases of applying mathematical arguments to real situations, great caution is needed to avoid serious fallacies.

PROBLEMS

1.   Consider the problem of drinking with the stranger in the bar, under the
assumptions made in the text.
  (i) Suppose he tosses the coin twice and gets "heads" each time.   What is
       your posterior probability that the coin is two-headed (a) for arbitrary π,
       (b) for $\pi = .1$?
  (ii) What, for arbitrary π, is this posterior probability if he tosses the coin n
       times and gets "heads" every time?
  (iii) What happens to the posterior probability of (ii) when n is very large
       (a) for $\pi < 1$, (b) for $\pi = 1$?

2.   Verify the entry in (1) for $\pi = .7$, $p = .3$.

3.   Suppose that the prior probability for a statement S is 1, and that an event A
is observed which is possible according to S.   Use Bayes' law to find $P(S|A)$.

4.   Let the prior probability of a statement S be π.   The ratio $\pi/(1 - \pi)$ is called
the *prior* odds ratio for S.   An event A is observed.   How is the *posterior* odds
ratio $P(S|A)/[1 - P(S|A)]$ related to the prior odds ratio?

5.   A detective, informed that a criminal had red hair, accused the nearest red-
haired person.   What would be a reasonable value to assign to P(accused has red
hair|accused is innocent)?

CHAPTER 5

RANDOM VARIABLES

5.1 DEFINITIONS

When a random experiment is performed, we are often not interested in all the details of the result, but only in the value of some numerical quantity determined by the result. For example, when deciding on the quality of a lot of mass-produced items by inspecting a random sample, we are interested only in the total number of defective items in the sample—which particular sampled items are good, and which are bad, is of no concern to us. Again, the dice player may be interested not in the number of points showing on each die separately but only in the total points showing on the two dice together. It is natural to refer to a quantity whose value is determined by the result of a random experiment as a *random quantity*.

EXAMPLE 1. *Number of boys in a two-child family.* If two-child families are classified according to the sex of the first and second child, four simple results are possible as listed in the first column of the tableau. Suppose we are interested in the random quantity "number of boys." The value of this quantity for each simple result is shown in the second column.

Simple result	Number of boys
first child male, second child male	2
first child male, second child female	1
first child female, second child male	1
first child female, second child female	0

EXAMPLE 2. *Number of matchings.* A magazine prints a row of three pictures of movie stars, say Alice, Barbara, and Charlotte. Also given are baby pictures of the three stars, and the reader is invited to match them. He may put the baby pictures in 3! = 6 different orders shown in

column one of the tableau. The quantity of interest is the "number of matchings." The value of this quantity corresponding to each simple result is shown in the second column.

Simple result	Number of matchings
Alice, Barbara, Charlotte	3
Alice, Charlotte, Barbara	1
Barbara, Alice, Charlotte	1
Barbara, Charlotte, Alice	0
Charlotte, Alice, Barbara	0
Charlotte, Barbara, Alice	1

A random quantity is represented in the model by a *random variable*. Just as a random quantity attaches a value to each simple result of a random experiment, the corresponding random variable attaches a value to each simple event. (In the language of mathematics, a random variable is thus a *function* of the simple event.) We shall consistently use capital letters to denote random variables.

EXAMPLE *1*. *Number of boys in a two-child family (continued)*. Let us represent the random quantity "number of boys" by the random variable B. In the notation of Example 1.5.2, the tableau of Example 1 becomes

Simple event	Value of B
MM	2
MF	1
FM	1
FF	0

EXAMPLE *2*. *Number of matchings (continued)*. Let us represent the random quantity "number of matchings" by the random variable M. If we use ABC to represent the arrangement "Alice, Barbara, Charlotte," etc., the tableau of Example 2 becomes

Simple event	Value of M
ABC	3
ACB	1
BAC	1
BCA	0
CAB	0
CBA	1

For general discussion, suppose that a random experiment has simple results $r_1, r_2, \ldots$, and that the corresponding values of a random quantity

q are $q(r_1)$, $q(r_2)$, ... It is not necessary for the numbers $q(r_1)$, $q(r_2)$, . . . all to be distinct, as both examples show. The set of distinct values of q is known as its *value set*. Thus, in Example 1 the value set of "number of boys" is $\{0, 1, 2\}$, while in Example 2 the value set of "number of matchings" is $\{0, 1, 3\}$. As other examples, the value set of the total number of points showing when two dice are thrown is $\{2, 3, . . . , 12\}$; when a random sample of size 25 is drawn from a lot consisting of 20 good and 30 defective items, the value set of the number of defective items in the sample is $\{5, 6, . . . , 25\}$.

Suppose that the random quantity q, which attaches value $q(r_1)$ to the simple result r_1, etc., is represented in the model by the random variable Z, which attaches the value $Z(e_1)$ to the simple event e_1, etc.

Random experiment		Model	
Simple result	Value of q	Simple event	Value of Z
r_1	$q(r_1)$	e_1	$Z(e_1)$
r_2	$q(r_2)$	e_2	$Z(e_2)$
.	.	.	.
.	.	.	.
.	.	.	.

The reader will notice that, in both examples, the second columns of the two tableaus are identical. This will be the case quite generally if the random variable Z is to represent the random quantity q. That is, if e_1 corresponds to r_1, etc., we must have

$$(1) \qquad\qquad Z(e_1) = q(r_1), \quad Z(e_2) = q(r_2), ...$$

Thus, the value set of Z will always coincide with that of q.

As we have pointed out, the numbers $q(r_1)$, $q(r_2)$, . . . need not be distinct. Therefore, the occurrence of a particular value of the random quantity q may be a composite result. Thus, in Example 1, the result "number of boys is one" is composed of the two simple results "first child male, second child female" and "first child female, second child male." This composite result is represented in the model by the composite event $\{MF, FM\}$, consisting of the two simple events for which B takes on the value 1. It is therefore natural to denote this composite event by the formula $B = 1$, as we have in fact already done in Example 1.5.2. Similarly, in Example 2, $M = 1$ denotes the composite event $\{ACB, BAC, CBA\}$ which represents the composite result "number of matchings is one." In general, if z is one of the possible values of a random variable Z, we shall denote by $Z = z$ the event consisting of all simple events e for which $Z(e) = z$. The event $Z = z$ represents in the model the result (which may be simple or composite) that the random quantity q takes on the value z.

Consider now the probability of the event $Z = z$, corresponding to the frequency with which the random quantity takes on the value z. It follows from Section 1.4 and the definition of the event $Z = z$ that $P(Z = z)$ is the sum of the probabilities of all those simple events e for which $Z(e) = z$. Therefore the value of $P(Z = z)$ will depend on the probabilities that are assigned to the simple events of the model. Thus, use in Example 1 of the probabilities

(2) $$P(\text{MM}) = \tfrac{1}{4}, \quad P(\text{MF}) = \tfrac{1}{4}, \quad P(\text{FM}) = \tfrac{1}{4}, \quad P(\text{FF}) = \tfrac{1}{4}$$

gives

$$P(B = 0) = \tfrac{1}{4}, \quad P(B = 1) = \tfrac{1}{4} + \tfrac{1}{4} = \tfrac{1}{2}, \quad P(B = 2) = \tfrac{1}{4}.$$

On the other hand, the product model suggested in Example 3.3.2 assigns to the simple events the probabilities

(3) $$P(\text{MM}) = (.514)^2 = .264, \; P(\text{MF}) = P(\text{FM}) = .514 \times .486 = .250,$$
$$P(\text{FF}) = (.486)^2 = .236,$$

and hence gives (as was already computed in Example 4.1.3),

$$P(B = 0) = .264, \; P(B = 1) = .250 + .250 = .500, \; P(B = 2) = .236.$$

PROBLEMS

1. What is the value set of the sum of points on (i) three dice, (ii) n dice?

2. When a random sample of size s is drawn from a lot containing 20 good and 8 defective items, what is the value set of the number of defective items in the sample when (i) $s = 5$; (ii) $s = 9$; (iii) $s = 12$; (iv) $s = 25$?

3. Solve the four parts of the preceding problem when the lot contains 8 good and 20 defective items.

4. In three tosses with a fair die, find the value sets of the following random variables:
    (i) the maximum of the numbers of points on the first two tosses;
    (ii) the maximum of the numbers of points on all three tosses;
    (iii) the product of the numbers of points on the first two tosses;
    (iv) the difference between the numbers of points on the first and second tosses.

5. In Example 2, what are the possible values of M
    (i) if there are four movie stars;
    (ii) if there are five movie stars?

6. Let D represent the difference between the number of heads and the number of tails when a coin is tossed n times. What is the value set of D?

7. If a random variable Z is defined over an event set ε consisting of n simple events, (i) what is the maximum number of elements that the value set of Z can contain; (ii) what is the minimum number?

8. For the random variable T of Example 1.5.3, list the simple events (1.5.2) making up each of the events $T = 2$, $T = 3$, $T = 4, \ldots, T = 12$.

9. In the model of Example 2.1.1, suppose that the marbles numbered 1, 2, 3 represent the Liberals, while 4, 5, 6, 7 represent the Conservatives. If D represents the number of Conservatives included in the delegation,

(i) list the simple events from (2.1.1) making up the event $D = 1$;

(ii) what value does D assign to the simple event (346)?

10. If M denotes the random variable described in part (i) of Problem 4, list the simple events making up the events (i) $M = 1$; (ii) $M = 2$.

11. Under the assumptions of Problem 1.5.6,

(i) find the value set of the random quantity "number of boys in the family";

(ii) if this random quantity is represented in the model by B, what value does B assign to the simple event FMF;

(iii) list the simple events making up the event $B > 0$;

(iv) find the probability $P(B > 0)$.

[Hint: (iii) The event $B > 0$ is the union of the events $B = 1$, $B = 2$, and $B = 3$.]

12. If two digits are produced by the random digit generator of Problem 1.5.8, and if Z represents the product of the two digits,

(i) list the simple events making up the event $Z = 12$;

(ii) find the probability of the event $Z = 12$;

(iii) list the simple events making up the event $Z < 5$;

(iv) find the probability of the event $Z < 5$.

13. Under the assumptions of Problem 1.5.9, let Z represent the number of correct answers.

(i) List the simple events making up the event $Z = 2$.

(ii) Find the probability $P(Z = 2)$.

(iii) Find the probability $P(Z \leq 1)$.

[Hint: (iii) "$Z \leq 1$" is read "Z is less than or equal to 1" and is the union of the events $(Z = 0)$ and $(Z = 1)$.]

14. If U denotes the random variable described in part (iii) of Problem 4, find (i) $P(U = 4)$; (ii) $P(U = 6)$; (iii) $P(U = 12)$.

15. If D denotes the random variable described in part (iv) of Problem 4, find (i) $P(D = -4)$; (ii) $P(D = 0)$; (iii) $P(D = 4)$.

16. For the total number T of points showing in two throws with a die, find the probability of the event $T = 4$ if the two throws are unrelated and

(i) the die is loaded and its probabilities are specified by (1.3.1);

(ii) the die is fair.

17. Let T represent the sum of the number of points on three unrelated throws with a fair die.

(i) If the simple events are labeled as in Problem 1.5.11, list the simple events making up the event $T = 5$.

(ii) Find the probability $P(T = 5)$.

18. In n tosses of a coin, let player A win 2 dollars each time the coin comes up heads and lose 2 dollars each time the coin comes up tails. Express player A's total winnings in the n tosses in terms of the random variable D of Problem 6.

19. List four random quantities that might be of interest when considering the performance in a course in which there was a midterm and a final.

5.2 DISTRIBUTION OF A RANDOM VARIABLE

The most important aspects of a random variable are its value set and the probabilities of it taking on its various possible values. As an illustration, consider the random variable T which represents the sum of the number of points showing on two dice. With the model of Example 1.5.3, the possible values t of T and their probabilities are given by

(1)

t	2	3	4	5	6	7	8	9	10	11	12
$P(T = t)$	$\frac{1}{36}$	$\frac{2}{36}$	$\frac{3}{36}$	$\frac{4}{36}$	$\frac{5}{36}$	$\frac{6}{36}$	$\frac{5}{36}$	$\frac{4}{36}$	$\frac{3}{36}$	$\frac{2}{36}$	$\frac{1}{36}$

(Problem 1.5.4).

A table like (1) which associates with each possible value z of a random variable Z the probability $P(Z = z)$ that the value z will occur, is known as the *distribution* of the random variable. (The distribution assigns a real number to each element in the value set and is therefore a function defined over the value set.) Since a random variable always assumes just one of its possible values, these values represent exclusive events that between them exhaust all the possibilities. It follows that the sum of the numbers $P(Z = z)$ over all values of z in its value set must be equal to 1. This fact provides a useful check when we are computing a distribution; for example, the probabilities displayed in the second row of (1) are seen to add up to 1.

EXAMPLE 1. *The delegation.* A very important class of problems, which we shall examine at length in Section 6.2, arises in the theory of sampling. In Example 2.1.1, we considered the drawing of a sample of three as a delegation from a seven-man council consisting of four Conservatives and three Liberals. The number of Conservatives in the delegation is a quantity whose value is determined by the result of the random selection of the sample. Therefore it is a random quantity, and we shall represent it in the model by the random variable D. The computation of Example 2.1.1 shows that $P(D = 3) = \frac{4}{35}$; we shall now obtain the complete distribution of D.

To facilitate the discussion, let us suppose that the Liberal members of the council are numbered 1, 2, 3 while the Conservative members are numbered 4, 5, 6, 7. The 35 possible samples are shown below, grouped according to the value of D.

$D = 0$:   123
$D = 1$:   124   125   126   127
        134   135   136   137
        234   235   236   237
$D = 2$:   145   146   147   156   157   167
        245   246   247   256   257   267
        345   346   347   356   357   367
$D = 3$:   456   457   467   567

If we regard the $\binom{7}{3} = 35$ samples as equally likely, the probability $P(D = d)$ that D takes on any specified value d can be found simply by counting the number of samples for which D has the value d, and dividing by 35. In this way, we find the following distribution for D, of which the last term agrees with the value found earlier.

d	0	1	2	3
$P(D = d)$	$\frac{1}{35}$	$\frac{12}{35}$	$\frac{18}{35}$	$\frac{4}{35}$

Notice that the probabilities add up to 1.

EXAMPLE 2. *Indicator.* A very simple but frequently useful kind of random variable is one whose value set is $\{0, 1\}$; such a random variable is called an *indicator*. Let I be an indicator random variable, and let E be the set of simple events for which I takes on the value 1. Then $\overline{E}$ must be the set of simple events for which I takes on its other possible value, 0. Thus $I = 1$ if and only if the event E occurs, and we say that I "indicates" the occurrence of E. For instance, suppose that $I = 1$ if and only if $D = 3$ in the previous example; then I indicates a delegation made up entirely of Conservative members. If the probability that $I = 1$ is p, then the distribution of I is

i	0	1
$P(I = i)$	$1 - p$	p

It is important to notice that quite different random variables may have the same distribution. This point is illustrated in Problems 16 and 17.

It is often instructive to present a distribution in a graphical form. Since $P(Z = z)$ takes on numerical values corresponding to the possible values z, we can do this by plotting the points $(z, P(Z = z))$ on graph paper. Another, slightly different, method that will have advantages for us later is illustrated for distribution (1) in Figure 1. Instead of plotting the point $(z, P(Z = z))$, one may draw a rectangle whose base is centered at z and whose area is equal to $P(Z = z)$. This graphical representation of a distribution is called a *histogram*. (Each bar of the histogram in Figure 1 has unit width, since the values of the random variable are consecutive integers, and height equal to the probability of the value to which it corresponds.)

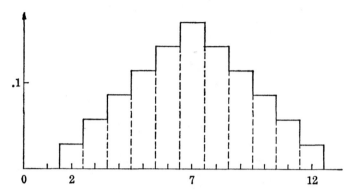

FIGURE 1. HISTOGRAM FOR POINTS ON TWO DICE

When dealing with experiments composed of several parts, one is frequently concerned with random quantities each of which relates to a different part. (Thus, in two throws of a die, one may be interested in the number of points on the first and the number of points on the second throw.) Let these be represented in the model by the random variables Z and W, and let E denote the event that Z takes on some particular value z, and F that W takes on some particular value w. Then, if a product model is being used, the probability that $Z = z$ and $W = w$ is the probability of the event (E and F), which by (3.2.13) equals $P(E)P(F)$.

. This argument shows quite generally that if Z and W correspond to different factors of a product model, we have

(2) $P(Z = z \text{ and } W = w) = P(Z = z)P(W = w)$ for all z, w.

Corresponding to the terminology for independent events mentioned in Section 3.2, it is customary to refer to random variables Z and W for which (2) holds as "independent."

Definition. Two random variables Z and W defined on any probability model are said to be *independent* if they satisfy (2).

In this terminology we have proved above that two random variables defined on different factors of a product model are independent. In fact, random variables defined on different factors will be the only independent random variables that we shall have occasion to discuss.

These remarks easily generalize to experiments with more than two parts. If the random variables $Z_1, Z_2, Z_3, \ldots$ represent random quantities relating to different parts and a product model is being used, then formula (3.2.14) shows that for any possible values z_1, z_2, z_3, ... we have

(3)
$$P(Z_1 = z_1 \text{ and } Z_2 = z_2 \text{ and } Z_3 = z_3 \text{ and } \ldots)$$
$$= P(Z_1 = z_1)P(Z_2 = z_2)P(Z_3 = z_3) \cdots .$$

Again extending the terminology for independent events, random variables Z_1, Z_2, Z_3, ... defined over any probability model are said to be *independent* if they satisfy (3) for all z_1, z_2, z_3, ... Thus in particular, random variables defined over different factors of a product model are independent.

PROBLEMS

1. Let B denote the number of boys in a three-child family. Assuming the 8 cases MMM, MMF, . . . , FFF to be equally likely, find the distribution of B.

2. In a surprise quiz, you are given two multiple-choice questions, one with three and the other with five possible answers. If you are completely unprepared and select one of the 15 possible answer combinations at random, find the distribution of the number Z of correct answers.

3. In two throws with a fair die, let D denote the difference between the number of points on the second and on the first die. Find the distribution of D.

4. For two throws with the loaded die whose probabilities are specified by formula (1.3.1) find the distribution of each of the following random variables:
    (i) the sum of points;
    (ii) the product of points.

5. Find $P(I = 1)$ when I is the indicator of the following event:
    (i) at least two sixes occur in three unrelated throws with a fair die;
    (ii) you will not be called upon for recitation in either of the classes of Problem 4.2.2.

6. Find $P(I = 0)$ when I is the indicator of the following event:
    (i) the reader of Problem 2.3.17 by purely random matching gets exactly two of the pictures right;
    (ii) at most three successes occur in four binomial trials with probability $\frac{1}{4}$ of success.

7. (i) If I is an indicator, can you find a relationship between I and I^2?
    (ii) If I is the indicator of an event E, show that $1 - I$ is also an indicator and find the event it indicates.
    (iii) If I_1, I_2 are indicators of two events E, F respectively, show that I_1I_2 is also an indicator, and find the event it indicates.

8. If I is an indicator with distribution $P(I = 1) = p$, $P(I = 0) = q$, find the distribution of the random variables (i) $2I$; (ii) $-I$; (iii) $3I + 2$.

9. Draw a histogram for the distribution of (i) the random variable D of Example

1; (ii) the random variable A of Example 1.4.1; (iii) the random variable B of Problem 1; (iv) the random variable S of Problem 1.7.10.

10. In two throws with the die of Problem 4, what is the probability that the number of points in both throws is even?

11. From a half-dozen eggs, of which two are rotten, two are selected at random. If D denotes the number of rotten eggs in the sample, (i) find $P(D = 0)$ and $P(D = 2)$; (ii) using the fact that the sum of the probabilities of all values is 1, find $P(D = 1)$.

12. From the city council of Example 1, a delegation of three is selected at random in April; in September it is necessary to send to Washington another delegation, of two members, and it is decided again to select it at random from the full council. What is the probability that both delegations consist entirely of Liberals? [Hint: Build a product model.]

13. Let I_1 and I_2 each indicate an event of probability $\frac{1}{2}$. What is the maximum possible value of $P(I_1I_2 = 1)$? The minimum?

14. If I_1, I_2 are two indicators, each having the distribution of the indicator I of Problem 8, and if I_1 and I_2 are independent, find the distribution of (i) $I_1 + I_2$; (ii) $I_2 - I_1$; (iii) I_1I_2.

15. Solve the three parts of the preceding problem if I_1 and I_2 are independent but have different distributions, namely if $P(I_1 = 1) = p_1$, $P(I_2 = 1) = p_2$.

16. In three tosses with a fair coin, let Z be the number of tails and W the number of heads.

(i) Show that Z and W have the same distribution.

(ii) Show that Z and W are different random variables.

[Hint: (i) Find the distribution of Z and that of W. (ii) Find a simple event e for which $Z(e)$ and $W(e)$ are not equal.]

17. If a model and two random variables Z and W are given by the following table

e	e_1	e_2	e_3	e_4	e_5
$P(e)$	$\frac{1}{8}$	$\frac{1}{8}$	$\frac{1}{8}$	$\frac{1}{4}$	$\frac{3}{8}$
$Z(e)$	3	6	6	3	8
$W(e)$	8	8	8	6	3

determine whether Z and W have the same distribution.

18. In the preceding problem, find the distribution of (i) $Z + W$; (ii) $W - Z$; (iii) ZW.

5.3 EXPECTATION

Like many of the ideas of probability, the notion of expectation arose in the study of gambling games, and was later found to have broader applications. As the motivating example, let us suppose that a gambler is to pay a casino a fixed price for the privilege of playing a chance game that may pay him varying amounts. If one disregards the casino's need to meet its overhead and make a profit, it would generally be agreed that a "fair" price for the game would be the amount the gambler might expect, on the average, to win. An example will illustrate the point.

EXAMPLE 1. *Throwing a die.* Suppose the gambler throws a fair die, after which the casino pays him as many dollars as there are points on the upper face. If he plays many times, the gambler may expect to win one dollar about $\frac{1}{6}$ of the time, two dollars $\frac{1}{6}$ of the time, and so forth. Thus his average winning per game would be about

$$\tfrac{1}{6}\cdot 1 + \tfrac{1}{6}\cdot 2 + \ldots + \tfrac{1}{6}\cdot 6 = \tfrac{1}{6}(1 + 2 + \ldots + 6) = \tfrac{21}{6} = \tfrac{7}{2}$$

dollars. This sum might be taken as a fair price to pay for the privilege of playing the game; a gambler paying this price would, in a long sequence of games, probably have an average winning per game near zero.

Study of the example will suggest how one may calculate for any random quantity the value which it may be expected to have on the average. So that we may refer to them easily, let us number the simple results, denoting them by $r_1, r_2, \ldots$, and denote the corresponding simple events by $e_1, e_2, \ldots$. In a sequence of n trials of the experiment, let $\#(r_1), \#(r_2), \ldots$ denote the numbers of occurrences of the results $r_1, r_2, \ldots$. Consider a random quantity that is represented in the model by the random variable Z. We shall now compute the sum of the n values taken on by the random quantity in these n trials. Whenever r_1 occurs, the value of the random quantity is $Z(e_1)$, so that the $\#(r_1)$ occurrences of r_1 contribute $Z(e_1)\#(r_1)$ to the sum. Similarly, the $\#(r_2)$ occurrences of r_2 contribute $Z(e_2)\#(r_2)$, and so on. Therefore the sum of all n values is

$$Z(e_1)\#(r_1) + Z(e_2)\#(r_2) + \ldots.$$

Dividing this total by n we see that the average of the observed values of the random quantity is

$$Z(e_1)f(r_1) + Z(e_2)f(r_2) + \ldots.$$

Our basic idea of probability as representing long-run frequency suggests that when n is large, the frequencies $f(r_1), f(r_2), \ldots$ should be close to the probabilities $P(e_1), P(e_2), \ldots$, so that the average of a great many observed values should be close to $Z(e_1)P(e_1) + Z(e_2)P(e_2) + \ldots$. This is what we would expect the random quantity to be on the average. It is called the *expected value* of Z, or the *expectation* of Z, and is denoted by $E(Z)$.

Definition. The expected value of the random variable Z is

(1) $$E(Z) = Z(e_1)P(e_1) + Z(e_2)P(e_2) + \ldots.$$

The expectation of Z in the model represents the long-run average value of the random quantity that Z represents, just as probability represents long-run frequency.

EXAMPLE *1. Throwing a die (continued).* Let us apply the definition to our motivating example. If X represents the number of points showing when a fair die is thrown, and we use the model of Example 1.3.1 which assigns probability $\frac{1}{6}$ to each of the six simple events, then $E(X)$ is seen to have the value $\frac{7}{2}$ calculated above.

Just as it is important to make a conceptual distinction between a *result*, which is something that may happen when a random experiment is performed, and the *event* that represents this result in the mathematical model for the experiment, it is important to distinguish between a *random quantity* and the *random variable* that represents this quantity in the model. However, in discussing specific applications it is often convenient to be somewhat careless, and to employ the same terms and notations for both. Thus, we may say that X "is" the number of points showing rather than that X "represents in the model" the number of points showing, or to speak of the "expected number of points" when in reality we mean the "expected value of the random variable representing the number of points." This short-cut usage is customary, and is exactly parallel to that discussed for results and events in Section 1.4.

EXAMPLE *2. The delegation.* What is the expected value of D, the random variable representing the number of Conservatives in the delegation of Example 2.1? The possible values of D are 0, 1, 2, 3 with probabilities $\frac{1}{35}$, $\frac{12}{35}$, $\frac{18}{35}$, $\frac{4}{35}$ respectively. Therefore

$$E(D) = 0 \cdot \tfrac{1}{35} + 1 \cdot \tfrac{12}{35} + 2 \cdot \tfrac{18}{35} + 3 \cdot \tfrac{4}{35} = \tfrac{12}{7} = 1.71.$$

EXAMPLE *3. Bridge bonus.* In the game of bridge, a rubber is won by the team that first wins two games. If the team wins a rubber by the score of two games to none (2–0), it gets a bonus of 700 points. If it wins the rubber 2–1, its bonus is 500 points. It sometimes happens that a rubber is interrupted when the score stands 1–0. What would be a fair bonus to give to the leading team in that case?

Let us suppose that each team has probability $\frac{1}{2}$ of winning each game, and that the games are unrelated, so that the binomial trials model with $p = \frac{1}{2}$ may be used. If the rubber had been continued, there would have been three possible outcomes, shown in the first two columns below. The third column lists the bonus won by the leading team in each case, while the last column gives the probabilities of the various possible outcomes.

Leading team on		Bonus	Probability
2nd game	3rd game		
Wins	—	700	$\frac{1}{2}$
Loses	Wins	500	$\frac{1}{4}$
Loses	Loses	−500	$\frac{1}{4}$

The expected bonus is

$$700 \cdot \tfrac{1}{2} + 500 \cdot \tfrac{1}{4} - 500 \cdot \tfrac{1}{4} = 350.$$

Actually, the rules of bridge provide a bonus of only 300 points. This may reflect the fact that the leading team (called "vulnerable") is placed at a disadvantage by the rules, so that its chance of winning the next game is somewhat less than $\frac{1}{2}$, and the binomial trials model is not appropriate.

EXAMPLE 4. *Gambling systems.* A gambler who has 7 dollars plays the following system. At the first toss of a coin, he bets 1 dollar on heads, and quits if he wins. If he loses, he bets 2 dollars on heads at the second toss, and quits if he wins. If he loses again, he bets his final 4 dollars on heads at the third toss. His system gives him probability $\frac{7}{8}$ of winning a dollar but what is the expected value of his final gain G? We distinguish four simple events, with the corresponding gains g and probabilities that $G = g$ as shown.

Event	H	TH	TTH	TTT
g	1	1	1	−7
$P(G = g)$	$\frac{1}{2}$	$\frac{1}{4}$	$\frac{1}{8}$	$\frac{1}{8}$

The expected gain is then

$$E(G) = 1 \cdot \tfrac{1}{2} + 1 \cdot \tfrac{1}{4} + 1 \cdot \tfrac{1}{8} - 7 \cdot \tfrac{1}{8} = 0.$$

Thus the gambler will on the average come out even: the game is "fair." Although he has a good chance to win a small amount, this is counterbalanced by the large loss he sustains if he loses. (It is a general fact, not recognized by many advocates of gambling systems, that in a game that is fair on each play, no system can give a positive expected gain.)

EXAMPLE 5. *Lottery.* The following model provides the basis for the important topic of sampling by variables which will be discussed in the second part of the book. Suppose that the amount of the prize in a lottery is to be determined by drawing a ticket from a box. The box contains N tickets, on each of which is written an amount. Let v_1 be the amount written on the first ticket, v_2 the amount on the second ticket, and so forth up to v_N. The tickets are thoroughly mixed and one is drawn at random. The prize is then equal to the amount written on the ticket that is drawn. (Most actual lotteries are of course conducted in a somewhat different way, but "lottery" will be a convenient label for the model described above.)

We shall assume that each ticket has the same probability $1/N$ of being drawn. Since the amount of the prize is determined by the result of the random draw, it is a random quantity; the random variable representing it in the model will be denoted by Y. The expected value of Y by formula (1) is

$$E(Y) = v_1 \cdot \frac{1}{N} + v_2 \cdot \frac{1}{N} + \ldots + v_N \cdot \frac{1}{N} = \frac{1}{N}(v_1 + v_2 + \ldots + v_N),$$

which is the arithmetic mean of the amounts written on the N tickets. We shall denote such an arithmetic mean by placing a bar over the symbol that denotes the values being averaged:

$$\bar{v} = \frac{v_1 + v_2 + \ldots + v_N}{N} \qquad \text{(read "v bar").}$$

The formula

$$(2) \qquad\qquad\qquad E(Y) = \bar{v}$$

asserts that the expected value of the prize is the arithmetic mean of the amounts shown on the tickets. As a numerical illustration suppose there are $N = 10$ tickets, showing amounts

$$v_1 = 5, \quad v_2 = v_3 = v_4 = 2, \quad v_5 = v_6 = \ldots = v_{10} = 0.$$

The expectation of Y is then

$$\bar{v} = (5 + 2 + 2 + 2 + 0 + \ldots + 0)/10 = 1.1.$$

If the amounts are in dollars, \$1.10 would be a fair price to pay for playing the lottery.

PROBLEMS

1. Throw a die 100 times, recording the numbers shown. Compare the average of these numbers with the expected value $\frac{7}{2}$ found in Example 1.

2. Let D be a digit drawn at random from $0, 1, \ldots, 9$. Find $E(D)$ and compare it with the average value of the digits in Table 3.4.1.

3. Find the expected number B of boys in a three-child family under the assumptions of Problem 2.1.

4. If X is the number of points on a loaded die with probabilities $P(1) = .21$, $P(2) = P(3) = P(4) = P(5) = .17$, $P(6) = .11$, determine (i) $E(X)$, (ii) $E(X^2)$.

5. Find the expected value of the random variable R of Problem 1.4.7.

6. Find the expected value of the random variable Z of Problem 2.2.

7. For two throws with the loaded die whose probabilities are specified by formula (1.3.1), find the expected value of

(i) the maximum of the two numbers of points;

(ii) the difference between the numbers of points on the second and first throw.

8. Find the expected number of pictures matched correctly in Problem 2.3.17.

9. Suppose that in training new workers to perform a delicate mechanical operation, the probabilities that the worker will be successful on his first, second and third attempt are $p_1 = \frac{1}{10}$, $p_2 = \frac{3}{10}$, $p_3 = \frac{6}{10}$ respectively. Find the expected number of successes the worker will have in these three attempts.

10. In Problem 2.17, find (i) $E(Z)$; (ii) $E(W)$; (iii) $E(Z + W)$; (iv) $E(ZW)$.

11. In Example 3, for what value of the probability that the vulnerable team will win, is the 300 point bonus fair? [Hint: Find the expected value of the bonus when the probability is p.]

12. A lottery of the type of Example 5 has first prize of \$1000, second prize \$500, and five \$100 prizes. Would you want to pay \$1 for a ticket if there were (i) 1000 tickets, (ii) 3000 tickets?

13. Study Example 4, and devise a system whereby a gambler with 31 dollars can arrange to win a dollar with probability $\frac{31}{32}$. With this system, what is his expected gain?

14. If Z denotes the number of plays the gambler of Example 4 will make before quitting, find $E(Z)$.

15. Give the value set and distribution of the random variable Y in the numerical illustration of Example 5.

16. Determine for each of the following two statements whether it is true:

(i) The expected value of a random variable X can never exceed all the values in the value set of X.

(ii) The expected value of a random variable X is always equal to one of the values in the value set of X.

17. Would you prefer to have half a chance to win \$2,000,000, or a tenth of a chance to win \$11,000,000 (tax free)? Discuss your preference in the light of the concept of expected winning.

18. (i) If each value of a random variable is doubled, what happens to its expected value?

(ii) If each value of a random variable is increased by three, what happens to its expected value?

19. In the famous Petersburg game, the gambler wins two roubles with probability $\frac{1}{2}$, four rubles with probability $\frac{1}{4}$, eight rubles with probability $\frac{1}{8}$, etc.

(i) Before making any calculations, how much would you be willing to pay to play this game?

(ii) Try to find a fair price for playing the game by using (1).

5.4 PROPERTIES OF EXPECTATION

Formula (1) of the preceding section

$$(1) \qquad E(Z) = Z(e_1)P(e_1) + Z(e_2)P(e_2) + \ldots$$

may be put into an alternative form which is often simpler to use. To introduce the idea of the simplification, let us recall Example 3.5. In that example, as so often happens, the random variable takes on the same value for two or more simple events. Thus, Y has the value 2 for three different tickets, which together contribute to $E(Y)$ the amount

$$2 \cdot \tfrac{1}{10} + 2 \cdot \tfrac{1}{10} + 2 \cdot \tfrac{1}{10} = 2 \cdot \tfrac{3}{10} = 2 \cdot P(Y = 2).$$

The single ticket with $Y = 5$ contributes $5 \cdot P(Y = 5)$ to the value of $E(Y)$ and the six tickets labeled 0 contribute nothing. Thus

$$E(Y) = 5 \cdot P(Y = 5) + 2 \cdot P(Y = 2) + 0 \cdot P(Y = 0)$$
$$= 5 \cdot \tfrac{1}{10} + 2 \cdot \tfrac{3}{10} + 0 \cdot \tfrac{6}{10} = 1.1.$$

The example suggests how formula (1) may be simplified in general. Let the distinct possible values of Z be denoted by $z_1, z_2, \ldots$. Consider first all of the simple events for which Z takes on the value z_1; let the set of these be E_1, so that $P(E_1) = P(Z = z_1)$. In the corresponding terms of (1), the first factor is always z_1. Their total contribution to (1) is therefore z_1 multiplied by the sum of the probabilities of the simple events in E_1, that is, $z_1 P(E_1)$. In just the same way, let E_2 denote the set of simple events for which Z takes on the value z_2. The contribution of these terms to (1) is then $z_2 P(E_2)$. By continuing in this way we see that (1) and therefore $E(Z)$ can be written as

$$E(Z) = z_1 \cdot P(E_1) + z_2 \cdot P(E_2) + \ldots,$$

or equivalently as

$$(2) \qquad E(Z) = z_1 P(Z = z_1) + z_2 P(Z = z_2) + \ldots.$$

When the probability distribution of Z is known, formula (2) is usually more convenient than formula (1). Furthermore, (2) shows that the expected value of a random variable depends only on the distribution of the random variable, that is, the values it can take on and the probabilities of taking on these values, and not on the random variable itself. (For an example of the distinction, see Problem 2.16.) For some purposes however, (1) is very convenient, as it does not require us to know the distribution of Z but only the model and the definition of Z.

EXAMPLE 1. *Gambling systems.* The expectation $E(G)$ of Example 3.4 can be computed by the use of (2) instead of (1). Since G takes on the values 1 and -7 with probabilities $P(G = 1) = \tfrac{7}{8}$ and $P(G = -7) = \tfrac{1}{8}$, we find $E(G) = 1 \cdot \tfrac{7}{8} - 7 \cdot \tfrac{1}{8} = 0$, as before.

EXAMPLE 2. *Indicators.* Let I be an indicator; that is, a random variable whose only values are 0 and 1. Then

$$E(I) = 0 \cdot P(I = 0) + 1 \cdot P(I = 1),$$

so that

$$E(I) = P(I = 1).$$

In words, the expected value of an indicator is the probability of the indicated event.

The expected value of a random variable Z is often used to represent the *center* of the distribution of Z. The interpretation of $E(Z)$ as a central value of the distribution is supported by two considerations.

(i) The expected value is exactly analogous to the physical concept of *center of gravity* of a mass distribution. Let us imagine that the z axis is a weightless rod with mass $P(Z = z_1)$ located at z_1, mass $P(Z = z_2)$ located at z_2, etc. Thus a unit of mass is distributed along the rod in the same way as the unit of probability is distributed along the z axis. The point at which the z axis would then be in balance is known as its center of gravity, and as we shall show below, this point is always at $E(Z)$. This is illustrated in Figure 1a for the distribution of the random variable D of Example 2.1, and in Figure 1b for the number B of heads that appear when three fair pennies are tossed (see Problem 5). In each case the center of gravity is marked by $\triangle$.

FIGURE 1.   CENTER OF GRAVITY

(ii) We shall say that the distribution of Z is *symmetric* about a point μ if it assigns equal probabilities to points equally distant from μ in both directions; that is, if

$$(3) \qquad P(Z = \mu + x) = P(Z = \mu - x) \qquad \text{for every } x.$$

(Thus in Figure 1b the distribution is symmetric about $\mu = 1.5$.) It can be shown (Problem 15) that then $E(Z) = \mu$, so that the expected value of a symmetrically distributed random variable is its "center of symmetry." The result is intuitively plausible from (i); for if the mass distribution is symmetric about μ, we would expect the rod to balance at μ.

In spite of its role as center of the distribution, $E(Z)$ need not be a typical, nor even a possible, value of Z (see for example Figure 1). Furthermore, if a distribution is highly asymmetric, it can happen that almost all of the probability lies on one side of the expected value. For example, if Z can take on only the two values 0 and 100 with probabilities $P(Z = 0)$

= .99 and $P(Z = 100) = .01$, then $E(Z) = 1$, so that 99% of the probability lies to the left of $E(Z)$.   Again, suppose one very rich family has a summer cottage in a village of poor fishermen.   If Z denotes the wealth of a family selected at random from this village, then $E(Z)$ will give a rather distorted picture of the prosperity of the village.

To avoid this difficulty, the *median* is sometimes proposed as an alternative value for the center of a distribution.   Roughly speaking, the median is the value which has half of the distribution on either side of it.

The equation

(4) $$E(Z) = z_1 P(E_1) + z_2 P(E_2) + \ldots$$

is valid (and useful) under somewhat more general conditions than those leading to (2).   Suppose that $E_1, E_2, \ldots$ are exclusive events such that every simple event belongs to one of the E's.   If $Z(e)$ has the same value z_1 for all simple events e in E_1, the same value z_2 for all simple events in E_2, and so on, then the argument leading to (2) shows (4) to hold even though $z_1, z_2, \ldots$ may not all be distinct.

For readers acquainted with elementary statics, we now give a proof of the relationship between expectation and center of gravity stated above.   We have to show that the sum of the torques tending to turn the rod about the point $E(Z)$,

$$[z_1 - E(Z)]P(Z = z_1) + [z_2 - E(Z)]P(Z = z_2) + \ldots,$$

is zero.   Since the probabilities of $(Z = z_1)$, $(Z = z_2)$, $\ldots$ add up to one, we have

$$E(Z) = E(Z)P(Z = z_1) + E(Z)P(Z = z_2) + \ldots.$$

Subtracting this equation from (2) gives the desired result.

PROBLEMS

1.   Use formula (2) to find the expected value of a ticket in the lottery of Problem 3.12(i) and 3.12(ii).

2.   Under the assumptions of Problem 2.11, find the expected number of rotten eggs in the sample.

3.   Find the expected number of keys that must be tried to open the door of Problem 1.5.7.

4.   Use (2) to find
   (i) the expectation of the random variable Z of Problem 2.2;
   (ii) the expected value of the random variable T whose distribution is given by (2.1).

5.   Calculate the distribution, and use (2) to obtain the expectation, of the number B of heads that appear when three fair pennies are tossed.   (See Figure 1b.)

6.   Find the expected value of
   (i) the random variable D of Problem 2.3;
   (ii) the maximum M of the numbers of points obtained in two throws with a fair die (cf. Problem 1.7.6).

7. Find the distributions of the random variables of parts (i) and (ii) of Problem 3.7, and then use formula (2) to find their expectations. Compare your results with those of Problem 3.7.

8. Find the expectations of the two random variables of Problem 2.4.

9. Find the expected number of customers preferring the new formula (i) in Problem 1.7.10; (ii) in Problem 1.7.12.

10. Find the expectations of the four random variables defined in parts (i)–(iv) of Problem 1.7.14.

11. Let I be an indicator random variable. Can you find any relationship between $E(I)$ and $E(I^2)$? Between I and I^2? Generalize to higher powers of I.

12. Check for each of the following distributions whether it is symmetric, and if so find the point of symmetry.
    (i) The distribution of T given by (2.1).
    (ii) The distribution of D in Example 2.1.
    (iii) The distribution of G in Example 3.4.
    (iv) The distribution of the number of keys that must be tried to open the door of Problem 1.5.7.
    (v) The distribution of B in Problem 5.

13. Let Z be a random variable which takes on the values $1, 2, \ldots, N$ each with probability $1/N$. Show that this distribution is symmetric about the value $\frac{1}{2}(N + 1)$ and hence that $E(Z) = \frac{1}{2}(N + 1)$. Use (2) to derive the identity

$$(5) \qquad\qquad 1 + 2 + \ldots + N = \frac{N(N + 1)}{2},$$

which will be used in Sections 6.8 and 12.4.

14. Find the expected gain for the gambling schemes of (i) Problem 4.3.15(i), (ii) Problem 4.3.16(i).

15. If Z is symmetric about μ, show that
    (i) the random variables $Z - \mu$ and $\mu - Z$ have the same distributions;
    (ii) $E(Z) = \mu$.
[Hint: (i) $P(\mu - Z = x) = P(Z - \mu = -x)$; apply (3). (ii) By (i), $E(Z - \mu) = E(\mu - Z)$.]

16. If Z and $-Z$ have the same distribution, show that Z is symmetric about zero.

5.5 LAWS OF EXPECTATION

In this section we shall develop several useful laws connecting the expected values of certain random variables which are defined on the same model. These laws often permit us to calculate expectations by expressing them in terms of other expectations that we already know, and thus to avoid the necessity of using formulas (1) or (2) of the preceding section. With their aid we shall obtain the expectations of several important random variables.

The simplest kind of random variable is one capable of assuming only a single value, say c, so that $Z(e) = c$, for all e in $\mathcal{E}$. Such a random variable, which does not vary at all, is called a *constant* random variable. Since it is always equal to c, it will clearly be equal to c on the average, and so its expected value must also be c. If we conveniently use c to denote a random variable that is constantly equal to the number c, this may be expressed by

$$(1) \qquad\qquad E(c) = c$$

and proved formally by reference to equation (2) of the preceding section. In fact, if c is the only possible value of Z, $P(Z = c) = 1$, so that

$$E(Z) = c \cdot P(Z = c) = c \cdot 1 = c.$$

If Z is any random variable, and c is any constant, we can obtain a new random variable by adding c to the value of Z. Let us denote the new random variable by $Z + c$. For example, if Z denotes the random amount a salesman will earn through commissions during a month and if c is his fixed monthly salary, then $Z + c$ will be his total monthly income. The expectation of the random variable $Z + c$ is related to that of Z by equation

$$(2) \qquad\qquad E(Z + c) = E(Z) + c.$$

In words, when a constant c is added to a random variable, the same constant is added to its expected value. This relation is just what one would expect: if Z is increased by c in each case, it will be increased by c on the average. A formal proof will be given following the proof of equation (4) below.

Relation (2) concerns the effect of adding a constant; analogously, multiplication by a constant gives

$$(3) \qquad\qquad E(cZ) = c \cdot E(Z).$$

This is again intuitively plausible since if the value of Z is multiplied by the constant c in each case, its average value will be multiplied by the same constant. (The proof of this is left as an exercise. See Problem 1.)

We now come to the most important law, the *addition law of expectation*: for any two random variables Z and W,

$$(4) \qquad\qquad E(Z + W) = E(Z) + E(W).$$

In words, the expectation of the sum of two random variables is the sum of their expectations. To get an intuitive understanding of the addition law, let us imagine a gambler who is paid an amount Z by the casino, depending on the roll of a die. In addition, the gambler makes a side-bet with a colleague, which pays him W. Then $Z + W$ is the total amount he receives on a play of the game. The expectation $E(Z + W)$ stands for the average of his total receipts in a long series of plays. Clearly the average value of his total winnings from both sources will equal the sum of his average winnings from the two sources separately. If we replace average by expectation, this statement becomes (4).

To obtain a formal proof of (4), we note that for the simple event e_1 the random variable $Z + W$ takes on the value $Z(e_1) + W(e_1)$, for e_2 the value $Z(e_2) + W(e_2)$, etc. Therefore, from formula (3.1),

$$E(Z + W) = [Z(e_1) + W(e_1)]P(e_1) + [Z(e_2) + W(e_2)]P(e_2) + \ldots.$$

By rearranging terms, the right-hand side may be rewritten

$$[Z(e_1)P(e_1) + Z(e_2)P(e_2) + \ldots] + [W(e_1)P(e_1) + W(e_2)P(e_2) + \ldots];$$

but this is of course equal to $E(Z) + E(W)$, which completes the proof.

If in (4) we take for W the constant random variable c, then (4) reduces to $E(Z + c) = E(Z) + E(c)$. By (1), this is equal to $E(Z) + c$, which proves (2).

It is easy to extend the addition law to more than two random variables. Thus, we may regard $Z + W + V$ as the sum of the random variable $(Z + W)$ and the random variable V. Therefore, by (4),

$$E(Z + W + V) = E(Z + W) + E(V).$$

Now we may apply (4) again to break up $E(Z + W)$ into $E(Z) + E(W)$, showing

$$E(Z + W + V) = E(Z) + E(W) + E(V).$$

More generally, if $Z_1, Z_2, \ldots, Z_n$ are any n random variables,

$$(5) \qquad E(Z_1 + Z_2 + \ldots + Z_n) = E(Z_1) + E(Z_2) + \ldots + E(Z_n).$$

This *general addition law* is useful when we can express a complicated random variable as the sum of several simple ones whose expectations are easier to compute.

EXAMPLE 1. *Dice.* Let T be the total number of points showing when two dice are thrown. If X_1 and X_2 denote the numbers on the two dice separately, then $T = X_1 + X_2$, and hence $E(T) = E(X_1) + E(X_2)$. By Example 3.1, $E(X_1) = E(X_2) = \frac{7}{2}$, so that $E(T) = 7$. The same answer can be obtained less simply by a direct evaluation of $E(T)$ (see Problem 4.4(ii)).

EXAMPLE 2. *Lottery.* Recall Example 3.5. We shall now modify it by supposing that two tickets are drawn from the box, without replacement. Let Y_1 denote the amount written on the first ticket drawn, while Y_2 denotes the amount on the second ticket drawn. The total amount paid in prizes is $Y_1 + Y_2$, and by (4) we have $E(Y_1 + Y_2) = E(Y_1) + E(Y_2)$.

To evaluate this sum, we shall use the model for ordered sampling developed in Section 2.3, with $s = 2$. According to the equivalence law, on each of the two draws each ticket has probability $1/N$ of appearing.

Therefore Y_1 and Y_2 each has the same distribution as the random variable Y considered in Example 3.5.   It follows from (3.2) that $E(Y_1) = E(Y_2) = \bar{v}$, and hence the expected total paid out is $E(Y_1 + Y_2) = \bar{v} + \bar{v} = 2\bar{v}$. (Notice that the same would be true if the sampling were with replacement, since in that case Y_1 and Y_2 would again have the same distribution as Y.)

An extension of the argument shows that if $s > 2$ tickets are drawn (with or without replacement), the expected total paid out is $s\bar{v}$.

EXAMPLE 3.   *Stratified sampling.*   In a survey of a three-block district, one dwelling is to be sampled from each block.   Some of the dwellings are apartment houses, as shown below:

Block	1	2	3
Number of apartment houses	2	5	1
Number of private residences	11	4	9

What is the expected value of the number A of apartment houses in the sample?

Let I_1 denote the number of apartment houses drawn from Block 1. Clearly, I_1 will be either 0 or 1, and in fact I_1 indicates that the dwelling drawn from Block 1 is an apartment house.   Similarly, let I_2 and I_3 denote the numbers of apartment houses drawn from Blocks 2 and 3.   Thus $I_1 + I_2 + I_3 = A$ is the total number of apartment houses in the sample.

By the addition law (5),

$$E(A) = E(I_1) + E(I_2) + E(I_3).$$

We know (Example 4.2) that the expected value of an indicator is just the probability of the indicated event.   Therefore $E(I_1)$ is the probability that the dwelling drawn from Block 1 is an apartment house, so that $E(I_1) = \frac{2}{13} = .154$.   Similarly, $E(I_2) = \frac{5}{9} = .556$ and $E(I_3) = \frac{1}{10} = .100$. Finally therefore

$$E(A) = .154 + .556 + .100 = .810.$$

The method used in this example is widely applicable.   Often the number of occurrences of some event may be represented as a sum of indicators, each of which specifies whether the event occurred on one occasion.   The reader may convince himself of the usefulness of the *method of indicators by* calculating $E(A)$ by the methods of Section 4.

EXAMPLE 4.   *Matching.*   A box contains five tickets numbered from 1 to 5. They are drawn out one by one.   What is the expected value of the number M of occasions when the number on the ticket matches the ordinal number of the draw (e.g., ticket number 2 obtained on second draw)?

Let $I_1, \ldots, I_5$ indicate that a match occurs on draws one, $\ldots$, five. (Thus $I_2 = 1$ if and only if the second ticket drawn bears number 2, etc.) Clearly $M = I_1 + \ldots + I_5$. Let us compute $E(I_2)$. Since each ticket has the same probability of appearing on the second draw (equivalence law of ordered sampling), $P(I_2 = 1) = \frac{1}{5} = E(I_2)$. Similarly $E(I_1) = E(I_3) = E(I_4) = E(I_5) = \frac{1}{5}$, and hence $E(M) = 1$.

The method applies equally well with any number of tickets; the expected number of correct matchings is always 1.

It is natural to ask whether there is a multiplication law of expectation analogous to the addition law: is the expected value of the product of two random variables equal to the product of their expectations? That this is not always true may be seen by a simple example. Let Z be the number of heads, and W be the number of tails, observed when a fair penny is tossed. Then $E(Z) = \frac{1}{2}$ and $E(W) = \frac{1}{2}$, so that $E(Z) \cdot E(W) = \frac{1}{4}$. But the product ZW is always equal to 0, so that $E(ZW) = 0 \neq \frac{1}{4}$.

There is however an important case in which the multiplication law does hold. Suppose that the random variables are defined on a product model with two factors, and that Z refers only to the first and W only to the second factor model. (For example, the random variables X_1 and X_2 of Example 1 refer to the throws of two different dice, for which a product model is used.)

We then have the following *multiplication law of expectation:*

(6) $E(ZW) = E(Z) \cdot E(W)$ if Z and W are defined on different factors.

The proof of (6) rests on the definition (3.1.4) of probabilities in a product model. Consider the value of the product ZW corresponding, for example, to the simple event (e_1 and f_2) of the product model. Since Z refers only to the first factor, Z in this case has the value $Z(e_1)$. Similarly W has the value $W(f_2)$, so that ZW has the value $Z(e_1)W(f_2)$. By (3.1.4), the probability of this event is p_1q_2. Thus the contribution to the expectation of ZW is $Z(e_1)W(f_2)p_1q_2$. The contributions of the other simple events are obtained analogously and hence, by the definition (3.1) of expectation, $E(ZW)$ equals

$$
\begin{aligned}
&Z(e_1)W(f_1)p_1q_1 + Z(e_1)W(f_2)p_1q_2 + \ldots \\
&+ Z(e_2)W(f_1)p_2q_1 + Z(e_2)W(f_2)p_2q_2 + \ldots \\
&+ \ldots .
\end{aligned}
$$

Taking out the factor $Z(e_1)p_1$ from the first row, $Z(e_2)p_2$ from the second row, etc., we find

$$
\begin{aligned}
E(ZW) &= [Z(e_1)p_1 + Z(e_2)p_2 + \ldots][W(f_1)q_1 + W(f_2)q_2 + \ldots] \\
&= E(Z)E(W).
\end{aligned}
$$

The multiplication law of expectation is in fact true for any independent random variables Z and W (Problem 21); however, we shall have occasion to use it only in the special case proved above.

PROBLEMS

1. Prove that $E(cZ) = cE(Z)$.

2. Let Y denote the amount you will win in a gambling game and suppose that $E(Y) = \$2.00$.
  (i) If you have to pay $2.50 to be allowed to play, what is your total expected gain or loss?
  (ii) State what law you have used in part (i).

3. How many points may we expect
  (i) when three dice are thrown;
  (ii) when n dice are thrown?

4. From a half-dozen eggs of which two are rotten, four are selected at random. Find the expected value of the number R of rotten eggs in the sample using the answer to Problem 4.2 by representing R as $R = 2 - D$ where D is the number of rotten eggs not included in the sample.

5. Prove that
$$E(Z - W) = E(Z) - E(W).$$
[Hint: Write $Z - W = Z + (-1) \cdot W$.]

6. Prove that
$$E(cZ + dW) = cE(Z) + dE(W)$$
where c and d are constants.

7. If $Z_1, Z_2, \ldots, Z_n$ are any n random variables having a common expectation ζ, and if $\overline{Z}$ denotes their average, prove that $E(\overline{Z}) = \zeta$.

8. Suppose that 10 percent of the parcels sent abroad do not reach their destination. A person wishing to send two dresses could send them either (a) in a single parcel or (b) in two separate parcels.
  (i) For both methods, find the value the sender can expect to reach the destination if the dresses are worth $10 and $15 respectively.
  (ii) For both methods find the probability that both dresses will reach their destination.
  (iii) For both methods find the probability that at least one of the dresses will reach its destination.
  (iv) For each of the criteria (i), (ii), (iii) determine which of the methods is preferable.
[Hints: (ii), (iii) Assume that different mailings are unrelated so that a product model is appropriate for the mailing of two packages.]

9. Use the method of indicators to find the expectation of the following random variables:

    (i) B of Problem 2.1;

    (ii) Z of Problem 2.2;

    (iii) D of Problem 2.11;

    (iv) the number of keys that must be tried in Problem 1.5.7.

10. Use (4) to find the expectation of the random variable W defined in Problem 1.7.14(iii).

11. Use the result of Problem 5 to find the expectation of the random variable D of Problem 1.7.14(iv).

12. In two throws with a die which may be loaded, let D denote the difference between the number of points on the second and on the first throw. What can you say about the expectation of D?

13. Let A be the number of aces in a five-card poker hand. Calculate $E(A)$ by the following two methods:

    (i) letting $I_1, \ldots, I_5$ indicate the appearance of an ace on the first, $\ldots$, fifth card dealt;

    (ii) letting J_1, J_2, J_3, J_4 indicate that the ace of spades, hearts, diamonds, clubs appears in the hand. (See Problem 2.2.17.)

14. If five dice are thrown, how many different faces may be expected to show? [Hint: Let I_1 indicate that at least one die shows an ace, etc.]

15. If three persons are seated at random at a lunch counter with six seats, how many persons may expect to have no one sitting next to them? [Hint: This problem may be worked by listing the $\binom{6}{3} = 20$ equally likely arrangements, and for each the number of isolated persons. Alternatively, one may let I_1 indicate that the first seat is occupied by an isolated person, etc.]

16. Under the assumptions of Problem 4.2.20, find

    (i) the expected rank assigned to Brand A by a single judge,

    (ii) the expected value of the sum of the ranks assigned to Brand A by the 5 judges,

    (iii) the expected average value of the ranks assigned to Brand A by the 5 judges.

17. Use the method of indicators to find the expected number of corner plots in the sample of Problem 4.2.15.

18. Use the method of indicators to find the expected number of successes in Problem 3.9.

19. (i) Find the distribution of $X_1 \cdot X_2$ in Example 1 by considering its value in each of the 36 equally likely cases.

    (ii) Use (i) to find $E(X_1 X_2)$ and verify that it equals $E(X_1) \cdot E(X_2) = (\frac{7}{2})^2$.

20. Use (6) to prove

$$P(E \text{ and } F) = P(E)P(F)$$

if E and F relate to different factors of a product model.

[Hint: Let Z indicate E and let W indicate F, and note that then ZW is also an indicator random variable.]

21. Prove that (6) holds whenever Z and W are independent.
[Hint: Compute the left-hand side of (6) using (4.4) where $E_1, \ldots$ are the events $(Z = z_1$ and $W = w_1)$, etc.]

5.6 VARIANCE

We noted in Section 4 that the expected value of a random variable can be used to specify the center of the distribution of the random variable. However, two distributions with the same expectation, and hence centered at the same point, may still be very different. Thus $E(Z) = 0$ if Z takes on the values -1, 0, 1 with probabilities $\frac{1}{4}$, $\frac{1}{2}$, $\frac{1}{4}$ respectively; but also $E(W) = 0$ if W takes on the values -5, 0, 5 with probabilities $\frac{1}{4}$, $\frac{1}{2}$, $\frac{1}{4}$ respectively (see Figure 1). A much more complete idea of a distribution is obtained if in addition to knowing its expectation, which tells us where it

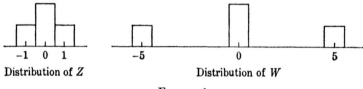

Distribution of Z Distribution of W

FIGURE 1.

is centered, we also know the extent to which the distribution is dispersed away from this center. In this section we shall study the quantity that is most commonly used to specify the dispersion of a distribution.

To simplify the writing, we shall denote the expected value $E(Z)$ of the random variable Z by the Greek letter ζ. It is customary to use, as a measure of dispersion, the expected value of $(Z - \zeta)^2$. This is known as the *variance* of Z, and is denoted by Var(Z).

Definition. The variance of the random variable Z is

$$(1) \qquad \qquad \text{Var}(Z) = E(Z - \zeta)^2.$$

In words, the variance of a random variable is the expectation of the square of the difference between Z and its expectation. (This measure of dispersion was introduced by K. F. Gauss (1777–1855).)

To see in what sense (1) measures the dispersion of the distribution of Z, note that if the distribution is widely dispersed, Z will with high probability differ greatly from ζ, so that $(Z - \zeta)^2$ will probably be large, and hence will be large "on the average." Thus if the distribution of Z is widely dispersed, Var(Z) will be large. (Variance is the analog of the physical concept of the *moment of inertia* of a mass distribution about its center of gravity.)

The reader may wonder why the difference $Z - \zeta$ is squared in defining variance. Suppose that instead we had tried to define a measure of dispersion by using the expected value of $Z - \zeta$ itself. Since $E(Z - \zeta) = E(Z) - \zeta$, we see that $E(Z - \zeta) = 0$, so that as our measure of dispersion we would always get zero. The reason for this is that $Z - \zeta$ is sometimes positive and sometimes negative, as Z happens to fall to the right or to the left of ζ, the positive and negative values just balancing out on the average. This difficulty is avoided by squaring, since $(Z - \zeta)^2$ can never be negative. There are other ways of making $Z - \zeta$ positive; for example, one might use $E|Z - \zeta|$ as a measure of dispersion, where the absolute value $|Z - \zeta|$ means the quantity $Z - \zeta$ always taken with positive sign. This measure was proposed by P. S. Laplace (1749–1827). Laplace's measure is also reasonable, but it turns out that Gauss' measure is easier to work with and has nicer mathematical properties.

EXAMPLE 1. *Throwing a die.* We shall calculate the variance of the number X of points showing when a fair die is thrown. In Example 3.1 the expected value of X was computed to be $E(X) = \frac{7}{2}$. The table below lists each possible value x of X, the corresponding probability, and the value of $(x - \frac{7}{2})^2$. (For example, when $x = 2$, $(x - \frac{7}{2})^2 = (2 - \frac{7}{2})^2 = \frac{9}{4}$.)

x	1	2	3	4	5	6
$P(X = x)$	$\frac{1}{6}$	$\frac{1}{6}$	$\frac{1}{6}$	$\frac{1}{6}$	$\frac{1}{6}$	$\frac{1}{6}$
$(x - \frac{7}{2})^2$	$\frac{25}{4}$	$\frac{9}{4}$	$\frac{1}{4}$	$\frac{1}{4}$	$\frac{9}{4}$	$\frac{25}{4}$

Using formula (1) and recalling the definition of expectation (formula 3.1), we find

$$\mathrm{Var}(X) = E(X - \tfrac{7}{2})^2 = \tfrac{25}{4} \cdot \tfrac{1}{6} + \tfrac{9}{4} \cdot \tfrac{1}{6} + \tfrac{1}{4} \cdot \tfrac{1}{6} + \tfrac{1}{4} \cdot \tfrac{1}{6} + \tfrac{9}{4} \cdot \tfrac{1}{6} + \tfrac{25}{4} \cdot \tfrac{1}{6}$$
$$= \tfrac{70}{24} = \tfrac{35}{12}.$$

As further illustrations, the reader may check that the distributions of Figure 1 have $\mathrm{Var}(Z) = \frac{1}{2}$ and $\mathrm{Var}(W) = \frac{25}{2}$.

There is an alternative to formula (1) which is sometimes more convenient for computing variances. Squaring out $(Z - \zeta)^2$ gives $Z^2 - 2Z\zeta + \zeta^2$. Using the laws of expectation, we then find $\mathrm{Var}(Z) = E(Z^2) - 2\zeta E(Z) + \zeta^2$. Since $E(Z) = \zeta$, we see that $2\zeta E(Z) = 2\zeta^2$, and hence

$$(2) \qquad\qquad \mathrm{Var}(Z) = E(Z^2) - \zeta^2.$$

Formula (2) is more convenient than (1) in that it dispenses with the differences $Z - \zeta$. Thus, in Example 1,

$$E(X^2) = 1 \cdot \tfrac{1}{6} + 4 \cdot \tfrac{1}{6} + 9 \cdot \tfrac{1}{6} + 16 \cdot \tfrac{1}{6} + 25 \cdot \tfrac{1}{6} + 36 \cdot \tfrac{1}{6} = \tfrac{91}{6}$$

so that by formula (2), $\mathrm{Var}(X) = \frac{91}{6} - (\frac{7}{2})^2 = \frac{35}{12}$; this agrees with the value found earlier with the aid of (1).

EXAMPLE 2. *Lottery.* As in Example 3.5 suppose that a box contains N

tickets, on which are written amounts $v_1, \ldots, v_N$. If Y denotes the amount written on a ticket drawn at random, we saw that $E(Y) = \bar{v}$, the arithmetic mean of $v_1, \ldots, v_N$. Let us now consider the variance of Y. By formula (1),

$$\mathrm{Var}(Y) = (v_1 - \bar{v})^2 \cdot \frac{1}{N} + \ldots + (v_N - \bar{v})^2 \cdot \frac{1}{N}$$

$$= [(v_1 - \bar{v})^2 + \ldots + (v_N - \bar{v})^2]/N.$$

This is the arithmetic mean of the N quantities $(v_1 - \bar{v})^2, \ldots, (v_N - \bar{v})^2$, which are the squared deviations of the amounts $v_1, \ldots, v_N$ from their arithmetic mean $\bar{v}$, and is called the *mean square deviation* of the numbers $v_1, \ldots, v_N$. We shall denote it by τ^2, so that

(3) $$\tau^2 = [(v_1 - \bar{v})^2 + \ldots + (v_N - \bar{v})^2]/N.$$

The formula

(4) $$\mathrm{Var}(Y) = \tau^2$$

will be important in Section 9.1 in the discussion of sampling by variables. If the v's have the numerical values assumed in Example 3.5, we find

$$\tau^2 = [(5 - 1.1)^2 + (2 - 1.1)^2 + (2 - 1.1)^2 + (2 - 1.1)^2$$
$$+ (0 - 1.1)^2 + \ldots + (0 - 1.1)^2]/10$$
$$= [(3.9)^2 + 3(0.9)^2 + 6(1.1)^2]/10 = 2.49.$$

There is an alternative way to compute τ^2 that is sometimes more convenient. Since $E(Y) = \bar{v}$ and $\mathrm{Var}(Y) = \tau^2$, it follows from (2) that $\tau^2 = E(Y^2) - \bar{v}^2$ and hence

(5) $$\tau^2 = \frac{1}{N} (v_1{}^2 + \ldots + v_N{}^2) - \bar{v}^2.$$

Using this method, we find for the numerical example above

$$E(Y^2) = (5^2 + 3 \cdot 2^2 + 6 \cdot 0^2)/10 = 3.7$$

so that $\tau^2 = 3.7 - (1.1)^2 = 2.49$ as before.

The mean square deviation defined above (which is also sometimes called the "population variance") is a measure of the extent to which the numbers $v_1, \ldots, v_N$ are unequal. As the average of squares it is always greater than or equal to zero; it is zero if and only if all the v's are equal.

EXAMPLE 3. *Indicators.* For later use we now obtain the variance of an indicator random variable. Since an indicator I takes on only the values 1 and 0, and since $1^2 = 1$ and $0^2 = 0$, it follows that $I^2 = I$ in all cases and hence that $E(I^2) = E(I)$. Thus, by formula (2),

$$\mathrm{Var}(I) = E(I) - [E(I)]^2 = E(I)[1 - E(I)].$$

If the probability of the event indicated by I is p (that is, $P(I = 1) = p$), we have

(6) $$\text{Var}(I) = p(1 - p).$$

PROBLEMS

1. Compute the variance of the random variable D of Example 3.2, by both formulas (1) and (2).

2. Compute the variance of the maximum M of the numbers of points obtained in two throws with a fair die (cf. Problem 4.6(ii)).

3. Compute the variance of
  (i) the bridge bonus of Example 3.3;
  (ii) the random variable B of Problem 2.1.

4. Compute the variance of the random variable
  (i) R of Problem 1.4.7;
  (ii) Z of Problem 3.14.

5. Compute the variance of
  (i) the random variable Z of Problem 2.2;
  (ii) the random variable T whose distribution is given by (2.1).

6. Compute the variance of the number of customers preferring the new formula in Problem 1.7.10 (cf. Problem 4.9(i)).

7. Compute the variance of the random variables
  (i) U defined in Problem 1.7.14(i);
  (ii) V defined in Problem 1.7.14(ii).

8. Compute the variance of
  (i) the random variable D of Problem 2.11;
  (ii) the random variable B of Problem 4.5.

9. Suppose that the lottery of Example 2 has an even number of tickets, half of which have the value 0 and half the value 1. Find $\text{Var}(Y)$.

10. In Problem 5.8 let Z denote the value of the dresses that reach their destination. Find $\text{Var}(Z)$ for both methods of mailing.

11. Find the variance of the value Y of a ticket drawn from the lottery of (i) Problem 3.12(i), (ii) Problem 3.12(ii).

12. Let J be a random variable which takes on the values -1, 0, 1 with probabilities p, q, r respectively. Find $E(J)$ and $\text{Var}(J)$.

13. Let D denote the number of rotten eggs of Problem 2.11 and let I_1 and I_2 indicate respectively that the first egg and second egg are rotten, so that $D = I_1 + I_2$. Find $\text{Var}(I_1)$ and $\text{Var}(I_2)$. By comparing $\text{Var}(I_1) + \text{Var}(I_2)$ with $\text{Var}(I_1 + I_2) = \text{Var}(D)$ computed in Problem 8(i), show that the variances of a sum is not always equal to the sum of the variances.

14. By applying the idea of the preceding problem to the number of correct answers of Problem 1.5.9 (the variance of which was computed in Problem 5(i)) show that the variance of a sum may be equal to the sum of the variances.

15. Compute Laplace's measure of dispersion $E|Z - \zeta|$ is Z is
  (i) the random variable D of Example 3.2;
  (ii) the random variable R of Problem 3.5.

16. Let a random variable X take on the values -1, 1, a with probability $\frac{1}{3}$ each. Find $\mathrm{Var}(X)$ for $a = 0, 2, 4, 6, 10$ and discuss what happens as a tends to infinity.

17. Let a random variable X take on the values 1, 2, 4 with probabilities $\frac{1}{2}$, $\frac{1}{4}$, $\frac{1}{4}$ respectively. Find $\mathrm{Var}(X)$, $\mathrm{Var}(X + 1)$ and $\mathrm{Var}(X + 2)$. What can you say about $\mathrm{Var}(X + c)$ for varying c? Do you think the same result is true for other random variables X?

5.7　LAWS OF VARIANCE

Variance is governed by laws analogous to the laws of expectation. Corresponding to (1)–(3) of Section 5, we have

$$(1) \qquad\qquad \mathrm{Var}(c) = 0$$

$$(2) \qquad\qquad \mathrm{Var}(Z + c) = \mathrm{Var}(Z)$$

$$(3) \qquad\qquad \mathrm{Var}(cZ) = c^2\,\mathrm{Var}(Z).$$

These are all intuitively plausible. If a random variable is always equal to the constant c, clearly c will also be its expectation. Since it therefore never differs from its expectation, its dispersion about this expectation will be 0. Again, if we add the constant c to the random variable Z, the expectation ζ of Z is also increased by c (this is just the law (5.2)). It follows that the difference $Z - \zeta$ is unchanged, and the variance is therefore also unchanged. Finally, multiplying a random variable by c will also multiply its expectation by c, so that the difference $Z - \zeta$ will be c times as large as before. Its square is then c^2 times as large, and it follows from (5.3) that the variance is multiplied by c^2. As a special case of (3) we obtain by putting $c = -1$

$$(4) \qquad\qquad \mathrm{Var}(-Z) = \mathrm{Var}(Z).$$

According to (3), multiplying a random variable by a constant multiplies its variance by the *square* of the constant. For example, if we convert from measuring in feet to measuring in inches, all measurements are multiplied by 12, so that in the new units the variance of the measurement is 144 times as large as before. For some purposes, it is more convenient to have a measure of dispersion that changes by the same factor as the scale, instead of its square. This has led to the introduction of the *standard deviation* of a random variable Z, denoted by $\mathrm{SD}(Z)$ and defined as the (positive) square root of the variance.

Definition. The standard deviation of the random variable Z is

(5) $$\mathrm{SD}(Z) = \sqrt{\mathrm{Var}(Z)}.$$

Consider now $\mathrm{SD}(cZ) = \sqrt{\mathrm{Var}(cZ)}$. It follows from (3) that this is equal to

$$\sqrt{c^2\,\mathrm{Var}(Z)} = c\sqrt{\mathrm{Var}(Z)} \quad \text{if } c \text{ is positive,}$$

and hence

(6) $$\mathrm{SD}(cZ) = c \cdot \mathrm{SD}(Z) \quad \text{if } c \text{ is positive.}$$

Thus, the standard deviation of a measurement in inches is just 12 times the standard deviation of the measurement in feet.

As was the case for expectation, the most important laws for variance are the addition laws. By analogy with the addition law of expectation, one might hope for an addition law of variance that would assert that "the variance of the sum of two random variables is the sum of their variances." The following examples show that such a law holds in some cases but not in others.

(a) Recall Example 5.1 where $T = X_1 + X_2$ is the total number of points showing when two dice are thrown. From the distribution of T given by (2.1) it is easy to calculate that $\mathrm{Var}(T) = \mathrm{Var}(X_1 + X_2) = \frac{35}{6}$ (Problem 4.4(ii)). We have seen in Example 6.1 that $\mathrm{Var}(X_1) = \mathrm{Var}(X_2) = \frac{35}{12}$, so that at least in this case the addition law holds.

(b) For any random variable Z it follows from (3) that $\mathrm{Var}(Z + Z) = \mathrm{Var}(2Z) = 4\,\mathrm{Var}(Z)$. Since this does not equal $\mathrm{Var}(Z) + \mathrm{Var}(Z) = 2\,\mathrm{Var}(Z)$, the addition law does not hold in this case.

To find the conditions under which the addition law holds, let us obtain a general formula for $\mathrm{Var}(Z + W)$. To simplify the writing we shall denote $E(Z)$ by ζ and $E(W)$ by η. From (5.4) $E(Z + W) = \zeta + \eta$; the difference between $Z + W$ and its expectation $\zeta + \eta$ is

$$(Z + W) - (\zeta + \eta) = (Z - \zeta) + (W - \eta).$$

The square of this expression is

(7) $$(Z - \zeta)^2 + 2(Z - \zeta)(W - \eta) + (W - \eta)^2.$$

Since the variance of $Z + W$ is by definition the expectation of (7),

(8) $$\mathrm{Var}(Z + W) = \mathrm{Var}(Z) + \mathrm{Var}(W) + 2E[(Z - \zeta)(W - \eta)].$$

The quantity $E[(Z - \zeta)(W - \eta)]$ appearing on the right-hand side is known as the *covariance* of Z and W, and is denoted by $\mathrm{Cov}(Z, W)$.

Definition. The covariance of the random variables Z and W is

(9) $$\mathrm{Cov}(Z, W) = E[(Z - \zeta)(W - \eta)].$$

In this notation, (8) may be written as

(10) $$\mathrm{Var}(Z + W) = \mathrm{Var}(Z) + \mathrm{Var}(W) + 2\,\mathrm{Cov}(Z, W),$$

and the desired *addition law for variance* is seen to be

(11) $\text{Var}(Z + W) = \text{Var}(Z) + \text{Var}(W)$ if $\text{Cov}(Z, W) = 0$.

A detailed discussion of the properties and meaning of covariance, which is a measure of the tendency of two random variables to vary together, is not required at this point and will be postponed to Section 7.2.   However we shall here note two simple properties.   By multiplying out the product indicated on the right side of (9), and then applying the addition law for expectation, we find

$$E[(Z - \zeta)(W - \eta)] = E[ZW - \zeta W - \eta Z + \zeta \eta]$$
$$= E(ZW) - \zeta E(W) - \eta E(Z) + \zeta \eta = E(ZW) - \zeta \eta.$$

Hence, (9) may be written in the alternative form

(12) $\text{Cov}(Z, W) = E(ZW) - E(Z)E(W)$,

which is analogous to the second form for variance given by (6.2).

Recalling (5.6), we see from (12) that if Z and W are defined on different factors of a product model, then

(13) $\text{Cov}(Z, W) = 0$.

Combining (11) and (13) gives the useful special form of the addition law

(14) $\text{Var}(Z + W) = \text{Var}(Z) + \text{Var}(W)$
                                     if Z, W are defined on different factors
                                     of a product model.

Extension of the argument (which is given in Section 7.2) leads to

(15) $\text{Var}(Z_1 + Z_2 + \ldots) = \text{Var}(Z_1) + \text{Var}(Z_2) + \ldots$
                                     if $Z_1, Z_2, \ldots$ are defined on different
                                     factors of a product model.

More generally, equations (13), (14), and (15) hold for any independent random variables (Problem 12).   However, we shall have occasion to use them only for the special case for which they are stated above.

EXAMPLE 1.   *Several dice.*   Suppose T is the total number of points showing when n dice are thrown.   Then, if $X_1, X_2, \ldots, X_n$ are the numbers showing on the n dice separately, we have $T = X_1 + X_2 + \ldots + X_n$.   If a product model is used to represent the experiment, the addition law (15) applies and gives

$$\text{Var}(T) = \text{Var}(X_1) + \text{Var}(X_2) + \ldots + \text{Var}(X_n).$$

By Example 6.1, each of these n terms equals $\frac{35}{12}$, and hence $\text{Var}(T) = 35n/12$.

We conclude the section by obtaining a useful identity.   Consider a random variable Z and an arbitrary constant a.   If $E(Z) = \zeta$ and if we put $X = Z - a$, it follows from (5.2) and (2) that

$$E(X) = \zeta - a \quad \text{and} \quad \text{Var}(X) = \text{Var}(Z).$$

By (6.2),

$$\text{Var}(X) = E(X^2) - [E(X)]^2 = E(Z - a)^2 - (\zeta - a)^2$$

and hence the expected squared difference between Z and a is

$$(16) \qquad E(Z - a)^2 = \text{Var}(Z) + (\zeta - a)^2.$$

Let us now apply the identity (16) to the lottery model of Example 6.2, replacing Z by Y. The variable Y takes on the values $v_1, \ldots, v_N$ each with probability $1/N$, so that

$$(17) \qquad E(Y - a)^2 = [(v_1 - a)^2 + \ldots + (v_N - a)^2]/N.$$

Since $E(Y) = \bar{v}$ and $\text{Var}(Y) = \tau^2$, it is seen from (16) that the left-hand side of (17) is equal to $\tau^2 + (\bar{v} - a)^2$. Multiplying through by N and using the expression (6.3) for τ^2, we find

$$(18) \qquad (v_1 - a)^2 + \ldots + (v_N - a)^2 = (v_1 - \bar{v})^2 + \ldots + (v_N - \bar{v})^2 + N(\bar{v} - a)^2.$$

This identity, which will be useful in proving several results later in the book, is also easily verified by direct computation (Problem 13). An important special case of (18) arises when $a = 0$, giving

$$(19) \qquad v_1^2 + \ldots + v_N^2 - N\bar{v}^2 = (v_1 - \bar{v})^2 + \ldots + (v_N - \bar{v})^2.$$

Relation (19) may be obtained also by comparing (6.3) and (6.5).

PROBLEMS

1. Note that the distributions of Figure 6.1 are those of two random variables, one of which is a constant multiple of the other. Check that (6) holds in this case.

2. Compute the variance of the random variable R of Problem 5.4 by the method suggested there for computing $E(R)$, and Problem 6.8(i).

3. Suppose a fair die is thrown and you receive 60 cents if the number of points showing is six.
  (i) If Y is the amount you will win, find $\text{Var}(Y)$.
  (ii) If you have to pay 10 cents to be allowed to play this game, use (i) and the laws of variance to find the variance of your total gain or loss.

4. (i) Prove that if Z and W are random variables defined on different factors of a product model, and if c and d are constants, then

$$\text{Var}(cZ + dW) = c^2 \text{Var}(Z) + d^2 \text{Var}(W).$$

  (ii) Use (i) to find $\text{Var}(Z - W)$.

5. If $Z_1, \ldots, Z_n$ are n unrelated random variables having a common variance σ^2, and if $\bar{Z}$ denotes their average, prove that $\text{Var}(\bar{Z}) = \sigma^2/n$.

6. Let I and J be the indicators of two events E and F. Show that

$$\text{Cov}(I, J) = P(E \text{ and } F) - P(E)P(F).$$

7. Find the covariance of the random variables Z and W defined in Problem 2.16.

8. Use the definition (6.1) and the laws of expectation to write out formal proofs of (1), (2), and (3).

9. If a random variable is multiplied by a negative constant, what happens to its standard deviation?

10. Find the variance of the random variable A of Example 5.3. [Hint: Use the method of indicators.]

11. Use the method of indicators to find the variance of the random variable (i) B of Problem 2.1; (ii) Z of Problem 2.2.

12. Prove that (13) and (14) hold whenever Z and W are independent. [Hint: Use the result of Problem 5.21.]

13. Prove the identity (18) by writing $v_1 - a$ as $(v_1 - \bar{v}) + (\bar{v} - a)$, and similarly for $v_2 - a, \ldots, v_N - a$, and then squaring out the terms of the left-hand side of (18).

14. Let Z be any random variable, and let ζ be its expectation. Show that the value of a which makes $E(Z - a)^2$ smallest is $a = \zeta$. [Hint: Note that the second term on the right-hand side of (16) is positive for all values of a different from ζ, and hence that for all $a \neq \zeta$ the left-hand side of (16) is greater than the first term of the right-hand side.]

15. Show that the value of a which makes $(v_1 - a)^2 + \ldots + (v_N - a)^2$ smallest is $a = \bar{v}$. [Hint: Apply the argument of the preceding problem to (18).]

16. Show that the variance of a random variable Z is always strictly positive (not only nonnegative) unless Z is constant. [Hint: If Z has at least two possible values, at least one of them must differ from $E(Z) = \zeta$. Apply (6.1).]

CHAPTER 6

SPECIAL DISTRIBUTIONS

6.1 THE BINOMIAL DISTRIBUTION

In Section 3.3 we introduced the concept of a sequence of binomial trials. These are trials on each of which there are just two possible results, conventionally called "success" and "failure." It is assumed that the various trials are unrelated, so that a product model is appropriate, and that the probability of success is the same, say p, on each trial. From these assumptions it was shown to follow that the probability of getting successes on b *specified* trials of a sequence of n trials, and failures on the remaining $n - b$ trials, is given by

$$(1) \qquad\qquad p^b q^{n-b}$$

where $q = 1 - p$ is the (constant) probability of failure on each trial.

However, in many cases we do not care on which particular trials the successes occurred, but are interested only in the total number of successes. Since this number is determined by the experimental results, it is a random variable, which we shall denote by B. Examples are the number of aces when a die is thrown ten times, the number of boys in a family of five children, the number of defectives among 20 items produced by a machine. We shall now obtain a formula for the distribution of B.

Let us begin by finding the probability that $B = 2$ when $n = 5$. In five trials, two successes occur in the following patterns of successes (S) and failures (F):

$$SSFFF, \quad SFSFF, \quad SFFSF, \quad SFFFS, \quad FSSFF,$$
$$FSFSF, \quad FSFFS, \quad FFSSF, \quad FFSFS, \quad FFFSS.$$

Since these are exclusive, the probability that $B = 2$ is the sum of the probabilities of these ten patterns. According to (1), the probability of each of these patterns is $p^2 q^3$, and hence $P(B = 2) = 10p^2 q^3$.

We could have determined the number of patterns without actually listing them. Each pattern is specified by indicating which two of the five trials are to be successes, and the number of ways of choosing two out of the five trials is just $\binom{5}{2} = 10$.

The argument extends easily to the general case. The number of patterns with b successes and $n - b$ failures is $\binom{n}{b}$. These patterns are the (exclusive) ways of getting just b successes, and the probability of each pattern is $p^b q^{n-b}$, as given by (1). Therefore the probability of getting just b successes in n binomial trials is

(2) $$P(B = b) = \binom{n}{b} p^b q^{n-b}.$$

Definition.   The number B of successes in n binomial trials each having success probability p is called a *binomial* random variable. Its distribution (2) will be referred to as the "binomial distribution (n, p)."

For illustration, consider once more the case $n = 5$. From formula (2) and Table A we find that the binomial distribution $(n = 5, p)$ is

(3)

b	0	1	2	3	4	5
$P(B = b)$	q^5	$5pq^4$	$10p^2q^3$	$10p^3q^2$	$5p^4q$	p^5

The term "binomial" derives from the fact that the probabilities (2) are the terms in the binomial formula for the expansion of $(q + p)^n$. The reader may check that when $(q + p)^5$ is expanded, the successive terms are those in (3) (Problem 21).

EXAMPLE 1.   *Random digits.*   What is the distribution of the number B of zeros produced when a random digit generator is operated 5 times? We can regard the five operations as $n = 5$ binomial trials with probability of success (i.e. getting a zero) $p = .1$. Therefore the desired distribution is obtained by putting $p = .1$ in (3):

b	0	1	2	3	4	5
$P(B = b)$	.5905	.3280	.0729	.0081	.0005	.0000

The six entries must of course add up to 1, which gives a check on the work.

The importance of the binomial distribution is attested by the publication of several extensive tables for it. The most notable are the tables published by the Ordnance Corps of the U.S. Army in 1952, which gives the binomial distribution to seven decimal places for all values of n up to 150, and for p at intervals of .01; and that published by Harvard University in 1955, which for these values of p gives the binomial distribution to five decimal places for selected values of n up to 1000.*

We give in Tables B and C at the back of the book a few examples of binomial distributions. Table B covers all values of n up to 15 for the five values $p = .05, .1, .2, .3,$ and $.4,$ while Table C is for $p = .5$ and $n \leqq 30.$

EXAMPLE 2. *Multiple choice examination.* An examination consists of eight multiple choice questions, each of which offers a choice between five answers of which only one is correct. To pass, it is necessary to answer at least three of the questions correctly. What is the probability that a student will pass who, being completely unprepared, for each of the questions selects one of the answers at random?

We can view the eight questions as eight unrelated trials; the probability of success (correct answer) is $p = \frac{1}{5} = .2$ for each. The desired probability is then the probability of at least three successes in eight binomial trials with $p = .2$. The probability of passing is therefore $P(B \geqq 3)$ where B has the binomial distribution $(n = 8, p = .2)$. Table B gives

$$P(B \geqq 3) = .1468 + .0459 + .0092 + .0011 + .0001 = .2031.$$

The reader will notice that no values of p greater than $\frac{1}{2}$ are given in our tables. Such values have been omitted, thereby cutting the size of Table B in half, because a binomial probability for $p > \frac{1}{2}$ can be reduced to one with $p < \frac{1}{2}$. To see how this is done, suppose we wish to find the probability of getting seven hits (and three misses) when firing ten times at a target, where it is assumed that the shots are unrelated and that the probability of a hit is .6 on each shot. If we call a hit a "success," the desired probability is $P(B = 7)$ for the binomial distribution $(n = 10, p = .6)$, which is not given in Table B. However, if we instead call a miss a "success," then the desired probability is $P(B = 3)$ for the binomial distribution $(n = 10, p = .4)$, which from Table B is seen to be .2150. By means of this device, which amounts to interchanging the roles of "success" and "failure," one sees that $P(B = b)$ for the binomial distribution (n, p) is equal to $P(B = n - b)$ for the binomial distribution $(n, 1 - p)$. The same identity may be proved from formula (2) (Problem 23).

In the special case when $p = \frac{1}{2}$, replacing p by $1 - p$ changes nothing, so that our identity in this special case states

$$P(B = b) = P(B = n - b) \quad \text{when} \quad p = \frac{1}{2}.$$

---

* *Table of the Cumulative Binomial Probabilities,* Ordnance Corps Pamphlet ORDP 20-1 (1952). 577 pages.
  Tables of the Cumulative Binomial Probability Distribution, Harvard University Press (1955). 503 pages.

That is, when $p = \frac{1}{2}$, B has a distribution which is symmetric about $n/2$ in the sense of Section 5.4 (Problem 18). Thus, when $p = \frac{1}{2}$ it suffices to give the values of $P(B = b)$ for $b \leq n/2$; this fact permits Table C to be half as large as would otherwise be necessary. For example, if we need $P(B = 7)$ for the binomial distribution ($n = 10$, $p = \frac{1}{2}$), this is the same as $P(B = 3)$ for that distribution, which from Table C is seen to be .1172.

EXAMPLE 3. *Quality control.* It is a characteristic of mass production that not all items coming off a production line will conform to specifications. Items that fail to conform may be called *defectives.* It is one of the tasks of *quality control* to seek to identify and eliminate the causes leading to defective items, but even after this has been done an occasional defective will appear as if by accident. The successive items coming off the line may in fact behave like binomial trials, with the appearance of a defective item constituting a "success." If this is the case, and if the probability of a defective is satisfactorily low, the process is said to be "in a state of (statistical) control."

A process which is in control may of course at any time lose this property if some part of the process deteriorates or breaks down. It is important to check, by regular inspection of the items produced, whether the process is still in control. To illustrate how this can be done, suppose that the process has been in control for some time and that observation during this period has shown the average frequency of defectives to be 5 percent. As a check, ten items of the day's production are inspected each day. If the process continues under control, it is reasonable to assume that the number B of defectives has the binomial distribution ($n = 10$, $p = .05$). The distribution of B is then given by the following table (taken from Table B);

(4)

b	0	1	2	3	4	5
$P(B = b)$	.5987	.3151	.0746	.0105	.0010	.0001

with the probabilities for $b > 5$ being negligible. Suppose now that it has been agreed to institute a careful physical check of the whole process on any day on which the sample contains three or more defectives. As long as the process remains in control, we see from the table that $P(B \geq 3) = .0116$. With this inspection procedure, we would therefore on the average institute the complete check unnecessarily on about one percent of the days. A more detailed discussion of the methods used for deciding such questions will be presented in Chapter 11.

EXAMPLE 4. *Size of an experiment.* It is known from past experience that on the average about 20 percent of the animals entering a certain experiment die before termination of the experiment. If we want to have probability at least .98 of having at least five animals complete the experiment, with how many animals should we start? It seems natural to regard the number B of surviving animals as the number of successes in n binomial trials with probability $p = .8$ of success, and we wish then to find the smallest number n of trials such that $P(B \geq 5) \geq .98$.

Let us first try $n = 7$. What is the probability of getting five or more successes on $n = 7$ trials with $p = .8$? Interchanging "success" and "failure" as discussed above, we see that the desired probability is the same as the probability of obtaining two or fewer successes with $n = 7$ and $p = .2$. From Table B we find this to be $.2097 + .3670 + .2753 = .8520$. Thus seven animals are not enough since the probability that five of them or more will survive is only 85%. A similar calculation shows that there is a probability $.9437$ of five or more survivors when eight animals are used, and a probability $.9804$ when nine are used. Thus nine is the smallest number of animals guaranteeing the desired result.

We will now find the expectation and variance of the binomial random variable B. To compute the expectation of the number B of successes in n binomial trials, we could multiply each possible value b of B by its probability $\binom{n}{b} p^b q^{n-b}$, and then add all these terms. There is however a much simpler method, based on the representation of B as a sum of indicators, an idea already employed in Section 5.5. The total number of successes in n trials is the number of successes in the first trial plus the number of successes in the second trial plus . . . plus the number of successes in the nth trial. Therefore, if $I_1, I_2, \ldots, I_n$ denote the numbers of successes in the first, second, . . . , nth trial respectively, then

$$(5) \qquad B = I_1 + I_2 + \ldots + I_n.$$

This equation can also be verified in a slightly different way. The number I_1 of successes in the first trial is 1 if the first trial is a success and 0 if the first trial is a failure. It is thus an indicator random variable, indicating success in the first trial. Similarly, I_2 indicates success in the second trial, and so forth. Since each indicator equals 1 if the corresponding trial succeeds and is otherwise 0, the number of successes is the number of indicators equal to 1, which is the sum of the indicators, as was to be proved. Now by Example 5.4.2, the expected value of each of the indicators $I_1, \ldots, I_n$ is p. It follows from (5) that $E(B)$ is the sum of n terms, each equal to p, so that

$$(6) \qquad E(B) = np.$$

This elegant, important, and useful formula asserts that in a sequence of binomial trials, the expected number of successes is the product of the number of trials and the success probability.

The representation (5) of B as a sum of indicators also gives us the variance of B. Since the indicators $I_1, \ldots, I_n$ are defined on different factors of a product model, it follows from the addition law for variance (5.7.15) that

$$\mathrm{Var}(B) = \mathrm{Var}(I_1) + \ldots + \mathrm{Var}(I_n).$$

The variance of each I was seen in Example 5.6.3 to be equal to $p(1 - p)$ $= pq$ so that

(7) $$\mathrm{Var}(B) = npq.$$

To illustrate formulas (6) and (7), let us recall Example 3, where $n = 10$ and $p = .05$. The number B of defectives in this example has expectation $E(B) = .5$ and variance $\mathrm{Var}(B) = .475$. The reader may check these values by computing $E(B)$ and $\mathrm{Var}(B)$ from (5.4.2) and (5.6.2), using the distribution (4).

At the end of Section 3.3, we briefly considered the extension of the binomial trials model to the case of unrelated trials with variable success probabilities. Suppose there are n unrelated trials, with success probabilities $p_1, p_2, \ldots$ and failure probabilities $q_1 = 1 - p_1, q_2 = 1 - p_2, \ldots$. Then the probability of the result SFSSF $\cdots$ is $p_1 q_2 p_3 p_4 q_5 \cdots$, and similarly for any other pattern of successes and failures. The resulting model is the *generalized binomial trials* model. (Also known as "Poisson binomial trials," after Siméon D. Poisson (1781–1840).) The distribution of the total number T of successes can in principle be obtained by computing and adding the probabilities of the different patterns with this number of successes, but this is cumbersome unless n is quite small.

Fortunately, there is a simple approximation, which is effective when $p_1, p_2, \ldots$ are not too different from each other. Let us denote the arithmetic mean of the success probabilities by $\bar{p} = (p_1 + p_2 + \ldots)/n$, and let B be the number of successes in n binomial trials with success probability $\bar{p}$. Since this sequence of trials has the same average chance of success as the sequence first considered, one may hope that the higher success probabilities on some of the trials will balance the lower probabilities on others, so that the distributions of T and B will be close. This turns out to be the case, and one may therefore use the binomial distribution $(n, \bar{p})$ to approximate the distribution of T.

EXAMPLE 5. *The marksmen.* Four marksmen of unequal skill have probabilities .05, .06, .13, .16 of hitting the bull's eye. If each fires once, what is the distribution of the total number T of hits on the bull's eye?

Here $n = 4$ and $\bar{p} = (.05 + .06 + .13 + .16)/4 = .1$, so the binomial approximation can be obtained from Table B. A comparison of the ap-

proximate with the correct values (Problem 24) shows the approximation
to be satisfactory:

t	0	1	2	3	4
$P(T = t)$ approx.	.6561	.2916	.0486	.0036	.0001
$P(T = t)$ correct	.6526	.2978	.0466	.0029	.0001

The approximation would have worked less well if success probabilities
that differed more widely had been used. (As is illustrated by Problem 25,
the error goes up like the sum of squares $(p_1 - \bar{p})^2 + (p_2 - \bar{p})^2 + \ldots$)

PROBLEMS

1. Use Table C to find the probabilities $P(B = b)$ for the following values of n
and b when $p = .5$: $n = 4$, $b = 2$; $n = 6$, $b = 3$; $n = 8$, $b = 4$; $n = 10$, $b = 5$;
$n = 20$, $b = 10$; $n = 30$, $b = 15$.

2. Use Tables B and C to find the probabilities $P(B = b)$ for the following values
of n, p, and b:
  (i) $n = 5$, $p = .2$, $b = 2$
  (ii) $n = 5$, $p = .8$, $b = 3$
  (iii) $n = 10$, $p = .6$, $b = 7$
  (iv) $n = 20$, $p = .5$, $b = 13$.

3. For $n = 6$, $p = .3$ find the probabilities of
  (i) at least three successes
  (ii) at most three successes.

4. Plot the histogram for the binomial distributions (i) $n = 15$, $p = .2$, (ii) $n = 30$,
$p = .5$.

5. Each student in a class of fourteen is asked to write down at random one of the
digits 0, 1, ... , 9. The instructor believes that digits 3 and 7 are especially
attractive. If the digits are really selected at random, how surprising would it be
if six or more students chose one of these "attractive" digits?

6. In the preceding problem how surprising would it be if fewer than three digits
were even?

7. In Example 3.3.8, what is the probability that eight or more patients receive
the same drug?

8. Compute the distribution of the number B of male children in a three-child
family, assuming $P(\text{Male}) = .514$.

9. (i) If $n = 5$, how small must p be before $P(B = 0) > P(B = 1)$?
  (ii) Solve the problem for an arbitrary value of n.

10. What is the most probable number of successes in eight binomial trials with
success probability $p = .1$?

11. In Table B, verify the entry $P(B = 2) = .2458$ when $n = 6$ and $p = .2$ from
formula (2).

12. In Example 3, suppose that the process has gone out of control, so that the probability of a defective has jumped from its usual value of .05 to the value .2. How likely is it that the inspection procedure will call for a check?

13. Find the probability that the frequency f of heads in n tosses with a fair coin lies between .4 and .6 inclusive for (i) $n = 5$, (ii) $n = 10$, and (iii) $n = 20$.

14. Use formulas (5.4.2) and (2) to check that $E(B) = np$ for the cases $n = 1, 2, 3$.

15. (i) Find formulas for the expectation and variance of the frequency B/n of success in n binomial trials.
   (ii) What can be said about the behavior of the variance as n becomes very large?

16. Find formulas for the expectation and variance of the number of failures in n binomial trials with success probability p.

17. In Example 4, if nine animals are used, (i) what is the expectation of the number of survivors?   (ii) What is its variance?

18. Using formulas (2.2.3) and (2), show that the binomial distribution is symmetric about $E(B) = n/2$ when $p = \frac{1}{2}$.

19. Find the expectation and variance of the number of successes in n unrelated trials when the probability of success is p_1 on the first trial, p_2 on the second trial, $\ldots , p_n$ on the nth trial.   Are these trials binomial?

20. Certain textbooks state that np is the value of B most likely to occur.   Criticize this statement.

21. Expand the binomial $(q + p)^5$ by multiplication and check the assertion made in the text about (3).

22. Use the fact that the probabilities (2) are the terms in the expansion of $(q + p)^n$ to show that these probabilities add up to one.

23. Use formula (2) to prove that $P(B = b)$ for the binomial distribution (n, p) is equal to $P(B = n - b)$ for the binomial distribution $(n, 1 - p)$.

24. Verify the entries in the last row of the tableau of Example 5.

25. Consider two unrelated trials with success probabilities $\bar{p} + \Delta$ and $\bar{p} - \Delta$. (i) Find the distribution of the number T of successes.   (ii) For any value t of T, find the maximum error if the exact distribution of T is replaced by the binomial distribution $(n = 2, \bar{p})$.

6.2 THE HYPERGEOMETRIC DISTRIBUTION

We have already on several occasions considered the problem of drawing a sample from a population containing two different kinds of items: red and white marbles, defective and nondefective fuses, Conservatives and Liberals, etc.   To have a general terminology, let us refer to the items of one kind as *special* and the remaining items as *ordinary*.   Then the number D of special items appearing in the sample is a random variable, and we shall now obtain its distribution.

Let the size of the population be N and the size of the sample s, and suppose that the population contains r special and $N - r$ ordinary items.

The distribution of D was already obtained in Example 5.2.1 for the case $N = 7$, $r = 4$, $s = 3$, if we regard the four Conservatives as the special items. A study of this example reveals the general method. An examination of the samples giving $D = 2$ in the example shows that the samples are arranged in a tableau of three rows and six columns, or $3 \cdot 6 = 18$ samples in all. The columns correspond to the $\binom{4}{2} = 6$ ways in which two Conservatives can be selected from the four available Conservative members: 45, 46, 47, 56, 57, 67. The rows correspond to the $\binom{3}{1} = 3$ ways in which one Liberal can be selected from the three available Liberals: 1, 2, 3.

Let us now carry through the argument in general. How many samples of size s can be formed to contain just d special, and hence $s - d$ ordinary, items? There are r special items available, so that the special items for the sample may be chosen in $\binom{r}{d}$ ways. Similarly, the $s - d$ ordinary items must be chosen from the $N - r$ such items in the population, which can be done in $\binom{N - r}{s - d}$ ways. Each choice of the special items may be combined with each choice of the ordinary items. Therefore the number of samples with $D = d$ is $\binom{r}{d}\binom{N - r}{s - d}$. If all $\binom{N}{s}$ samples are assumed to be equally likely, this gives

$$(1) \qquad P(D = d) = \frac{\binom{r}{d}\binom{N - r}{s - d}}{\binom{N}{s}}.$$

Definition. The number D of special items in a random sample of size s from a population of size N that contains r special items is called a *hypergeometric* random variable. Its distribution (1) will be referred to as the "hypergeometric distribution (N, r, s)." (The term has its origin in the fact that the quantities (1) appear in a series of that name studied in analysis.)

As an illustration, let us use this formula to derive the hypergeometric distribution of Example 5.2.1. Putting $N = 7$, $r = 4$, $s = 3$ in (1), we obtain the following results, which agree with those obtained earlier by direct enumeration.

d	$\binom{4}{d}$	$\binom{3}{3 - d}$	$P(D = d)$
0	1	1	$\frac{1}{35}$
1	4	3	$\frac{12}{35}$
2	6	3	$\frac{18}{35}$
3	4	1	$\frac{4}{35}$
			1

The hypergeometric probabilities (1) depend on four quantities: N, r, s, and d. It would be essentially impossible to publish a table covering all values that might arise in practice. A table* has been published giving the probabilities to six decimal places for populations of size $N \leq 50$, and certain selected higher values.

To obtain the expected value of D, one might use formula (5.4.2), writing $E(D)$ as the sum of the products of the possible values d and the corresponding probabilities (1), and then simplifying the result. (This method was illustrated in Example 5.3.2.) A much simpler derivation may be based on the method of indicators, used in the preceding section for the binomial random variable. Recalling from Section 2.3 that a random sample can be obtained by drawing an ordered random sample, we may think of D as the number of special items in an ordered sample of size s. In this model, D will have the same distribution (and hence the same expectation) as before.

The total number of special items included in s draws is the number of special items obtained on the first draw plus the number of special items obtained on the second draw plus . . . plus the number of special items obtained on the sth draw. Therefore, if $I_1, I_2, \ldots, I_s$ denote the number of special items obtained on the first, second, . . . , sth draw respectively, then

(2) $$D = I_1 + I_2 + \ldots + I_s.$$

As in the binomial case, this equation can also be seen in a slightly different way. The number I_1 of special items on the first draw is 1 or 0 as the first draw produces a special or an ordinary item. It is thus an indicator random variable, indicating success on the first draw. Similarly I_2 indicates a special item on the second draw, and so forth. Since each indicator equals 1 if the corresponding draw produces a special item and is otherwise 0, the number of special items in the sample is the number of indicators equal to 1, which is the sum of the indicators, as was to be shown. To compute $E(D)$ it is now only necessary to find the expectation of each of the I's.

By Example 5.4.2, we have $E(I_1) = P(I_1 = 1)$, which is the probability that the first item drawn is special. Since all N items are equally likely to be drawn on the first draw and since r of them are special, it follows that

$$E(I_1) = \frac{r}{N}.$$

By the same argument, $E(I_2) = P(I_2 = 1)$ is the probability that the second item is special. By the equivalence law of ordered sampling (2.3.5), on each draw the probability of getting a special item is r/N and therefore $E(I_2) = \ldots = E(I_s) = r/N$. Hence by applying the addition law of expectation to (2) we find

(3) $$E(D) = s \cdot \frac{r}{N}.$$

* *Tables of the Hypergeometric Probability Distribution*, by Gerald J. Lieberman and Donald B. Owen. Stanford University Press (1961). 726 pages.

We shall now develop an approximation for the hypergeometric distribution which is accurate when the sample constitutes only a small fraction of the population, so that the *sampling fraction* s/N is small. Suppose that a sample of s items is drawn *with replacement* from the population of N items of which r are special. On each draw the chance of getting a special item is r/N, just as it was for sampling without replacement, but now the s draws are unrelated. Therefore they form a sequence of s binomial trials, and the number of special items in the sample has the binomial distribution with $n = s$ and $p = r/N$.

As we have remarked in Example 3.3.3, when the sampling fraction s/N is small, it should make little difference whether sampling is done with or without replacement. This suggests that the distribution of D, the number of special items when sampling without replacement, will be approximately the same as the binomial distribution with $n = s$ and $p = r/N$, *provided s/N is small.* For example, if in fact D has the hypergeometric

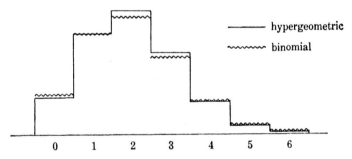

————— hypergeometric

~~~~~~ binomial

FIGURE 1. BINOMIAL APPROXIMATION TO HYPERGEOMETRIC

distribution ($N = 100$, $r = 20$, $s = 10$), we might as an approximation treat D as if it had the binomial distribution ($n = 10$, $p = .2$). Figure 1 shows the two histograms to be in reasonably good agreement. It is common practice to use the binomial tables for the computation of hypergeometric probabilities when the sampling fraction is small.

The expected value of the approximating binomial distribution with $n = s$ and $p = r/N$ is of course $np = s \cdot r/N$, which agrees exactly with $E(D)$. Thus the exact and approximate distributions are centered at the same place. The reason for this agreement is that whether we sample with or without replacement, the number of special items in the sample is the sum

$$I_1 + I_2 + \ldots + I_s$$

of the indicators of special items on the s draws. With either sampling method, each indicator has the same expectation r/N.

The variance of the approximating binomial distribution, from formula (1.7) with $n = s$ and $p = r/N$, is

(4)
$$s \cdot \frac{r}{N} \cdot \left(1 - \frac{r}{N}\right).$$

Will this again agree exactly with $\text{Var}(D)$? We should not expect this to happen, because in deriving (1.7) use was made of the unrelatedness of the indicators, which does not hold for sampling without replacement. The formula for $\text{Var}(D)$ requires a correction factor, which turns out to be $(N - s)/(N - 1)$. This gives for the variance of a hypergeometric random variable

$$(5) \qquad \text{Var}(D) = \frac{N - s}{N - 1} \cdot s \cdot \frac{r}{N}\left(1 - \frac{r}{N}\right).$$

This formula will be proved in Section 7.2.

We note that the correction factor is always less than 1 (unless $s = 1$ in which case there is no difference between sampling with and without replacement) but that if N is much larger than s, it is quite close to 1. That the two formulas for the variance are in close agreement when s/N is small corresponds to the fact noted earlier that in this case it makes little difference whether the sampling is with or without replacement.

In the special case when $s = N$, in which the "sample" consists of the entire population, we see that $\text{Var}(D) = 0$. This is as it should be, since D has the constant value r if we take the entire population into the sample.

PROBLEMS

1. Use (1) to check the distribution of the random variable D of Problem 5.2.11.

2. Compute the distribution of the number of red cards in a poker hand, using Table A and the value $\binom{52}{5} = 2{,}598{,}960$. Give the probabilities to three decimals and graph the histogram.

3. A sample of size $s = 5$ is drawn from a population of $N = 20$ items. For what values of r is $P(D = 0) > P(D = 1)$?

4. A batch of 20 items contains five defectives. Find the probability that a sample of four will contain more than one defective.

5. Suppose that the sample of four houses of Example 4.2.3 is drawn at random from all 13 houses rather than by the method of stratified sampling. Find the distribution of
    (i) the number of corner houses included in the sample;
    (ii) the number of houses from the south side of the street included in the sample.

6. Suppose that in Example 3.3.8, so as to avoid the possibility that most of the ten patients receive the same drug, drug A is assigned to five of the patients at random with the other five receiving drug B. If five of the patients have a light case of the disease and five a severe case, what is the probability that at least four of the severe cases will receive drug A?

7. Using formula (1), solve (i) Problem 2.2.5, (ii) Problem 2.2.9.

8. A lot of ten items contains five defectives. Plot the histogram of the distribution of the number D of defectives in a sample of four, and that of the binomial approximation to this distribution.

9. Work the preceding problem if the number of defectives is four instead of five.

10. Work Problem 8 if the lot size is 20 and the sample size is 5 when the number of defectives in the lot is (i) 10, (ii) 8, (iii) 6.

11. A sample of ten items is drawn from a lot of 1000, which contains 50 defective items. Use the binomial approximation to find approximately the probability that the sample will contain two or more defectives.

12. Solve the preceding problem when the lot contains 100 instead of 50 defective items.

13. The 200 students in a class are divided at random into 20 sections of 10 each. Suppose the class consists of 160 undergraduates and 40 graduates, and let G denote the number of graduates in Section 1. Use the binomial approximation to find approximately the probability that $G \geq 3$.

14. Derive formulas for the expectation and variance of the fraction D/s of special items, when the distribution of D is given by (1).

15. A sample of size s is drawn from a population of N items, of which r are special. Derive formulas for the expectation and variance of the number of special items left in the population after the sample has been drawn. What is the distribution of this number?

16. A box contains nine marbles, three red, three white, three blue.
   (i) What is the probability that a sample of four will contain marbles of only two colors?
  (ii) How many colors do you expect in such a sample?
 (iii) What is the distribution of the number of red marbles?

17. A stratified sample is obtained by drawing unrelated samples from four blocks as follows:

Block	1	2	3	4
Private dwellings	8	10	13	7
Apartment houses	4	3	2	5
Sample size	2	2	3	2

Find the expectation and variance of the number of apartment houses included in the sample. [Hint: Let A_1 be the number of apartment houses drawn from block 1, etc.]

18. In the preceding problem, find the probability that the sample includes at least one apartment house from each block.

19. In Problem 17, find the probability that no apartment house is drawn.

20. The random variable D of equation (2) may be represented as a sum of indicators in a different way. Number the special items from 1 to r, and let J_1 indicate that the first special item is included in the sample, etc.
   (i) Prove that $D = J_1 + J_2 + \ldots + J_r$.
  (ii) Use this representation to obtain an alternative proof of (3).

21. Find the expected number of corner houses in the samples of Problems 4.2.5(i) and 4.2.5(ii).

22. Find the expected number of freshmen in the samples of Problems 4.2.6(i) and 4.2.6(ii).

23. Find the expected number of freshmen in the samples of Problems 4.2.16(i) and 4.2.16(ii).

6.3 STANDARD UNITS

One of the most remarkable facts in probability theory, and perhaps in all of mathematics, is that the histograms of a wide variety of different distributions are very nearly the same when the right units are used on the horizontal axis. We cannot present the theoretical reasons underlying this fact, but we can and shall support it by some computational evidence, and then show how it may be used to calculate, with little effort, approximate values for certain probabilities that would be very cumbersome to compute exactly. Let us begin by considering an example.

EXAMPLE *1.   The sum of points on several dice.*   We obtained in Problem 1.5.4 the distribution of the number, say T_2, of points showing when two dice are thrown. By an extension of the method used there, it is possible in principle to find the distribution of the number T_n of points showing when n dice are thrown, for any value of n, although the work gets heavier as n is increased. We present the histograms for $n = 1, 2, 4$, and 8 in Figure 1 (here T_1 denotes the number of points on a single die).
  A study of this figure will make clear the following points. (a) As the number n of dice is increased, the distribution moves off to the right. This is not surprising since it is easily seen (Problem 5.5.3) that $E(T_n) = 7n/2$. The distribution of T_n is centered at $E(T_n)$, which gets arbitrarily large as n gets sufficiently large. (b) The distribution tends to get more spread out as n gets larger. This is also reasonable, since we know (Example 5.7.1) that $SD(T_n) = \sqrt{\frac{3.5}{12}} \cdot \sqrt{n}$. The standard deviation is a measure of the spread of the distribution and as n becomes large, so does $\sqrt{n}$, and hence so does $SD(T_n)$. We note also that the minimum and maximum values of T_n are n and $6n$ respectively, so that the range of T_n is $6n - n = 5n$, which also becomes large with n. (c) The histogram is a rectangle when $n = 1$, and roughly the shape of a triangle when $n = 2$. As n increases, the histogram appears to smooth out, though for $n = 8$ it is already so low and spread out that its shape cannot be seen very clearly.
  In order to bring the distribution into sharper focus, we shall make a change in the units of the graphs. To overcome the tendency of the distribution to move off to the right, let us take $E(T_n)$ as the new origin on the horizontal axis. This can be achieved by considering the random variable $T_n - E(T_n)$, whose expected value is 0, so that its histogram is centered at 0. Similarly, the tendency to spread out may be overcome by using

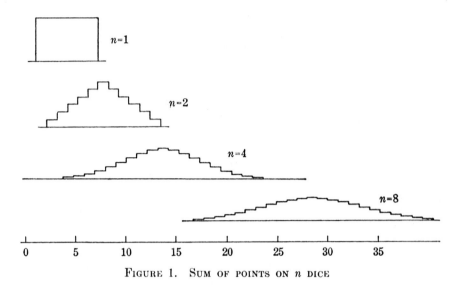

FIGURE 1. SUM OF POINTS ON n DICE

$\mathrm{SD}(T_n)$ as the new unit of scale, or equivalently by considering the random variable

$$T_n^* = \frac{T_n - E(T_n)}{\mathrm{SD}(T_n)}.$$

The change of origin and scale that we have suggested for T_n may be applied to any random variable Z, and leads to the consideration of the "standardized" variable

(1) $$Z^* = \frac{Z - E(Z)}{\mathrm{SD}(Z)},$$

which is also known as the random variable Z "reduced to standard units." It is seen that

$$E(Z^*) = \frac{E[Z - E(Z)]}{\mathrm{SD}(Z)} = 0$$

and

$$\mathrm{Var}(Z^*) = \frac{\mathrm{Var}[Z - E(Z)]}{[\mathrm{SD}(Z)]^2} = \frac{\mathrm{Var}(Z)}{\mathrm{Var}(Z)} = 1,$$

so that a random variable, when it has been reduced to standard units, has expectation 0 and variance 1.

To see how to construct the histograms of standardized variables, let us recall from Section 5.2 that a histogram is a row of contiguous bars, one for each possible value of the random variable. Each bar is centered at the value it represents, and its area is equal to the probability of that value. As an example consider the number B of successes in two binomial trials with success probability $p = \frac{1}{9}$. Since

$$E(B) = \tfrac{2}{9} \quad \text{and} \quad SD(B) = \sqrt{2\cdot\tfrac{1}{9}\cdot\tfrac{8}{9}} = \tfrac{4}{9},$$

the distributions of B and B^* are given in the following table.

b	0	1	2
b^*	$\dfrac{0 - (\frac{2}{9})}{\frac{4}{9}} = -\dfrac{2}{4}$	$\dfrac{1 - (\frac{2}{9})}{\frac{4}{9}} = \dfrac{7}{4}$	$\dfrac{2 - (\frac{2}{9})}{\frac{4}{9}} = \dfrac{16}{4}$
Probability	$\tfrac{64}{81}$	$\tfrac{16}{81}$	$\tfrac{1}{81}$

The three bars of the histogram of B^* are centered at $-\tfrac{2}{4}$, at $\tfrac{7}{4}$, and at $\tfrac{16}{4}$ respectively, as shown in Figure 2. The distance between successive centers is

$$\tfrac{7}{4} - (-\tfrac{2}{4}) = \tfrac{9}{4} \quad \text{and} \quad \tfrac{16}{4} - \tfrac{7}{4} = \tfrac{9}{4}.$$

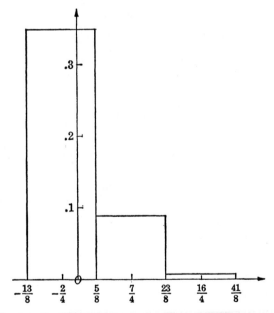

FIGURE 2. HISTOGRAM OF A STANDARDIZED BINOMIAL

Each bar must therefore have width $\tfrac{9}{4}$ extending half this amount on either side of the center. For example, the bar corresponding to the value $b = 0$ covers the interval from

$$-\tfrac{2}{4} - \tfrac{1}{2}\cdot\tfrac{9}{4} = -\tfrac{13}{8} \quad \text{to} \quad -\tfrac{2}{4} + \tfrac{1}{2}\cdot\tfrac{9}{4} = \tfrac{5}{8}.$$

Similarly, the second bar of the histogram covers the interval from $\tfrac{5}{8}$ to $\tfrac{23}{8}$, and the third one the interval from $\tfrac{23}{8}$ to $\tfrac{41}{8}$.

Let us next consider the height of the first bar. This is determined by the fact that

$$\text{area} = \text{height} \times \text{width}$$

and that the area (by definition of a histogram) must equal the probability, which is $\frac{64}{81}$. Since the width is $\frac{9}{4}$, we must have

$$\frac{9}{4} \times \text{height} = \frac{64}{81}$$

or

$$\text{height} = \frac{64}{81} \cdot \frac{4}{9} = \frac{256}{729}.$$

Similarly, the height of the second bar is found to be $\frac{84}{729}$ and that of the last bar $\frac{4}{729}$.

Let us now generalize this result. Suppose that Z is a random variable whose possible values are consecutive integers. To simplify the notation, let

$$E(Z) = \zeta \quad \text{and} \quad \text{SD}(Z) = \sigma.$$

If z is any possible value of Z, the next value is $z + 1$, and the corresponding values of Z^* are

$$\frac{z - \zeta}{\sigma} \quad \text{and} \quad \frac{z + 1 - \zeta}{\sigma}.$$

The distance between these two values is

(2)
$$\frac{z + 1 - \zeta}{\sigma} - \frac{z - \zeta}{\sigma} = \frac{1}{\sigma}.$$

(This agrees with the value $\frac{9}{4}$ between successive values in the example, since there $\text{SD}(Z) = \frac{4}{9}$.) Since this is the distance between any two successive values, it follows as in the example that each bar of the histogram must have width $1/\sigma$, extending half this amount on either side of the center. The bar of the histogram corresponding to z therefore has as its base the interval

$$\left(\frac{z - \zeta}{\sigma} - \frac{1}{2} \cdot \frac{1}{\sigma}, \frac{z - \zeta}{\sigma} + \frac{1}{2} \cdot \frac{1}{\sigma} \right) \quad \text{or} \quad \left(\frac{z - \zeta - \frac{1}{2}}{\sigma}, \frac{z - \zeta + \frac{1}{2}}{\sigma} \right).$$

Since the area of the bar is $P(Z = z)$ and its width is $1/\sigma$, its height must satisfy the equation

$$\text{height} \cdot \frac{1}{\sigma} = P(Z = z),$$

so that

$$\text{height} = \sigma \cdot P(Z = z).$$

In more general notation, we have proved that the bar of the histogram of Z^* corresponding to the value $Z = z$ has as its base the interval with end points

(3)
$$\frac{z - E(Z) - .5}{\text{SD}(Z)} \quad \text{and} \quad \frac{z - E(Z) + .5}{\text{SD}(Z)}$$

while its height is

(4)
$$\mathrm{SD}(Z) \cdot P(Z = z)$$

as shown in Figure 3.

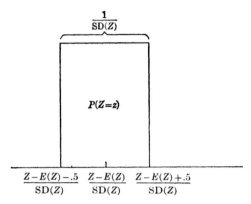

$$\frac{1}{\mathrm{SD}(Z)}$$

$P(Z=z)$

| $\dfrac{Z-E(Z)-.5}{\mathrm{SD}(Z)}$ | $\dfrac{Z-E(Z)}{\mathrm{SD}(Z)}$ | $\dfrac{Z-E(Z)+.5}{\mathrm{SD}(Z)}$ |

FIGURE 3. SINGLE BAR OF HISTOGRAM OF Z^*

Application of these formulas gives the histograms of T_1^*, T_2^*, T_4^*, and T_8^* shown in Figure 4. Comparing this with Figure 1, it is seen that in the new units it is much easier to perceive the shapes of the histograms and that they become smoother as n is increased.

In calculating a standard deviation, it is necessary to extract the square root of a variance. While there are arithmetic methods for extracting square roots, it is frequently quicker to use a table such as Table D. This table gives the square roots of the integers from 1 to 99, and also (as $\sqrt{10n}$) the square roots of the integers 10, 20, 30, ... , 990. Some examples will indicate how the table is used.

(a) $\sqrt{17} = 4.1231$

(b) $\sqrt{470} = 21.679$

(c) $\sqrt{1700} = \sqrt{17 \times 100} = \sqrt{17} \times \sqrt{100} = \sqrt{17} \times 10 = 41.231$

(d) $\sqrt{.047} = \sqrt{470/10{,}000} = \sqrt{470}/100 = .21679$

(e) To find $\sqrt{17.3}$, we must use interpolation. The table gives $\sqrt{17} = 4.1231$ and $\sqrt{18} = 4.2426$. Since 17.3 is three-tenths of the way from 17 to 18, it is natural to take $\sqrt{17.3}$ to be three-tenths of the way from 4.1231 to 4.2426. To obtain $\sqrt{17.3}$, we therefore start with $\sqrt{17}$ and add to it three-tenths of the difference between $\sqrt{18}$ and $\sqrt{17}$, that is

$$\sqrt{17.3} = 4.1231 + .3(4.2426 - 4.1231) = 4.1231 + .3(.1195)$$
$$= 4.1590.$$

(The correct value is 4.1593.)

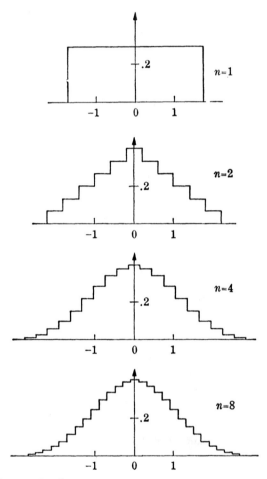

FIGURE 4.   STANDARDIZED SUM OF POINTS ON n DICE

(f) The computation of $\sqrt{1.53}$ would seem to call for interpolation between $\sqrt{1}$ and $\sqrt{2}$; however $\sqrt{n}$ is changing too rapidly here for the interpolation to be accurate.   Since $\sqrt{1.53} = \sqrt{153}/10$, we may instead interpolate between $\sqrt{150} = 12.247$ and $\sqrt{160} = 12.649$ to obtain the result $\sqrt{1.53} = 12.368/10 = 1.2368$.   (The correct value is 1.2369.)

PROBLEMS

1. Construct the histogram for B^* where B is the number of successes in 10 binomial trials with probability of success $p = .5$.

2. Let B have the binomial distribution with $n = 100, p = \frac{1}{2}$. Find the beginning and end point of the base of the bar in the histogram of B^* corresponding to the values (i) $B = 50$, (ii) $B = 47$, (iii) $B = 52$.

3. Construct the histogram for D^* where D is the random variable defined in Example 5.2.1.

4. Let D have the hypergeometric distribution with $N = 200$, $r = 30$, $s = 20$. Find the beginning and end point of the base of the bar in the histogram of D^* corresponding to the values (i) $D = 2$, (ii) $D = 3$, (iii) $D = 5$.

5. Let T be the sum of points on two throws with a fair die. Find the beginning and end point of the base of the bar in the histogram of T^* corresponding to the values (i) $T = 2$; (ii) $T = 4$; (iii) $T = 6$.

6. Find the heights of the bars in parts (i)–(iii) of the preceding problem.

7. Let M be the number of matchings of Example 5.1.2. Construct the histogram of (i) M; (ii) M^*.

8. Use Table D to find the square roots of (i) 900, (ii) 47,000, (iii) .000173, (iv) .963.

6.4   THE NORMAL CURVE AND THE CENTRAL LIMIT THEOREM

In Figure 4 of the preceding section we saw that the histograms of the standardized variables T_n^* (where T_n is the sum of the number of points on n dice) appear to smooth out and take on a definite shape as n increases. In more advanced treatments of probability theory it is proved in fact that as n increases indefinitely, the histogram of T_n^* tends to the smooth curve pictured in Figure 1, which is known as the *normal curve*. Some of its properties will be discussed later in this section, but we first give another example of distributions tending to the normal curve.

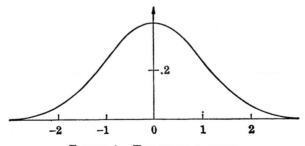

FIGURE 1.   THE NORMAL CURVE

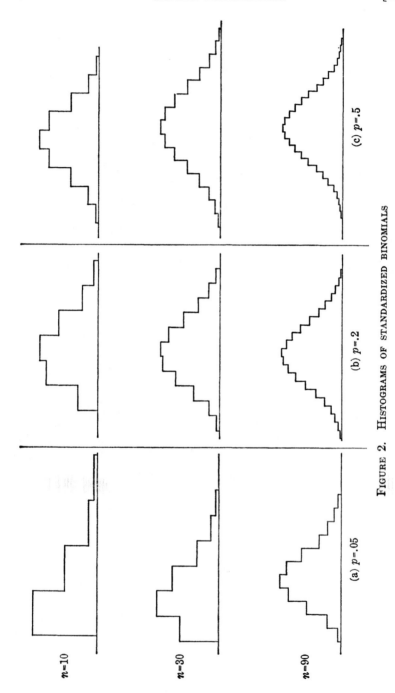

FIGURE 2. HISTOGRAMS OF STANDARDIZED BINOMIALS

EXAMPLE *1*. *Binomial distributions.* In Section 1 we defined the binomial random variable B as the number of successes in n binomial trials each having success probability p. Since $E(B) = np$ and $\text{Var}(B) = npq$, where $q = 1 - p$, the variable B in standard units is

(1) $$B^* = \frac{B - np}{\sqrt{npq}}.$$

We show in Figure 2 the histograms of B^* for increasing n (10, 30, 90) and for three different values of p (.05, .2, .5). The figure shows that for each fixed value of p, as n is increased the histogram of B^* looks more and more like the normal curve. Notice that the approach to the normal curve is considerably slower for $p = .05$ than for $p = .5$. Generally, the approach to the normal curve is slower the further p is from .5; if p is close either to 0 or to 1, n must be quite large before the approximation is acceptable. As a rough rule-of-thumb, the approximation is fairly good provided npq exceeds 10.

The behavior of T_n^* and B^* illustrates a quite general phenomenon. Whenever a random variable Z is the sum of a large number of independent random variables, all of which have the same distribution,† then $Z^* = [Z - E(Z)]/\text{SD}(Z)$ will have a histogram close to the normal curve. (Thus $T_n = X_1 + \ldots + X_n$ is the sum of independent random variables representing the numbers of points on the n dice, while $B = I_1 + \ldots + I_n$ is the sum of n independent indicator random variables.) If $Z = Z_1 + \ldots + Z_n$, where the Z's are independent and have the same distribution, it can be shown that as n increases without limit, the histogram of Z^* tends to the normal curve. This result is known as the *central limit theorem.*

Since the central limit theorem tells us that certain histograms are close to the normal curve, it often justifies the use of areas under the normal curve as approximations to the corresponding areas of these histograms, and hence as approximations to certain probabilities. The details of this procedure will be explained in the next section, but first let us introduce a table of areas under the normal curve.

It is customary to denote by $\Phi(z)$ the area under the normal curve, above the horizontal axis, and to the left of the vertical line at z. Thus $\Phi(z)$ is the area of the shaded region of Figure 3. A short table of $\Phi(z)$ for positive values of z is given as Table E. Table E may be supplemented by several remarks about Φ.

(a) The total area under the normal curve equals 1. This is not surprising since the curve is a limit of histograms each having total area equal to 1.

---

† Actually this result requires a mild restriction on the distribution, which is, however, automatically satisfied when the value set of the distribution is finite, as is assumed throughout this book.

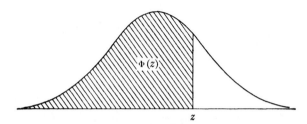

FIGURE 3. AREA UNDER NORMAL CURVE

(b) According to Table E, $\Phi(3.90) = 1.0000$. The value of $\Phi(3.90)$ is not exactly 1, but it is 1 to four decimals of accuracy. In fact $\Phi(3.90) = .999952$ to six decimals.

(c) An immediate consequence of (a) is that

(2)   Area to the right of $z = 1 - \Phi(z)$.

For example, the area to the right of $z = 1.2$ is $1 - \Phi(1.2) = 1 - .8849 = .1151$.

(d) The normal curve is symmetric about 0. Since the total area under the curve is 1, it follows that the area to the left of 0 and that to the right of 0 are both $\frac{1}{2}$, which checks the entry $\Phi(0) = .5000$. More generally, it follows from the symmetry of the curve that the area to the left of $-z$, which is $\Phi(-z)$, is equal to the area to the right of z, which by (2) is equal to $1 - \Phi(z)$, so that

(3) $\Phi(-z) = 1 - \Phi(z),$

as illustrated in Figure 4a. This explains why it is not necessary to extend Table E to negative values of z. To obtain for example $\Phi(-1.3)$ one computes $\Phi(-1.3) = 1 - \Phi(1.3) = 1 - .9032 = .0968$.

(e) It is often required to find the area under the normal curve *between* two vertical lines. Such areas can be obtained as the difference of two values of Φ. The area between z and z', where $z < z'$, is the area to the left of z' minus the area to the left of z. This is illustrated in Figure 4b, where the shaded region has area $\Phi(z') - \Phi(z)$. For example, the area between .4 and 1.3 is

$$\Phi(1.3) - \Phi(.4) = .9032 - .6554 = .2478.$$

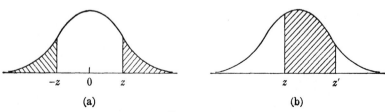

(a)   (b)

FIGURE 4.

As another example, let us find the area between $-.35$ and 1.24. The area to the left of 1.24 is $\Phi(1.24) = .8925$; that to the left of $-.35$ is $1 - \Phi(.35) = .3632$; thus, the desired area is $.8925 - .3632 = .5293$.

(f) Table E gives values of $\Phi(z)$ only for values of z that are multiples of one hundredth. Other values may be obtained by interpolation. Let us illustrate the method by finding a value for $\Phi(.364)$. The table gives $\Phi(.36) = .6406$ and $\Phi(.37) = .6443$. Since $.364$ is four-tenths of the way from $.36$ to $.37$, it is natural to take $\Phi(.364)$ to be four-tenths of the way from $.6406$ to $.6443$. To obtain $\Phi(.364)$, we therefore start with $\Phi(.36)$ and add to it four-tenths of the difference between $\Phi(.37)$ and $\Phi(.36)$, that is,

$$\Phi(.364) = .6406 + .4(.6443 - .6406) = .6406 + .4(.0037)$$
$$= .6406 + .0015 = .6421.$$

(g) In applications one often wishes to find the value of z corresponding to a given value of $\Phi(z)$. Such values can of course be obtained by interpolating backwards in Table E, but for convenience we give at the bottom of Table E a small auxiliary table of this "inverse" function. The auxiliary table also shows how rapidly $\Phi(z)$ approaches 1 as z increases.

PROBLEMS

1. Find the area under the normal curve to the left of (i) $.87$, (ii) -1.46, (iii) 1.072, (iv) $-.156$.

2. Find the area under the normal curve to the right of (i) $.04$, (ii) -3.97, (iii) $.423$, (iv) -1.006.

3. Find the area under the normal curve between (i) $.41$ and 1.09, (ii) $-.26$ and 2.13, (iii) -1.41 and $-.07$, (iv) -1.237 and 1.237.

4. Find the area under the normal curve outside the interval (i) $(-1.82, 1.82)$, (ii) $(-.37, .37)$, (iii) $(.08, 2.15)$, (iv) $(-.91, -.16)$, (v) $(-1.15, .09)$, (vi) $(-2.01, 1.89)$.

5. Find the value of z such that the area under the normal curve to the left of z is (i) $.9732$, (ii) $.2912$, (iii) $.6780$, (iv) $.1960$.

6. Find the value of z such that the area under the normal curve to the right of z is (i) $.8186$, (ii) $.0071$, (iii) $.85$.

7. Find the value of z such that the area between $-z$ and z is (i) $.6826$, (ii) $.5$, (iii) $.08$.

8. Find a value of z such that the area between (i) z and $2z$ is $.1$; (ii) z and $2z$ is $.12$; (iii) z and $3z$ is $.1$.

6.5 THE NORMAL APPROXIMATION

Let us now consider some examples of the use of the normal curve in obtaining approximate values of probabilities.

EXAMPLE 1. *The sum of points on three dice.* What is the probability that the number T_3 of points showing on three dice is equal to 15? The desired probability is the area of the corresponding bar of the histogram of T_3^*. From Example 3.1 we know that $E(T_3) = 3(\frac{7}{2}) = 10.5$ and $SD(T_3) = \sqrt{3(\frac{35}{12})} = 2.958$. According to (3.3), this bar covers the interval from $(15 - 10.5 - .5)/2.958 = 1.352$ to $(15 - 10.5 + .5)/2.958 = 1.690$. The normal approximation to the probability is the area under the normal curve above this interval, or $\Phi(1.690) - \Phi(1.352) = .9545 - .9118 = .0427$, as illustrated in Figure 1a. For comparison, the true value of $P(T_3 = 15)$ is $\frac{10}{216} = .0463$. The approximation is thus in error by $.0427 - .0463 = -.0036$; it is too small by 8% of the true value. For many purposes this degree of accuracy would suffice.

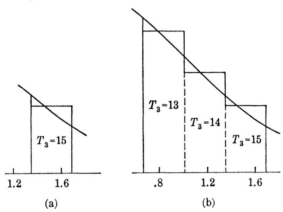

FIGURE 1. NORMAL APPROXIMATION TO DISTRIBUTION OF T_3

The method extends readily to the sum of the probabilities of several consecutive values. As an example, let us compute the probability $P(13 \leq T_3 \leq 15)$ that T_3 lies between 13 and 15 inclusive. By the addition law, this is

$$P(T_3 = 13) + P(T_3 = 14) + P(T_3 = 15),$$

which is the sum of the areas of the adjacent bars covering the intervals (12.5, 13.5), (13.5, 14.5), and (14.5, 15.5). Together, these bars cover the interval (12.5, 15.5). When we go over to standard units, this becomes the interval from $(12.5 - 10.5)/2.958 = .676$ to $(15.5 - 10.5)/2.958 = 1.690$, as shown in Figure 1b. The normal curve area above this interval is

$$\Phi(1.690) - \Phi(.676) = .9545 - .7505 = .2040$$

which may be compared with the exact value

$$P(13 \leq T_3 \leq 15) = \frac{46}{216} = .2130.$$

The approximation is low by .0090 which is 4% of the true value.

Let us now generalize the foregoing results. Suppose Z is any random variable taking on consecutive integers as its possible values, and let a and b be any of these values, with $a < b$. We are interested in obtaining an approximation for the probability

$$P(a \leqq Z \leqq b) = P(Z = a) + P(Z = a + 1) + \ldots + P(Z = b).$$

This probability is the sum of the areas of the bars in the Z histogram that cover the interval $(a - .5, b + .5)$. The corresponding bars in the Z^* histogram extend from

$$[a - .5 - E(Z)]/\text{SD}(Z) \quad \text{to} \quad [b + .5 - E(Z)]/\text{SD}(Z).$$

The desired approximation is therefore

$$(1) \qquad P(a \leqq Z \leqq b) \doteq \Phi\left(\frac{b + .5 - E(Z)}{\text{SD}(Z)}\right) - \Phi\left(\frac{a - .5 - E(Z)}{\text{SD}(Z)}\right)$$

EXAMPLE 2. *Male births.* Of the 5000 births from New York State represented in Figure 1.2.3c, 2641 or 52.8% were boys. A twenty-year record of over two million births from New York State gives 51.4% boys. Is the higher frequency of boys in our 5000 cases evidence that they really differ in this regard from the general experience? Let us see how likely would be so large or larger a number of boys, if in fact $p = P(\text{Male}) = .514$.

To compute $P(B \geq 2641)$ when the number B has the binomial distribution with $n = 5000$, $p = .514$ exactly from formula (1.2) would be tedious, and no binomial table covers so large a value of n. To obtain an approximation for this probability, we apply (1) with B instead of Z and $E(Z) = 5000 \times .514 = 2570, \text{SD}(Z) = \sqrt{5000 \times .514 \times .486} = 35.34$, $a = 2641$ and $b = 5000$. Substitution in (1) gives for the desired probability the approximate value

$$\Phi\left(\frac{5000.5 - 2570}{35.34}\right) - \Phi\left(\frac{2640.5 - 2570}{35.34}\right) = \Phi(68.77) - \Phi(1.995)$$

$$= 1 - .9770 = .0230.$$

This probability is small enough to cast some doubt on the hypothesis that these 5000 births are in accord with the general experience in New York State. (We shall consider the interpretation of such "significance probabilities" in Section 11.2.)

When in (1) the value of b is the largest value that Z can take on, so that the left-hand side of (1) can be written as $P(Z \geqq a)$, as was the case in the preceding example, a simplification of (1) is possible. The value of $\Phi\left(\dfrac{b + .5 - E(Z)}{\text{SD}(Z)}\right)$ (in the preceding example it was $\Phi(68.77)$), is then usually so close to 1 that one may simplify (1) to

(2) $$P(a \leqq Z) = 1 - \Phi\left(\frac{a - .5 - E(Z)}{\mathrm{SD}(Z)}\right)$$

Similarly, when a is the smallest possible value of Z, $\Phi\left(\dfrac{a - .5 - E(Z)}{\mathrm{SD}(Z)}\right)$ will be nearly 0, and we get

(3) $$P(Z \leqq b) = \Phi\left(\frac{b + .5 - E(Z)}{\mathrm{SD}(Z)}\right).$$

While many random variables have a representation as a sum of independent terms with the same distribution, this is by no means true of all random variables. The central limit theorem, however, has many extensions to sums of independent random variables not all having the same distribution, and also to certain sums of dependent variables. Examples of the first of these possibilities are found in Problem 20 and in Sections 12.6 and 12.7; the second possibility is illustrated by the following two examples.

EXAMPLE 3. *Hypergeometric.* The hypergeometric random variable D, introduced in Section 2, is seen by formula (2.2) to be the sum of s indicators; these all have the same distribution but are not independent. It is shown in more advanced treatments of probability theory that the histogram of D, expressed in standard units, is close to the normal curve provided Var(D) is not too small. Figure 2 shows the histogram of D^* for the hypergeometric distribution ($N = 20$, $r = 8$, $s = 7$), with the normal curve superimposed.

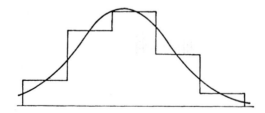

FIGURE 2. NORMAL APPROXIMATION TO HYPERGEOMETRIC DISTRIBUTION

Experience has shown that a great many variable quantities encountered in nature are distributed in a shape closely resembling the normal. For example, if we measure the heights of a large number of men and plot the frequencies with which various heights are observed, we obtain an *empirical histogram* whose shape is nearly of the normal form. We may give a theoretical argument to explain this phenomenon. The height of any individual is the resultant of the action of a large number of more or less unrelated genetic and environmental effects, so that the central limit theorem would lead us to expect a normal form.   It should not however be thought that all distributions encountered in nature are normal.   For example, in certain regions of central Africa, the distribution of heights of adult males would resemble Figure 3a: one factor (whether the man is or is not a pygmy) in this case dominates the other factors.   Another non-normal distribution is that of individual income: a few individuals have

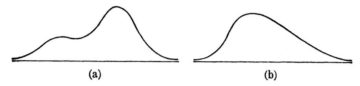

(a)   (b)

FIGURE 3.   DISTRIBUTIONS NOT RESEMBLING THE NORMAL CURVE

incomes far larger than the rest, so that the distribution (Figure 3b) has a long "tail" to the right.

PROBLEMS

1. Suppose B has the binomial distribution ($n = 100, p = .2$).   Calculate the normal approximations to (i) $P(B = 20)$, (ii) $P(B < 18)$, (iii) $P(B \leqq 17)$, (iv) $P(24 < B \leqq 29)$, (v) $P(B > 29)$.

2. Suppose B has the binomial distribution ($n = 10, p = .6$).   Calculate the normal approximations to (i) $P(B = 1)$, (ii) $P(B = 3)$, (iii) $P(B = 5)$ and compare them with the exact values.

3. Let B be the number of successes in 100 binomial trials with probability $p = \frac{1}{5}$ of success.   Use the normal approximation to find a number c such that the probability $P(B < c)$ is approximately (i) .95, (ii) .8, (iii) .5.   [Hint: (i) use the approximate formula

$$P(B < c) = \Phi\left(\frac{c - 20.5}{4}\right)$$

and the fact that $\Phi(1.645) = .95$ to get an equation for c.]

4. Let B be the number of successes in 300 trials with probability $\frac{3}{4}$ of success. Use the method of the preceding problem to find a number c such that $P(B < c)$ is approximately (i) .95, (ii) .9, (iii) .75.

5. Under the assumptions of the preceding problem find c such that $P(B > c)$ is approximately (i) .95, (ii) .7.

6. Under the assumptions of Problem 3 find a number c such that $P(|B - 20| \leqq c)$ is approximately (i) .95, (ii) .8, (iii) .5.   [Hint: $|B - 20| \leqq c$ means that $20 - c \leqq B \leqq 20 + c$.   Use formula (4.3).]

7. A college has invited 800 guests to a charter-day picnic.   For the purpose of ordering the picnic baskets, it is assumed that the decisions of the guests are unrelated, and that each has a probability $\frac{2}{3}$ of accepting the invitation.   With this assumption, how many baskets should be ordered if the college wishes to be (approximately) 99% certain that there will be a basket for each guest who comes?

8. If A is the number of apartment houses in the stratified sample of Problem 2.17, calculate the normal approximation to $P(A \geq 5)$.

9. A random digit generator is operated n times. Find the approximate probability that the frequency $f(\text{even})$ of even digits satisfies $.4 \leq f(\text{even}) \leq .6$, when (i) $n = 50$, (ii) $n = 250$, (iii) $n = 50,000$.

10. Suppose D has the hypergeometric distribution (2.1) with $N = 800$, $r = 200$, $s = 40$. Calculate the normal approximation to (i) $P(D = 10)$, (ii) $P(D < 8)$, (iii) $P(D \leq 12)$, (iv) $P(10 \leq D \leq 13)$, (v) $P(D > 9)$.

11. Suppose D has the hypergeometric distribution (2.1) with $N = 1000$, $r = 300$, $s = 50$. Use the normal approximation to find a number c such that the probability $P(D \leq c)$ is approximately (i) .9, (ii) .5, (iii) .05. [Hint: see Problem 3.]

12. Suppose D has the hypergeometric distribution (2.1) with $N = 1200$, $r = 600$, $s = 80$. Use the normal approximation to find a number c such that the probability $P(D \leq c)$ is approximately (i) .9, (ii) .5, (iii) .05. [Hint: See Problem 3.]

13. A city with 10,000 voters is holding an election involving two candidates A and B. To make a forecast, a random sample of size s is taken two days before the election. If D denotes the number of voters who express a preference for A, the poll will predict A as the winner if $D/s > \frac{1}{2}$ and B as the winner if $D/s < \frac{1}{2}$. If actually 51% of the voters prefer A, what is the probability that the poll will predict the election correctly if (i) $s = 50$, (ii) $s = 100$, (iii) $s = 200$? [Hint: Use the normal approximation and assume that every voter is willing to express his preference correctly, that all voters will cast a ballot, and that no one will change his mind between the poll and the election.]

14. Work the three parts of the preceding problem if the proportion of voters prefering A is 52% rather than 51%.

15. Work Problem 13 if the number of voters is 100,000 instead of 10,000 and if (i) $s = 100$, (ii) $s = 200$, (iii) $s = 500$, (iv) $s = 1000$.

16. In Problem 13, how large does s have to be so that the probability of correctly predicting the winner is (i) .9, (ii) .95, (iii) .99?

17. Solve the preceding problem if the number of voters instead of being 10,000 is (i) 50,000, (ii) 100,000, (iii) 500,000.

18. A bag contains a two-headed penny and a two-tailed penny. A penny is chosen at random from the bag and is then tossed five times.

  (i) Obtain the distribution of the number H of heads and plot its histogram.
  (ii) Calculate $E(H)$ and $\text{Var}(H)$, and show that the normal approximation is not satisfactory in this case.
  (iii) Would the normal approximation be better if the coin were tossed many times rather than only five times?
  (iv) Would the normal approximation be better if the coin were returned to the bag and a fresh draw were made before each toss?

[This problem uses the concepts and results of Sections 4.3 and 4.4.]

19. Suppose the possible values of the random variable Z are $0, c, 2c, 3c, \ldots$. If b is one of these values, how should (3) be modified to give a normal approximation for $P(Z \leq b)$?

20. Consider the generalized binomial trials model of Section 6.1.

    (i) Find $E(T)$ and $\text{Var}(T)$.

    (ii) For $n = 4$ and $p_1 = .2$, $p_2 = .4$, $p_3 = .6$, $p_4 = .8$, calculate the distribution of T and compare it with its normal approximation.

    (iii) Why does the binomial approximation of Section 6.1 work poorly in this example?

6.6 THE POISSON APPROXIMATION FOR $np = 1$

As was pointed out in Section 4, the normal approximation to the binomial distribution does not work very well if p is near 0 unless n is quite large. Since $q = 1 - p$ is then close to 1, the rule of thumb given in Example 4.1 suggests that the normal approximation could be relied upon for small p only if np is greater than or equal to 10. We would not, for example, expect it to work well when $p = .01$ and $n = 100$, in which case $np = 1$. Fortunately there is available a quite different method, the *Poisson approximation*, which gives good results when p is near 0. (By the device of relabeling failures as successes, the same method can also be used for values of p near 1.)

To explain the Poisson approximation, let us at first fix the expected number of successes, $E(B) = np$, at the value 1. Table 1 shows five different binomial distributions, corresponding to $n = 2, 5, 10, 20,$ and 40, with p respectively $\frac{1}{2}$, $\frac{1}{5}$, $\frac{1}{10}$, $\frac{1}{20}$, and $\frac{1}{40}$, so that in each case $np = 1$.

TABLE 1. BINOMIAL DISTRIBUTIONS WITH $np = 1$

	$n = 2$ $p = \frac{1}{2}$	$n = 5$ $p = \frac{1}{5}$	$n = 10$ $p = \frac{1}{10}$	$n = 20$ $p = \frac{1}{20}$	$n = 40$ $p = \frac{1}{40}$	$\ldots$	Poisson approximation: n very large $p = 1/n$
$b = 0$	.2500	.3277	.3487	.3585	.3632		.3679
1	.5000	.4096	.3874	.3774	.3725		.3679
2	.2500	.2048	.1937	.1887	.1863		.1839
3		.0512	.0574	.0596	.0605		.0613
4		.0064	.0112	.0133	.0143		.0153
5		.0003	.0015	.0022	.0026		.0031
6			.0001	.0003	.0004		.0005
7							.0001

If the first two rows of the table are compared, it is apparent that the values of $P(B = 0)$ and $P(B = 1)$ come closer together as n is increased and correspondingly p is decreased. When $n = 2$, $P(B = 1)$ is twice as large as $P(B = 0)$, but by the time $n = 40$, the ratio of $P(B = 1)$ to $P(B = 0)$ is only $.3725/.3632 = 1.026$. This suggests that, when n is very large, $P(B = 1)/P(B = 0)$ might be very close to 1, so that to a good approximation the same value could be used for both $P(B = 1)$ and $P(B = 0)$.

It is easy to verify this suggestion. It follows from the formula for binomial probabilities (1.2) and the fact that $\binom{n}{0} = 1$ and $\binom{n}{1} = n$, that

$$P(B = 0) = q^n \quad \text{and} \quad P(B = 1) = npq^{n-1}.$$

Since for the present we are only dealing with the case $np = 1$, this means that

(1)
$$\frac{P(B = 1)}{P(B = 0)} = \frac{1}{q}.$$

If p is close to 0, then $q = 1 - p$ is close to 1, and therefore so is $1/q$. This proves the suggestion that the ratio (1) will be near 1 if p is near 0. Therefore, if n is large and $p = 1/n$ is correspondingly small, we have

$$P(B = 1) \doteq P(B = 0)$$

where $\doteq$ means that the two expressions are approximately equal.

In a precisely similar way, it can be shown that $P(B = 2)/P(B = 1)$ will be close to $\frac{1}{2}$ when n is large and p small and $np = 1$ (see Problem 1). Since $P(B = 2)$ is about half as large as $P(B = 1)$, which in turn is close to $P(B = 0)$, we have the approximation

$$P(B = 2) \doteq \tfrac{1}{2} P(B = 0).$$

Continuing in the same way, we find

$$P(B = 3) \doteq \tfrac{1}{3} \cdot \tfrac{1}{2} P(B = 0) = \tfrac{1}{6} P(B = 0)$$
$$P(B = 4) \doteq \tfrac{1}{4} \cdot \tfrac{1}{3} \cdot \tfrac{1}{2} P(B = 0) = \tfrac{1}{24} P(B = 0)$$

and so forth. But of course the probabilities of all possible values of B must add up to 1, so that

$$1 = P(B = 0) + P(B = 1) + P(B = 2)$$
$$+ P(B = 3) + P(B = 4) + \ldots$$

(2)
$$\doteq P(B = 0) + P(B = 0) + \tfrac{1}{2} P(B = 0)$$
$$+ \tfrac{1}{6} P(B = 0) + \tfrac{1}{24} P(B = 0) + \ldots$$
$$= P(B = 0)[1 + 1 + \tfrac{1}{2} + \tfrac{1}{6} + \tfrac{1}{24} + \ldots]$$

where in the last step the quantity $P(B = 0)$ has been factored out.

The factor $1 + 1 + \tfrac{1}{2} + \tfrac{1}{6} + \tfrac{1}{24} + \ldots$ can be calculated numerically to any desired degree of accuracy. The work is shown in Table 2, where the factor is computed to four decimal places, giving the value 2.7183. We have added up only eight terms, and could have continued to add as many more as we wished, but the remaining ones are so small that they do not amount to much even when taken all together. (This fact is proved rigorously in theoretically more advanced books.) The number we have computed to four decimals is a very important one in mathematics, and is denoted by the letter e, in honor of the Swiss mathematician Leonhard Euler (1707–1783).

TABLE 2.   CALCULATION OF $1 + 1 + \frac{1}{2} + \frac{1}{6} + \frac{1}{24} + \ldots = e$

$$
\begin{aligned}
1 &= 1.0000 \\
1 &= 1.0000 \\
\tfrac{1}{2} &= .5000 \\
1/2\cdot3 = \tfrac{1}{6} &= .1667 \\
1/2\cdot3\cdot4 = \tfrac{1}{24} &= .0417 \\
1/2\cdot3\cdot4\cdot5 = \tfrac{1}{120} &= .0083 \\
1/2\cdot3\cdot4\cdot5\cdot6 = \tfrac{1}{720} &= .0014 \\
1/2\cdot3\cdot4\cdot5\cdot6\cdot7 = \tfrac{1}{5040} &= .0002 \\
\hline
&2.7183
\end{aligned}
$$

Now let us substitute this result in (2), to find

$$1 \doteq P(B = 0)2.7183.$$

Dividing, we get for $P(B = 0)$ the approximate value $1/2.7183 = .3679$. That is to say, if n is large, so that $p = 1/n$ is small, the probability that there will be no successes is approximately .3679, regardless of the precise values of n and $p = 1/n$. This approximation is already reasonably good when $n = 40$ and $p = \frac{1}{40}$, where the correct value is .3632, and the larger n is, the better the approximation will be. Since $P(B = 1) \doteq P(B = 0)$, the same value .3679 may be used to approximate $P(B = 1)$; since $P(B = 2) \doteq \frac{1}{2}P(B = 0)$, the value $\frac{1}{2}(.3679) = .1839$ may be used to approximate $P(B = 2)$, and so forth. The approximate values are shown as the last column of Table 1.

EXAMPLE 1.   *Triplets.*   It is stated that the chance of triplets in human births is $1/10,000$. What is the probability of observing at least 4 sets of triplets in a record of 10,000 human births?

Let us regard the births as unrelated, so that the number B of sets of triplets has the binomial distribution corresponding to $n = 10,000$ and $p = 1/10,000$. Since $np = 1$ and n is very large, we may use the Poisson approximation shown in the last column of Table 1. Thus

$$
\begin{aligned}
P(B \geqq 4) &= P(B = 4) + P(B = 5) + P(B = 6) + \ldots \\
&\doteq .0153 + .0031 + .0005 + .0001 \\
&= .0190.
\end{aligned}
$$

There is a little less than a 2% chance of observing so many sets of triplets.

PROBLEMS

1.   (i) Show that $P(B = 2) = \dfrac{n(n - 1)}{2}\, p^2 q^{n-2}$.   [Hint: Problem 2.2.19.]

(ii) In the case $np = 1$, show that the preceding formula reduces to $P(B = 2)$
$= \frac{1}{2}\left(1 - \frac{1}{n}\right)q^{n-2}.$

(iii) Use result (ii) to establish that when $np = 1$,

$$P(B = 2)/P(B = 1) = \frac{1}{2}\left(1 - \frac{1}{n}\right) \cdot \frac{1}{q}.$$

(iv) Use result (iii) to explain why $P(B = 2) \doteq \frac{1}{2}P(B = 1)$, when n is large and $p = 1/n$ is small.

2. By steps parallel to those of Problem 1, explain why $P(B = 3) \doteq \frac{1}{3}P(B = 2)$ when n is large and $p = 1/n$ is small.

3. Calculate the Poisson approximation for the following binomial probabilities:
(i) $P(B > 4)$ when $n = 300$ and $p = 1/300$,
(ii) $P(3 \leq B \leq 5)$ when $n = 80$ and $p = 1/80$,
(iii) $P(B \leq 1)$ when $n = 50$ and $p = 1/50$.

4. A machine is claimed on the average to produce only one defective in a hundred. If the claim is true, how surprising would it be to find five or more defectives in a lot of 100?

5. Refine the computations in Table 2 to obtain a five-decimal value for e.

6. Under the assumptions of Example 1, determine the largest c for which $P(B \geq c) \geq .5$.

7. A lot of 2000 items contains 20 defectives. If a sample of 100 is selected at random, use the Poisson approximation to find the probability that the sample will contain at least 2 defectives. [Hint: As an intermediate step use the binomial approximation to the hypergeometric distribution.]

8. In Example 1, use the Poisson approximation to find the probability of observing no triplets in 20,000 births. [Hint: Let the number of triplets be $B = B_1 + B_2$ where B_1 and B_2 denote the numbers of triplets in the first and second set of 10,000 births respectively. Then $P(B = 0) = P(B_1 = 0)P(B_2 = 0)$.]

9. Under the assumptions of the preceding problem, find (i) $P(B = 1)$, (ii) $P(B = 2)$, (iii) $P(B = 3)$.

10. Let B denote the number of successes in n binomial trials with success probability p. Use the method of Problems 8 and 9 to find (i) $P(B = 0)$, (ii) $P(B = 1)$, (iii) $P(B = 2)$ when $np = 3$ and n is large.

6.7 THE POISSON APPROXIMATION: GENERAL CASE

For simplicity of exposition, in the preceding section we considered only the case $np = 1$, but the method is applicable to any fixed value of $E(B) = np$. Suppose we consider the binomial distributions with np fixed at the value

(1) $np = \lambda.$

We shall be interested in obtaining an approximation to these distributions when p is close to 0 and hence n is very large. Just as in the case

$np = 1$, we have
$$P(B = 0) = q^n \quad \text{and} \quad P(B = 1) = npq^{n-1}$$
and hence
$$\frac{P(B = 1)}{P(B = 0)} = \frac{np}{q}.$$

Since q is close to 1, and np is equal to λ, we have for large n
$$\frac{P(B = 1)}{P(B = 0)} \doteq \lambda$$
or

(2) $$P(B = 1) \doteq \lambda P(B = 0).$$

By an exactly analogous argument we find
$$\frac{P(B = 2)}{P(B = 1)} \doteq \tfrac{1}{2} \lambda$$

(3) $$\frac{P(B = 3)}{P(B = 2)} \doteq \tfrac{1}{3} \lambda$$

$$\frac{P(B = 4)}{P(B = 3)} \doteq \tfrac{1}{4} \lambda, \quad \text{etc.}$$

To see how these relations can be used to obtain approximate probabilities, suppose that $\lambda = 2$. Then it is found by exactly the same method as for the case $\lambda = 1$ in the preceding section that

(4) $P(B = 2) \doteq \tfrac{1}{2}4P(B = 0), \quad P(B = 3) \doteq \tfrac{1}{6}8P(B = 0),$

$$P(B = 4) \doteq \tfrac{1}{24}16P(B = 0)$$

and so forth. Adding the probabilities and factoring out the common factor $P(B = 0)$ gives, in analogy to (6.2),
$$1 = P(B = 0)[1 + 2 + \tfrac{1}{2}4 + \tfrac{1}{6}8 + \ldots].$$

The factor in brackets can be calculated to any desired degree of accuracy. Taking only the first 12 terms gives to it the value 7.3891 and hence gives $P(B = 0) \doteq .1353$. The probabilities of the other values can now be computed from (2) and (4).

This method can be used to obtain the Poisson approximation corresponding to any given value of λ. Table F shows the results for 20 different values of λ ranging from .1 to 10. The Poisson approximation applies not only to the binomial random variable B, but also to a more general random variable T to be defined below, and for this reason the headings of Table F are in terms of T rather than B. A much more extensive table is available giving to eight decimal places the probabilities for values of λ ranging from .0000001 to 205.*

As an illustration of the use of Table F, let us find $P(B \leq 3)$ when B has the binomial distribution corresponding to $n = 120$ and $p = .04$.

* *Tables of the Individual and Cumulative Terms of Poisson Distribution*, D. Van Nostrand Co., Inc. (1962). 202 pages.

Here $E(B) = np = 4.8$, which is too small to permit safe use of the normal approximation. But, since n is large and p is small, we may use the Poisson approximation. By adding the four entries for $\lambda = 5$ and $T = 0, 1, 2, 3$ from Table F, we find that $P(B \leq 3)$ is about .2650 when $E(B) = 5$. Similarly $P(B \leq 3)$ is about .4335 when $\lambda = E(B) = 4$. As $E(B) = 4.8$ in the present case, it is necessary to interpolate to find $P(B \leq 3) \doteq .2987$. (The correct value, computed from the binomial formula, is .2887.)

As another illustration, let us find an approximation for $P(B = 60)$ when $n = 60$ and $p = \frac{19}{20}$. In this case, p is near 1, while the Poisson approximation requires that p be near 0. We can make the method applicable by looking at failures rather than successes. The chance of a failure is $\frac{1}{20}$, and 60 successes means 0 failures. The desired probability is therefore the same as $P(B = 0)$ when $n = 60$ and $p = \frac{1}{20}$. For this problem $np = 3$, and the result .0498 may be read directly from Table F. (The correct value computed from the binomial formula is .0461.)

We have developed the Poisson approximation for use with the binomial distribution, but it is also applicable in certain cases of the generalized binomial trials considered at the end of Section 6.1. As was pointed out there, if the success probabilities $p_1, p_2, \ldots$ of n unrelated trials are not too different from each other, the total number T of successes has a distribution very close to the binomial distribution $(n, \bar{p})$, where $\bar{p}$ is the arithmetic mean of the success probabilities. If in addition $p_1, p_2, \ldots$ are all small, they will of necessity be close to each other; also, $\bar{p}$ will then be small, and the Poisson distribution with

$$\lambda = n\bar{p} = p_1 + p_2 + \cdots + p_n$$

will be reasonably close to the binomial distribution $(n, \bar{p})$. By combining these two approximations, it is seen that the distribution of T can reasonably be approximated by the Poisson approximation with this value of λ.

EXAMPLE 1. *Effect of training.* Suppose that in the training of new workers to perform a delicate mechanical operation, the probabilities that the worker will be successful on his first, second and third attempt are $p_1 = .03$, $p_2 = .06$, $p_3 = .11$ respectively, and that the attempts may be considered to be unrelated. The distribution of the number T of successes among these three attempts, which was obtained in Problem 3.3.15, is shown below. We also give the binomial approximation with $\bar{p} = .20/3 = .0667$, and the Poisson approximation corresponding to $\lambda = .03 + .06 + .11 = .20$.

t	0	1	2	3	≥ 4
$P(T = t)$	.8115	.1772	.0111	.0002	.0000
Binomial approx.	.8130	.1742	.0124	.0003	.0000
Poisson approx.	.8187	.1637	.0164	.0011	.0001

As was the case in Example 1.5, the binomial approximation to the generalized binomial is excellent. The Poisson approximation is also good, but the value of $\bar{p}$ is a little too large for best results. Generally speaking, when dealing with a generalized binomial with small $p_1, p_2, \ldots$, the binomial approximation $(n, \bar{p})$ will work better than the Poisson approximation $(\lambda = n\bar{p})$. Why then ever use the latter? Essentially for the same two reasons that the Poisson approximation to the binomial distribution itself is so useful. First, the Poisson tables require only two entries (λ, t) rather than the three (n, p, b) or $(n, \bar{p}, b)$ needed by the binomial; this permits a much finer coverage by the table in the same number of pages, easier interpolation, etc. Second, the Poisson approximation requires knowledge only of the product $\lambda = np$ or $\lambda = n\bar{p}$, rather than the separate values of n and p or of the separate probabilities $p_1, p_2, \ldots$. Two typical examples will illustrate this point.

EXAMPLE 2. *Telephone traffic.* A telephone company anticipates that there will be, on the average, five lines in use at any given moment between two communities during the peak hours. How many lines should be built so that there is less than one chance in twenty a subscriber will find all lines in use when he makes a call?

Let T be the number of subscribers who are using their telephones for a call between the two communities at a given moment. Each subscriber is a "trial" having only a small chance of "success," i.e. of making a call at that moment. Under ordinary circumstances these trials, although certainly not having the same success probability, will be unrelated, so that the Poisson approximation can be used for the distribution of T. This will of course not be the case if some event causes many subscribers to call the other community at the same time.

From Table F we see that with $\lambda = 5$, $P(T \geq 9) = .0680$ while $P(T \geq 10) = .0317$. Thus, under the assumption that the number of lines in use at a given moment follows the Poisson distribution, the company will want to provide ten lines.

EXAMPLE 3. *Suicides.* During the week following the suicide of a film star, a newspaper columnist notes that in his city there were 12 suicides as compared with an average figure of 8 per week. He attributes the extra deaths to the suggestive effect of the star's suicide. Is this explanation convincing?

Let T be the number of suicides committed in the city during a given week. Considering each inhabitant of the city as a trial, the assumptions underlying the Poisson approximation appear not unreasonable. From Table F with $\lambda = 8$, we find $P(T \geq 12) = .1118$. Thus, if we assume the number of suicides per week to follow the Poisson approximation, there would be 12 or more suicides in about one ninth of the weeks. The observed number could well be explained as a chance event, without the necessity of invoking a suggestion effect.

Notice that, in the two preceding examples, it was not necessary to know the value of n or of the probabilities $p_1, p_2, \ldots$.

In Section 6 we mentioned that when $\lambda = np = 1$, the probability $P(T = 0)$ is equal to $1/e$ where e is the number 2.7183 . . . calculated to four decimal places in Table 6.2. More generally, it is shown in more advanced treatments of the subject that in the Poisson approximation for arbitrary λ, one has $P(T = 0) = 1/e^\lambda$.

Each of the columns of Table F consists of a sequence of positive numbers which (except for rounding errors) add up to 1. It is natural to think of such a sequence as defining a probability distribution for a random variable whose possible values are $s = 0, 1, 2, \ldots$. This Poisson random variable, unlike those we have previously considered, has an infinite number of possible values. An interesting feature of the Poisson distribution is that its expected value and variance are equal. This fact can be seen intuitively by considering the formula for binomial variance,

$$\text{Var}(B) = np(1 - p) = E(B)(1 - p).$$

If p is very small, $1 - p$ will be very close to 1, and $\text{Var}(B)$ will nearly equal $E(B)$. Thus, if T is a Poisson random variable with $E(T) = \lambda$, then we have $\text{Var}(T) = \lambda$ and hence $\text{SD}(T) = \sqrt{\lambda}$.

Table F gives the Poisson distribution for $\lambda \leqq 10$; for larger values of λ the normal approximation gives good results. That is, if λ is as large as 10, one may apply formula (5.3) to get the approximation

$$(5) \qquad\qquad P(T \leqq t) \doteq \Phi\left(\frac{t + .5 - \lambda}{\sqrt{\lambda}}\right).$$

PROBLEMS

1. Verify the first approximate equation of (3) by the method of Problem 6.1.

2. Check the value 7.3891 derived in the text from (4).

3. Compare the Poisson approximation with the correct binomial probability for the following cases:

   (i) $P(B = 3)$ when $n = 8$ and $p = .05$,
   (ii) $P(B = 9)$ when $n = 10$ and $p = .95$,
   (iii) $P(1 \leqq B \leqq 4)$ when $n = 10$ and $p = .1$,
   (iv) $P(B = 2)$ when $n = 9$ and $p = .05$.

4. Use the Poisson approximation to find the following binomial probabilities:
   (i) $P(B \geqq 3)$ when $n = 800$, $p = .005$,
   (ii) $P(B > 3)$ when $n = 12$, $p = \frac{1}{36}$,
   (iii) $P(B < 144)$ when $n = 150$, $p = .98$.

5. Work Example 2 if the average number of lines in use is (i) 4, (ii) 6, (iii) 8.

6. Solve the three parts of the preceding problem if the chance that a subscriber will find all lines in use when he makes a call is to be less than one in fifty.

7. In Example 3, suppose that the average number of suicides per week (instead of 8) had been 6 and the number in the particular week in question 11 (instead of 12). How convincing is the explanation in this case?

8. Solve Problem 5.7 under the assumption that each guest has a probability of .99 of accepting the invitation.

9. If a table such as Table F is subject on the average to one erroneous digit out of 1000, how likely is it that Table F is error free?

10. If you buy a lottery ticket in 100 lotteries, in each of which your chance of winning a prize is $\frac{1}{200}$, what is the (approximate) probability that you will win a prize (i) at least once, (ii) exactly once, (iii) exactly twice?

11. Compute the (approximate) probabilities of the preceding problem if the probability of winning a prize is $\frac{1}{50}$ in 20 of the 100 lotteries and $\frac{1}{800}$ in the remaining 80 lotteries.

The remaining problems relate to the material of the section that is in small print.

12. Duplicate the analysis of Example 1 if there are four attempts with probabilities $p_1 = .01$, $p_2 = .02$, $p_3 = .03$, $p_4 = .05$. [See Problem 3.3.16.]

13. If T has the Poisson distribution with $\lambda = 10$, compare the normal approximation (5) of $P(T \leq 7)$ with the correct value given by Table F.

14. Given that T has the Poisson distribution with $\lambda = .1$, check numerically from Table F that $E(T)$ and $\text{Var}(T)$ are both equal to λ.

15. Obtain a formula for $E(T^2)$ when T has a Poisson distribution, using the fact that $\text{Var}(T) = E(T) = \lambda$.

16. Let Z_1 and Z_2 be independent random variables having Poisson distributions with $E(Z_1) = .1$ and $E(Z_2) = .3$.

   (i) The distribution of Z_1 is a good approximation to the distribution of the number B_1 of successes in 100 binomial trials with success probability $p = .001$; the distribution of Z_2 is a good approximation to the distribution of the number B_2 of successes in 300 binomial trials with success probability $p = .001$. What is the distribution of the random variable $B_1 + B_2$? By considering the Poisson approximation to this distribution, conjecture the distribution of $Z_1 + Z_2$.

   (ii) Check your conjecture from Table F. [Hint: $P(Z_1 + Z_2 = 0) = P(Z_1 = 0)P(Z_2 = 0)$; $P(Z_1 + Z_2 = 1) = P(Z_1 = 0)P(Z_2 = 1) + P(Z_1 = 1)P(Z_2 = 0)$; etc.]

   (iii) Carry out the work of parts (i) and (ii) for the case $E(Z_1) = .2$ and $E(Z_2) = .3$.

17. Show that in the generalized binomial trials model

$$\frac{P(T = 1)}{P(T = 0)} = \frac{q_1}{p_1} + \frac{q_2}{p_2} + \cdots + \frac{q_n}{p_n}.$$

6.8 THE UNIFORM AND MATCHING DISTRIBUTIONS

In this section we present two simple distributions which further illustrate the concept of this chapter and which will have applications in Part II.

(a) *The uniform distribution.* Suppose that a box contains N tickets which bear the labels $1, 2, \ldots, N$. A ticket is chosen at random and the number X written on the ticket is observed. If the phrase "chosen at random" means that each ticket has the same probability $1/N$ of being chosen, then the random variable X has probability $1/N$ of assuming each of its possible values $1, 2, \ldots, N$, so that its probability distribution is

(1)

x	1	2	$\cdots$	N
$P(X = x)$	$\dfrac{1}{N}$	$\dfrac{1}{N}$	$\cdots$	$\dfrac{1}{N}$

If the events $X = 1, \ldots, X = N$ constitute the simple events of the model, the model is uniform in the sense of Section 1.5. Correspondingly the distribution of X is called the *uniform distribution* (on the integers $1, \ldots, N$). It is also known as the rectangular distribution since the histogram of X has the form of a rectangle.

We already encountered this distribution when considering the experiment of throwing a fair die. The number X of points showing when a fair die is thrown has the uniform distribution for $N = 6$. As another illustration, let X be a digit produced by a random digit generator (Example 1.2.2); then $X + 1$ has the uniform distribution for $N = 10$. The following examples illustrate some other situations in which a uniform distribution may arise.

(i) When a calculation is carried to five decimal places and then rounded off to four decimals, the digit X appearing in the fifth place must be one of $0, 1, 2, \ldots, 9$. In many types of calculation it appears that these ten digits occur about equally often; that is, we may suppose that $X + 1$ has the uniform distribution on the integers $1, \ldots, 10$. This model is often employed in studying the accumulation of rounding errors in high-speed computing machines.

(ii) Students of industrial accidents may conjecture that because of a fatigue effect accidents are more likely to occur late than early in the day, or in the week. A skeptic maintains that no such effect exists, and that an accident is equally likely to occur on Monday, $\ldots$, Friday. If the days of the work week are numbered 1 to 5, he would assume a uniform model for the number X of the day on which an accident occurs. By observing the numbers X for several unrelated accidents, one may investigate the adequacy of the uniform model in comparison with a "fatigue effect" model. A similar approach can be used when studying the possibility of a "birth order effect" for rare maladies which some geneticists think more likely in later than in earlier children of a family; in the study of a possible seasonal effect on earthquakes; etc. Methods for dealing with such problems will be discussed in Part II.

What is the expectation of the uniform random variable X defined by (1)? From (5.4.2) and (1), we have

$$E(X) = 1 \cdot \frac{1}{N} + 2 \cdot \frac{1}{N} + \ldots + N \cdot \frac{1}{N} = (1 + 2 + \ldots + N)/N.$$

It is shown in elementary books on algebra that

(2) $$1 + 2 + \ldots + N = \tfrac{1}{2}N(N + 1)$$

from which it follows that

(3) $$E(X) = \tfrac{1}{2}(N + 1)$$

This formula shows, for example, that the expected number of points showing when a fair die is thrown, is $\tfrac{7}{2}$ (see Example 5.3.1). For an alternative proof of (2) and (3), see Problem 1. In a similar way, the algebraic formula for the sum of squares,

(4) $$1^2 + 2^2 + \ldots + N^2 = \tfrac{1}{6}N(N + 1)(2N + 1)$$

shows that

(5) $$E(X^2) = \tfrac{1}{6}(N + 1)(2N + 1)$$

and hence that

(6) $$\mathrm{Var}(X) = \frac{N^2 - 1}{12}.$$

Again, the laws of expectation provide an alternative proof of (4) and (5) (see Problem 2). The formulas (3) and (6) will prove useful when we discuss the Wilcoxon test in Section 12.4.

(b) *The matching distribution.* As in case (a), consider again N tickets labeled $1, 2, \ldots, N$, but now suppose that all N tickets are drawn at random one at a time and are laid out in a row. Whenever the label on the ticket agrees with the number of its position in the row, we say that a *match* has occurred. Let M denote the number of such matches; then the distribution of M is called the *matching* distribution.

For illustration, suppose that $N = 4$ and that the tickets are laid out

$$3 \quad 2 \quad 1 \quad 4.$$

Here the ticket labeled 2 is in the second position and the ticket labeled 4 is in the fourth position; hence two matches have occurred and $M = 2$.

What is the distribution of M? If "drawn at random" means that the $N!$ possible orderings of the N tickets are equally likely (see Section 2.4), it is only necessary to count the number of orderings giving each value of M to find the desired distribution. Again using $N = 4$ for illustration, we know from Table 2.3.1 that there are $4! = 24$ possible orderings. They are shown in the tableau below, grouped according to the value of M. (Here each instance of a match is indicated by italics.)

$M = 0$: 2143 2341 2413 3412 3421 3142 4321 4312 4123
$M = 1$: *1342* *1423* *3241* *4213* 2431 4132 2314 3124
$M = 2$: *1243* *1432* *1324* *4231* *3214* 2134
$M = 4$: *1234*

Thus, when $N = 4$, the distribution of M is

m	0	1	2	3	4
$P(M = m)$	$\frac{9}{24}$	$\frac{8}{24}$	$\frac{6}{24}$	0	$\frac{1}{24}$

In general, there will always be exactly one ordering which gives a match in every place, so that $P(M = N) = 1/N!$. Since it is not possible to have all but one place matched, $P(M = N - 1) = 0$. The entire distribution for any particular value of N can in principle be worked out by listing all cases as we have done for $N = 4$ (Problem 7), but the work rapidly becomes prohibitive as N increases. In Problem 10 we present a short-cut method by means of which the entries in Table 1 were obtained. This table gives the number $C(m, N)$ of orderings of N tickets which have exactly m matches, for $N \leqq 8$.

TABLE 1. $C(m, N)$

	$N = 1$	2	3	4	5	6	7	8
$m = 0$	0	1	2	9	44	265	1854	14833
1	1	0	3	8	45	264	1855	14832
2		1	0	6	20	135	924	7420
3			1	0	10	40	315	2464
4				1	0	15	70	630
5					1	0	21	112
6						1	0	28
7							1	0
8								1
$N! = (N)_N = 1$		2	6	24	120	720	5040	40,320

EXAMPLE *1*. *Baby pictures.* A magazine prints the photographs of five movie stars, and also (in scrambled order) a baby picture of each. If a reader correctly matches at least three of them, can he justly claim to have demonstrated an ability to recognize resemblances? It might be argued that the correct matchings were just luck—after all, even if the reader matched at random, he might happen to get three or more right. But how likely would this be?

From Table 1 for $N = 5$, we see that the probability of getting three or more correct matches by luck is $(10 + 1)/120 = 0.092$. That is, about one out of every 11 persons who matches at random will do as well or better as the reader who makes three correct identifications.

As was shown in Example 5.5.4, the expected value of M is for all N equal to one,

$$(7) \qquad\qquad E(M) = 1.$$

It is a remarkable fact that we expect exactly one match, regardless of the number of tickets. The variance of M turns out to be also equal to one for all N (Problem 7.2.20),

$$(8) \qquad\qquad \mathrm{Var}(M) = 1.$$

If the sampling of the N tickets had been done with replacement, the number M of matches would have the binomial distribution corresponding to N trials with success probability $1/N$ on each trial. For large N we would therefore by Sections 6 and 7 expect the distribution of M to be approximately the Poisson distribution corresponding to $\lambda = N(1/N) = 1$. It is another remarkable fact, which unfortunately we cannot prove here, that the matching distribution is well approximated by this Poisson distribution even though the sampling is done without replacement. The reader may compare the distribution of M for $N = 8$ with the Poisson distribution with $\lambda = 1$ (Problem 9).

PROBLEMS

1. Let the distribution of X be given by (1).
  (i) Apply formula (5.4.2) to obtain expressions for $E(X^2)$ and $E(X + 1)^2$. By forming the difference of these two expressions, show that

$$E(X + 1)^2 - E(X^2) = [(N + 1)^2 - 1]/N = N + 2.$$

  (ii) Use the fact that

$$E(X + 1)^2 - E(X^2) = E[(X + 1)^2 - X^2] = 2E(X) + 1$$

  and the result of (i) to find $E(X)$.
  (iii) By comparing the expression (5.4.2) of $E(X)$ with its value found in (ii), prove (2).

2. (i) Using the results and methods of the preceding problem, prove (5) by taking the difference of $E(X + 1)^3$ and $E(X^3)$.
  (ii) Use the result of (i) to prove (4).

3. (i) If X has the distribution (1), what are the possible values of $Z = 2X - 1$ and what are the probabilities of these values?
  (ii) By comparing the expression for $E(Z)$ obtained from (5.4.2) with the value $2E(X) - 1$ obtained from (3), prove that the sum of the first N odd integers is

$$(9) \qquad 1 + 3 + 5 + \ldots + (2N - 1) = N^2.$$

4. By applying the method of Problem 3 to $E(Z^2)$ instead of $E(Z)$, prove that

$$(10) \qquad 1^2 + 3^2 + 5^2 + \ldots + (2N - 1)^2 = N(4N^2 - 1)/3.$$

5. Let X_1, X_2 be independent random variables, each having the distribution (1), and let Y be the larger of X_1 and X_2.

(i) Show that $P(Y \leq y) = y^2/N^2$ for $y = 1, 2, \ldots, N$.

(ii) Show that $P(Y = y) = (2y - 1)/N^2$ for $y = 1, 2, \ldots, N$.

6. Let X_1, X_2 be independent random variables, each having the distribution (1). Find the distribution of $Y = X_1 + X_2$.

7. Check the value $C(2, 5)$ of Table 1 by enumeration.

8. Suppose that the magazine of Example 1 asks its readers to match the pictures of eight (instead of five) movie stars. What is the probability that by purely random matching a reader will get at least four right?

9. Compare the distribution of M for $N = 8$ with the Poisson distribution for $\lambda = 1$.

10. (i) Show that $C(2, 5) = \binom{5}{2} C(0, 3)$ by counting the number of choices of the two tickets which are to provide the matchings, and the number of orderings of the remaining tickets which will not lead to any additional matchings.

(ii) Obtain analogous formulas for $C(1, 5)$ and $C(3, 5)$, and from these formulas compute $C(1, 5)$, $C(2, 5)$, $C(3, 5)$ using the first four columns of Table 1.

(iii) Find $C(0, 5)$ by using the fact that

$$C(0, 5) + C(1, 5) + C(2, 5) + C(3, 5) + C(5, 5) = 5!$$

and compare the values $C(0, 5), \ldots, C(3, 5)$ obtained with the fifth column of Table 1.

11. Use the method of the preceding problem to extend Table 1 to the case $N = 9$.

12. Show that for all values of N, $C(N, N) = 1$, $C(N - 1, N) = 0$, and

$$C(N - 2, N) = \binom{N}{2}.$$

6.9 THE LAW OF LARGE NUMBERS

In this section we shall derive a law which, in a sense, justifies the model for probability built in Chapter 1. The justification consists in showing that this model predicts the phenomenon of long-run stability which was the conceptual basis for our notion of probability.

We begin by putting into a quantitative form the notion, introduced in Section 5.6, that a widely dispersed random variable Z must have a large variance. Let us again denote $E(Z)$ by ζ, and consider an interval centered at ζ, say $(\zeta - c, \zeta + c)$ where c is any positive number. Of the possible values of Z, some may fall outside this interval and some inside it. Denote the possible values of Z outside (or on the end points of) the interval by $z_1, z_2, \ldots$ and those inside it by $w_1, w_2, \ldots$. By the definitions (5.6.1) of variance and (5.4.2) of expectation, the variance of Z can then be written in the form

$$\text{Var}(Z) = E(Z - \zeta)^2$$

(1)
$$= (z_1 - \zeta)^2 P(Z = z_1) + (z_2 - \zeta)^2 P(Z = z_2) + \ldots$$
$$+ (w_1 - \zeta)^2 P(Z = w_1) + (w_2 - \zeta)^2 P(Z = w_2) + \ldots.$$

Since the values $z_1, z_2, \ldots$ all lie outside the interval $(\zeta - c, \zeta + c)$ (or on its end points) each of the quantities $(z_1 - \zeta)^2$, $(z_2 - \zeta)^2$, $\ldots$ is at least as large as c^2. We shall therefore not increase the right side of (1) if we replace $(z_1 - \zeta)^2$, $(z_2 - \zeta)^2$, $\ldots$ each by c^2. The terms in the last row of (1) are all nonnegative, so we shall again decrease (or at least not increase) the right side of (1) if we eliminate these terms. Consequently

(2)
$$\text{Var}(Z) \geqq c^2 P(Z = z_1) + c^2 P(Z = z_2) + \ldots.$$

Factoring out c^2, the right-hand side can be written as $c^2 [P(Z = z_1) + P(Z = z_2) + \ldots]$. Since $z_1, z_2, \ldots$ are the possible values of Z outside the interval $(\zeta - c, \zeta + c)$, the probabilities in brackets add up to the probability that Z falls outside the interval, or $P(|Z - \zeta| \geqq c)$. This proves the famous *Chebyshev inequality*, that

(3)
$$\text{Var}(Z) \geqq c^2 P(|Z - \zeta| \geqq c)$$

for any random variable Z and any positive number c.

This inequality gives quantitative expression to the idea that wide dispersion means large variance. If the distribution of Z is widely dispersed, there exists a large number c such that a substantial probability lies outside $(\zeta - c, \zeta + c)$; that is, such that $P(|Z - \zeta| \geqq c)$ is substantial. Since c is large, c^2 will be very large, and the right side of (3) must be large. But by (3), the variance of Z is then still larger.

The inequality holds for any random variable Z. Let us apply it to the particular random variable B/n, where B has the binomial distribution (n,p) introduced in Section 1. The expectation and variance of B/n are $E(B/n) = p$ by (1.6) and (5.5.3), and $\text{Var}(B/n) = pq/n$ by (1.7) and (5.7.3). Substituting these values in (3) and dividing by c^2, we get

(4)
$$P\left(\left|\frac{B}{n} - p\right| \geqq c\right) \leqq \frac{pq}{nc^2}$$

for any positive number c and any binomial distribution (n,p).

The inequality (4) has an important consequence which serves to reinforce our concept of probability as long-run frequency. We began (Section 1.2) by presenting empirical evidence for the fact that when a long sequence of unrelated trials is performed under similar conditions, the frequency of a specified result (a "success") tends to be stable. Conceptually, the probability p of success represents in the model this stable frequency. The binomial trials model (Section 3.3) was constructed to represent a sequence of unrelated trials under constant conditions, and in this model

B/n represents the frequency of success.   If the model is realistic, it should then turn out that within the model B/n is in some sense close to p, at least if n is large.

The inequality (4) expresses a sense in which B/n is close to p when n is large enough.   For let c be a small number, so that the interval $(p - c, p + c)$ is a narrow interval centered at p.   Since c is small, c^2 will be very small and hence pq/c^2 will be very large.   In spite of this, by taking n large enough, we can make pq/nc^2 as small as we please, and by (4), $P(|B/n - p| \geq c)$ will then be still smaller; that is, by taking n large enough we can make sure that B/n is very unlikely to differ from p by as much as c.   In other words, if the number of trials is large enough, it is unlikely that the frequency B/n of success will differ much from the probability p of success.   This fact, which was discovered by James Bernoulli (1654–1705), is known as the *law of large numbers*.

It is important to realize that the law of large numbers does not "prove" experimental long-run frequencies to be stable.   The real world is not under the sway of mathematical arguments; rather, mathematical models are realistic only to the extent that their conclusions correspond to observation.   Long-run stability is an empirical fact.   The law of large numbers merely asserts that our model for probability is sufficiently realistic to agree with this fact.   A model for probability that led to an opposite conclusion would presumably not be sufficiently realistic to be of much use.

While the consequences of (4) are of great conceptual interest and theoretical value, the bound provided by the right side is usually very crude.   (This is only to be expected in view of the wholesale reductions involved in passing from (1) to (2).)   As an illustration, consider the special case $p = \frac{1}{2}$.   Table 1 provides a comparison of the actual probability $P(|B/n - \frac{1}{2}| \geq c)$ with the bound (4), which in the present case is $pq/nc^2 = 1/4nc^2$, for several values of n and c.   The last column of Table 1 shows the normal approximation (6.5.1) for $P(|B/n - p| \geq c)$.   We see that the normal approximation is good, but that the bound (4) is often far larger than the true value.   Thus, B/n tends to be actually much closer to p than is guaranteed by (4).

TABLE 1.   ILLUSTRATION OF THE BOUND FOR $p = \frac{1}{2}$

c	n	$P\left(\left\lvert\dfrac{B}{n} - \dfrac{1}{2}\right\rvert \geq c\right)$	$\dfrac{1}{4nc^2}$	normal approx.
.1	50	.203	.5	.203
	100	.0567	.25	.0574
	200	.00568	.0125	.00582
.05	100	.368	1	.368
	200	.179	.5	.179
	400	.0510	.25	.0512
	800	.00518	.125	.00522

The inequality (4) and its consequences may easily be generalized to deal with averages rather than frequencies. Recall that, in Section 5.3, the expectation of a random variable Z was motivated as corresponding to the long-run average value of many unrelated observations on the quantity that Z represents. Let these observations be represented by random variables $Z_1, Z_2, \ldots, Z_n$ which are defined on different factors of a product model, and each of which has expectation ζ and variance σ^2. Let the average of these random variables be denoted by

$$\overline{Z} = (Z_1 + Z_2 + \ldots + Z_n)/n.$$

It is then easy to see that $E(\overline{Z}) = \zeta$ and $\mathrm{Var}(\overline{Z}) = \sigma^2/n$ (Problem 5.7.5). If we now apply (3) to $\overline{Z}$, we find

$$(5) \qquad\qquad P(|\overline{Z} - \zeta| \geq c) \leq \sigma^2/nc^2.$$

This is a generalization of (4) to which it specializes when $Z_1, Z_2, \ldots$ are indicators (see (1.5)). It follows from (5) that it is unlikely that $\overline{Z}$ will be far from its expected value ζ, if n is sufficiently large. This theorem supports the common practice of averaging many unrelated observations on a quantity when one wishes to determine it with high precision.

PROBLEMS

1. For $p = \frac{1}{2}$, $n = 30$ and three suitable values of c, compare the right side of (4) with the corresponding probabilities from Table C.

2. Verify the entry .0574 of Table 1, using Table E.

3. Let Z be a random variable, with $E(Z) = 0$ and $\mathrm{Var}(Z) = 1$, whose distribution is adequately described by the normal curve. Use Table E to find the value of c for which the right side of (3) is largest.

4. Try to find values of n and c for which the bound provided by the right side of (4) is close to the value on the left side, for $p = \frac{1}{2}$.

5. (i) Find a distribution for Z and a value of c for which the two sides of (3) are equal.
(ii) Describe all distributions of Z for which this can occur.
[Hint: Equality requires that nothing is thrown away in passing from (1) to (2).]

6. (i) Show that $pq = p(1 - p)$ is equal to $\frac{1}{4} - (p - \frac{1}{2})^2$.
(ii) Use (i) to prove that $pq \leq \frac{1}{4}$ for all values of p.
(iii) Use (ii) to prove the *uniform* law of large numbers

$$P\left(\left|\frac{B}{n} - p\right| \geq c\right) \leq 1/4nc^2$$

in which the right side no longer depends on p.

7. Consider n unrelated trials with success probabilities $p_1, p_2, \ldots, p_n$.
(i) If $\overline{p} = (p_1 + \ldots + p_n)/n$, and S denotes the number of successes in the n trials, show that

$$P\left(\left|\frac{S}{n} - \bar{p}\right| \geq c^2\right) \leq 1/4nc^2.$$

(ii) What does this inequality tell us about the behavior of the success frequency S/n for large n?

[Hint (i): In (3), use the variance for S found in Problem 1.19, and apply the inequality obtained in part (ii) of the preceding problem.]

8. If Z^* denotes the random variable Z reduced to standard units as defined in (3.1), show that

(6) $$P(|Z^*| \geq d) \leq 1/d^2.$$

[Hint: In (5), replace c by $d\sigma/\sqrt{n}$.]

9. Let Z have the uniform distribution on the integers $1, \ldots, 9$. Compare the left and right sides of (3) for $c = 1, 2, 3$ and 4. [Hint: Use the results of Section 6.8.]

10. Obtain the entries for the last two columns of Table 1 if p (instead of being equal to $\frac{1}{2}$) has the values (i) $\frac{1}{3}$, (ii) $\frac{1}{4}$.

11. Prepare a table analogous to that asked for in the preceding problem, for the hypergeometric random variable D with distribution (2.1) when $N = 100,000$, $r = 50,000$, $c = .1$ and $.05$ and $s = 100, 200, 500$.

6.10 SEQUENTIAL STOPPING

In deriving the binomial distribution for the number B of successes in n binomial trials (Section 1), we assumed that the number n of trials was fixed in advance, regardless of how the trials would turn out. When the trials are performed *sequentially*, with the outcome of each trial known before the next trial is performed, it is possible to let the number of trials depend on the results obtained. A medical experiment may, for example, be abandoned if there are too many failures; the length of the world series in baseball depends on the outcomes of the successive games; the number of attempts needed to reach a person by telephone depends on the number of times the line is found busy; and so on. In such situations, the number of trials becomes a random variable, and the number of successes no longer has a binomial distribution. A model for sequential binomial trials will depend not only on the success probability p for single trials, but also on the *stopping rule* which specifies when observation is discontinued.

EXAMPLE 1. *World Series.* In matches between two contestants or teams, the rules often specify that the number of games played to determine the winner will depend on the outcomes. The baseball world series, for example, continues until one of the two teams has won four games. (Other examples occur in bridge, tennis, yacht races, etc.) It is of interest in such a series to know how many games will be played, and what is the chance that a given team will win.

Let us treat the successive games as a series of binomial trials, in which Team A has fixed probability p of winning any one game (which we shall call a "success"). The stopping rule is: continue the trials until a total of four successes or four failures has occurred and then stop. The record of play can be represented by a sequence of the letters S and F, where for example FSSFSFS would mean that Team A won the second, third, fifth and seventh games and hence the series. Under the assumptions made, the probability of this pattern is $qppqpqp = p^4q^3$. There are in fact 70 possible patterns, which may be grouped according to the numbers N_S and N_F of successes and failures, as shown in tableau (1):

	n_S	n_F	$P(N_S = n_S$ and $N_F = n_F)$
	4	0	p^4
	4	1	$4p^4q$
	4	2	$10p^4q^2$
(1)	4	3	$20p^4q^3$
	3	4	$20p^3q^4$
	2	4	$10p^2q^4$
	1	4	$4pq^4$
	0	4	q^4

Here, for example, the second line reflects the fact that four patterns (FSSSS, SFSSS, SSFSS, SSSFS) give the result ($N_S = 4$ and $N_F = 1$), and each pattern has probability p^4q. Notice that in this case the four patterns correspond to the four possible positions of the one failure among the first four trials. Similarly, the coefficient 10 in $P(N_S = 4, N_F = 2)$ represents the $\binom{5}{2} = 10$ ways of placing two F's among the first five symbols, and so forth.

From tableau (1) it is possible to obtain answers to various questions related to the series. For example, the number $N = N_S + N_F$ of games played has the distribution shown below, which is also given numerically for several values of p. As one might expect, the series tends to be longer when the teams are evenly matched than when one team is superior to the other.

n	$P(N = n)$	$p = .5$	$p = .6$	$p = .7$
4	$p^4 + q^4$	.1250	.1552	.2482
5	$4pq(p^3 + q^3)$	.2500	.2688	.3108
6	$10p^2q^2(p^2 + q^2)$	.3125	.2995	.2558
7	$20p^3q^3$	.3125	.2765	.1852

The probability that Team A wins can also be obtained from (1), but even more simply by another method. Imagine a different rule of play, under which the teams always play seven games with victory going to the team that wins four or more. It is clear that this rule would always result in the same outcome as that actually used, since the team winning four or more out of seven games is also the first team to win four. Under the new rule, however, the number of trials is fixed at $n = 7$, so that the number B of games won by Team A has the binomial distribution ($n = 7$, p). Thus, Table B gives as the probability of Team A winning the series for several values of p:

p	.5	.6	.7	.8	.9	.95
P(Team A wins series)	.5	.7102	.8740	.9667	.9973	.9998

EXAMPLE 2. *Sequentially stopped families.* Parents sometimes have a preference with regard to the sexes of their children, and may then base the decision whether to have another child on the sexes of their present children. If the sexes of successive children are regarded as binomial trials with fixed probability p of a boy on any given birth, the exercise of such a decision amounts to a sequential stopping rule. What influence would parental favoritism for children of one sex or the other have on the frequencies of the sexes in the population? In particular, could the observed slight excess of boys be explained by a favoritism for boys?

In an attempt to throw some light on these questions, let us consider the extreme case in which the parents are insistent on having a son, but otherwise want as small a family as possible. Then their rule will be: continue having children until a son is born, and then stop. Since we do not wish to consider families of indefinitely large size, let us fix a maximum number of children in the family, say t. Then possible family patterns, the associated number B of boys, G of girls, and C of children, and the probabilities of the patterns, are shown in tableau (2):

	Pattern	b	g	c	Probability
	M	1	0	1	p
	FM	1	1	2	qp
(2)	FFM	1	2	3	q^2p
	$\cdots$				
	FF$\cdots$FM	1	$t-1$	t	$q^{t-1}p$
	FF$\cdots$FF	0	t	t	q^t

What will be the frequency of boys in a large population of families generated according to (2)? Suppose there are k such families, the first family having B_1 boys, G_1 girls and $C_1 = B_1 + G_1$ children; the second family having B_2 boys, G_2 girls, and $C_2 = B_2 + G_2$ children; and so forth.

The frequency of boys is the total number of boys divided by the total number of children, or

$$f(\text{boys}) = \frac{B_1 + \ldots + B_k}{C_1 + \ldots + C_k} = \frac{(B_1 + \ldots + B_k)/k}{(C_1 + \ldots + C_k)/k}.$$

By the law of large numbers, if k is large, $(B_1 + \ldots + B_k)/k$ will probably be close to $E(B)$, the expected number of boys in a single family. Similarly $(C_1 + \ldots + C_k)/k$ will probably be close to $E(C)$. Therefore, it is probable that $f(\text{boys})$ will be close to $E(B)/E(C)$.

From (2) it is seen that $E(B) = 1 - q^t$. It can be shown (Problem 5) that $E(C) = (1 - q^t)/p$, so that $E(B)/E(C) = p$. This equation shows that the frequency of boys in a large population of families generated by our rule is likely to be close to p, the probability of a boy on a single birth. Thus, even an extreme stopping rule does not alter the frequency of boys in a large population.

The fact that even a stopping rule that favors boys in such an extreme way does not alter the frequency of boys suggests that this frequency would also remain constant with less extreme rules. An intuitive argument can be given for this conjecture. Suppose that the probability of a boy is p, and that the sexes of children are unrelated. Suppose that each family determines its size in its own way, which may differ from one family to another. Then regardless of how family sizes are decided, the frequency of males in the population will be near p. To see this, consider the analogous problem of a group of gamblers who are tossing a penny for which the probability of heads is p. The first gambler tosses it, using the stopping rule of the first family (identify "heads" with "boy"). When he has stopped, he passes the penny to the second gambler, who uses the stopping rule of the second family, etc. From the point of view of the coin, which is indifferent to who is doing the tossing, there is one long sequence of tosses, and the long-run frequency of heads will be close to the probability p of heads on a single toss. Analogously, the frequency of males in the population will be near the probability of a male on a single birth, regardless of the stopping rules used by the various families.

The situation is different, however, if the chance of a male varies from family to family, as some biologists believe. In this case, paradoxically, the effect of the assumptions leading to (2), which were motivated by a preference for boys, increase the frequency of girls. This can be seen intuitively by noting that under the assumed rules, girl-prone families will tend to be larger than boy-prone families. (This phenomenon is illustrated in Problem 7.)

PROBLEMS

1. Two teams continue playing until one of them has won (i) two games, (ii) three games. For each of these cases obtain a tableau analogous to (1).

2. If the probability of team A winning a single game is $p = .7$, find the probability of team A winning the series under the assumptions (i) and (ii) of the preceding problem, and compare your results with the corresponding one of Example 1.

3. Discuss the relation of Example 2 with the gambling system of Problem 5.3.13.

4. Using the fact that the probabilities in the last column of (2) must add up to one, show that

$$1 + q + q^2 + \ldots + q^{t-1} = (1 - q^t)/p.$$

5. In a family generated according to (2), let C denote the total number of children, and let I_1 indicate that $C \geq 1$, I_2 indicate that $C \geq 2, \ldots, I_t$ indicate that $C = t$. Show that

  (i) $C = I_1 + I_2 + \ldots + I_t$,

  (ii) $E(I_1) = 1$, $E(I_2) = q, \ldots, E(I_t) = q^{t-1}$,

  (iii) $E(C) = (1 - q^t)/p$.

[Hint (iii): Combine (ii) with the result of Problem 4.]

6. Give a formula for the expected number of girls in a family generated according to (2). [Hint: $G = C - B$.]

7. Suppose that a population consists of 1000 families with $p = .4$ and 1000 families with $p = .6$. Find the expected number of boys and of girls in this population if

  (i) each family has three children,

  (ii) each family is generated according to (2) with $t = 4$.

[Hint: the number of boys in the population is the sum of the numbers of boys in the 2000 families; use the addition theorem of expectation.]

8. Calculate the expected number of children in the family (2) when $p = \frac{1}{2}$ and $t = 2, 4, 8, 16$. What do you think happens to $E(C)$ when t becomes indefinitely large? Suggest an approximation for $E(C)$ in model (2) if t is large.

CHAPTER 7

MULTIVARIATE DISTRIBUTIONS

7.1 JOINT DISTRIBUTIONS

We have defined a random quantity as a quantity whose value is determined by the result of a random experiment. Actually, an experiment usually will determine the values of many different quantities. For example, when two dice are thrown the experiment determines the number of points on the first die, the number of points on the second die, the total number of points, the number of dice showing an odd number of points, etc. All of these quantities are random quantities determined by one and the same trial of the experiment. Again, when a random sample of items is drawn from a population, one usually observes two or more quantities for each item in the sample. Thus, for each child in a sample drawn from a school, the investigator may record both age and IQ; when a sample is drawn from a lot of ball bearings, the inspector may measure both diameter and hardness of each bearing in the sample. In many cases one is interested in the joint behavior of two or more random quantities, relating to the same experiment, which are represented in the model by random variables.

Example *1*. *Two dice*. In a throw of two dice let F denote the maximum of the numbers of points on the two dice and G the number of dice showing an even number of points. Let us list all possible pairs of values F and G and calculate the probability for each pair of values, assuming the dice to be fair. These probabilities can be found by listing the 36 equally likely results, as was done in Example 1.5.3, and can be exhibited conveniently in a table, whose rows correspond to the possible values of G and columns correspond to the possible values of F, and whose entries are the probabilities.

Thus the result ($G = 0$ and $F = 2$) is impossible, since if the maximum of the numbers showing on the two dice is two (i.e. $F = 2$), at least one of

f / g	1	2	3	4	5	6	Total
0	$\frac{1}{36}$	0	$\frac{3}{36}$	0	$\frac{5}{36}$	0	$\frac{9}{36}$
1	0	$\frac{2}{36}$	$\frac{2}{36}$	$\frac{4}{36}$	$\frac{4}{36}$	$\frac{6}{36}$	$\frac{18}{36}$
2	0	$\frac{1}{36}$	0	$\frac{3}{36}$	0	$\frac{5}{36}$	$\frac{9}{36}$
Total	$\frac{1}{36}$	$\frac{3}{36}$	$\frac{5}{36}$	$\frac{7}{36}$	$\frac{9}{36}$	$\frac{11}{36}$	$\frac{36}{36} = 1$

the dice must show an even number of points ($G > 0$). The entry $\frac{2}{36}$ at ($G = 1$ and $F = 2$) is obtained by noting that the only results with just one die showing an even number and with the maximum of the two numbers being 2, are (1, 2) and (2, 1).

The table above shows the *joint distribution* of the random variables G and F. In general, the joint distribution of two random variables Z and W gives, for each pair of possible values z of Z and w of W, the probability

$$(1) \qquad\qquad P(Z = z \text{ and } W = w)$$

of these values occurring on the same trial of the experiment. If Z can take on the values $z_1, z_2, \ldots$ and W the values $w_1, w_2, \ldots$, the joint distribution can be exhibited in a table as follows.

	w_1	w_2	$\ldots$
z_1	$P(Z = z_1 \text{ and } W = w_1)$	$P(Z = z_1 \text{ and } W = w_2)$	$\ldots$
z_2	$P(Z = z_2 \text{ and } W = w_1)$	$P(Z = z_2 \text{ and } W = w_2)$	$\ldots$

(2)

EXAMPLE 2. *Double lottery.* An important general class of problems is obtained by generalizing the lottery model (Examples 5.3.5, 5.5.2, 5.6.2). In this model, a ticket is chosen at random from N tickets bearing the numbers $v_1, \ldots, v_N$. If the number on the chosen ticket is denoted by Y, then the distribution, expectation, and variance of Y can be expressed in terms of the numbers $v_1, \ldots, v_N$.

Let us now suppose that each ticket bears two numbers, say v and w. Thus, on the first ticket are written the numbers v_1 and w_1, on the second ticket the numbers v_2 and w_2, etc. A ticket is drawn at random, and as before Y denotes the v number on the ticket while the w number on the same ticket is denoted by Z. Then Y and Z are jointly distributed random variables. In fact, $P(Y = v \text{ and } Z = w)$ is just the fraction of the N tickets which bear the numbers v and w. This model arises whenever we draw an item at random from a population and are concerned with two characteristics of the items. This is illustrated by the following example (other applications are made in Section 9.4).

EXAMPLE 3. *Distribution of boys and girls.* Suppose that 100 families are classified according to the number of male and female children as follows.

Boys \ Girls	0	1	2	3	4	Total
0	7	11	4	4	2	28
1	6	12	12	6	1	37
2	7	9	8	4	0	28
3	3	2	1	1	0	7
Total	23	34	25	15	3	100

A family is drawn at random and we observe the number B of boys and G of girls in the family. Here we may represent each of the 100 families by a ticket on which is written the number of boys and girls in that family. The random variables B and G correspond to the variables Y and Z of the general discussion above. The joint distribution of B and G may be obtained by dividing each entry of the table by 100. Thus, $P(B = 2$ and $G = 1) = \frac{9}{100} = .09$, since just 9 of the 100 equally likely families have these values for B and G.

EXAMPLE 4. *Items of three types.* We considered in Section 6.2 the problem of drawing a random sample of s from a population of N items of two types, called ordinary and special. The number D of special items in the sample is a random variable having the hypergeometric distribution. Let us now see how this distribution generalizes when the items are of three types and we are interested in the numbers of each type that appear in the sample. These numbers will be jointly distributed random variables.

To be specific, suppose that a lot of $N = 20$ manufactured items consists of 2 having major defects, 5 having minor (but no major) defects, and 13 which are without defects. As a check on the lot, a sample of $s = 4$ is drawn at random. The numbers D_1 of items with major defects and D_2 with minor defects have a joint distribution which can be read from the table on page 220.

The table shows the number of different samples having d_1 items with major and d_2 with minor defects, and the probability $P(D_1 = d_1$ and $D_2 = d_2)$ is found by dividing the appropriate entry of the table by the total number of samples, which is $\binom{20}{4} = 4845$.

d_1 \ d_2	0	1	2	3	4	Total
0	715	1430	780	130	5	3060
1	572	780	260	20	0	1632
2	78	65	10	0	0	153
Total	1365	2275	1050	150	5	4845

To see how the table was computed, let us find the entry for $d_1 = 0$ and $d_2 = 2$. The number of ways of selecting two items with minor defects for the sample from the five items with minor defects in the lot is $\binom{5}{2} = 10$. The remainder of the sample must consist of two nondefective items, and these can be chosen in $\binom{13}{2} = 78$ ways so that the total number of samples with $D_1 = 0$ and $D_2 = 2$ is $10 \cdot 78 = 780$.

As another example consider the entry $d_1 = 1$ and $d_2 = 2$. We must now choose two items with minor defects, $\binom{5}{2} = 10$ possibilities; one item with major defects, $\binom{2}{1} = 2$ possibilities; and one nondefective item, $\binom{13}{1} = 13$ possibilities. Altogether, there are therefore $10 \cdot 2 \cdot 13 = 260$ possible samples with $D_1 = 1$ and $D_2 = 2$.

In a situation in which the joint distribution of two random variables Z and W is given, it may be of interest to find the distribution of each random variable separately, the so-called *marginal distributions* of Z and W. How can we, from (2), obtain the marginal distribution of Z, that is the probabilities $P(Z = z_1)$, $P(Z = z_2)$, . . . ? The event $Z = z_1$ occurs when Z takes on the value z_1 and W any one of the values $w_1, w_2, \ldots$, so that

$$(Z = z_1) = [(Z = z_1 \text{ and } W = w_1) \text{ or } (Z = z_1 \text{ and } W = w_2) \text{ or}. . .].$$

Since the events on the right-hand side of this equation are exclusive, it follows that

(3) $P(Z = z_1) = P(Z = z_1 \text{ and } W = w_1) + P(Z = z_1 \text{ and } W = w_2) + \ldots .$

The probability $P(Z = z_1)$ is therefore obtained from (2) by adding all the probabilities in the row corresponding to z_1. Similarly $P(Z = z_2)$ is the sum of the probabilities in the row corresponding to z_2, etc. The same argument also shows that $P(W = w_1)$ is the sum of the probabilities in the column corresponding to w_1 and analogously for the other values of W.

The term "marginal distribution" derives from the fact that the sum of the probabilities of each row and column is frequently shown in the margin of the table, as was done in the examples above. Thus in Example 1, we read off the marginal distribution of G from the right-hand margin of the table as

$$P(G = 0) = \tfrac{9}{36}, \quad P(G = 1) = \tfrac{18}{36}, \quad P(G = 2) = \tfrac{9}{36}$$

and similarly that of F from the bottom margin of the table. Analogously, in Example 3, the marginal distribution of the number G of girls is seen to be

$$P(G = 0) = .23, \quad P(G = 1) = .34, \quad P(G = 2) = .25,$$
$$P(G = 3) = .15, \quad P(G = 4) = .03.$$

The marginal probability distribution of W gives us the probabilities that W takes on the values $w_1, w_2, \ldots$. What can we say about these probabilities if we are given the value of Z? The probability that $W = w$ given that $Z = z$ is the conditional probability $P(W = w|Z = z)$ (see Section 4.2). Thus for example, if z is any given value of Z, the conditional probability that $W = w_1$, given $Z = z$, is

$$P(W = w_1|Z = z) = \frac{P(W = w_1 \text{ and } Z = z)}{P(Z = z)}.$$

It is easy to check (Problem 5) that the conditional probabilities

(4) $$P(W = w_1|Z = z), \quad P(W = w_2|Z = z), \ldots$$

add up to one, and thus constitute a distribution.

Definition. The distribution which assigns to $w_1, w_2, \ldots$ the probabilities (4) is called the *conditional distribution of W given $Z = z$.*

This distribution has of course an expectation, known as the *conditional expectation* of W given $Z = z$, and denoted by $E(W|Z = z)$.

Illustrations of conditional distributions are provided by each of the rows and columns of Examples 1, 3, and 4. In Example 1, for instance, the conditional distribution of G given that $F = 3$ is obtained by dividing the entries of the column $f = 3$ by the total of that column:

$$P(G = 0|F = 3) = \tfrac{3}{5}, \quad P(G = 1|F = 3) = \tfrac{2}{5}, \quad P(G = 2|F = 3) = 0.$$

The joint distribution of Z and W is particularly simple if Z and W are independent (Section 5.2), i.e. if

(5) $$P(Z = z \text{ and } W = w) = P(Z = z) \cdot P(W = w) \qquad \text{for all } z, w.$$

In this case, the conditional distribution of W, given that $Z = z$, is the same as the marginal distribution of W, for all z (Problem 12).

The concepts of this section extend in a natural way to more than two variables. For example, if Z, W, and V are any three random variables, their joint distribution is given by the probabilities

(6) $$P(Z = z, \quad W = w, \quad \text{and} \quad V = v)$$

that Z, W, and V take on any particular values z, w, and v. The marginal distribution of Z is obtained by summing the probabilities (6) over all possible pairs of values w and v; the joint marginal distribution of Z and W is obtained by summing the probabilities (6) over all possible values v. The conditional distribution of V given $Z = z$ and $W = w$ is given by the probabilities

$$P(V = v_1 | Z = z \text{ and } W = w), \quad P(V = v_2 | Z = z \text{ and } W = w), \ldots$$

etc.

PROBLEMS

1. Let Z and W be the minimum and maximum of the numbers of points showing when two fair dice are thrown.   Find the joint distribution of Z and W.

2. Let S and D denote the sum and (absolute) difference of the numbers of points showing when two fair dice are thrown.   Find the joint distribution of S and D.

3. A sample of five items is drawn at random from a population of 25 items, of which six have major defects and four have minor defects (but no major ones). Find the joint distribution of the numbers of items in the sample having major defects and having minor defects.

4. Let Z and W denote the number of ones and the number of twos appearing when three fair dice are thrown.   Find the joint distribution of Z and W.

5. Find the marginal distribution of the random variable D_2 of Example 4 from formula (6.2.1) and check this against the marginal distribution of D_2 read from the table of Example 4.

6. Check that the probabilities (4) add up to one.   [Hint: Use formula (4.4.5).]

7. Find the following conditional distributions:
  (i) the distribution of F given $G = 1$ in Example 1;
  (ii) the distribution of B given $G = 3$ in Example 3.

8. Find the conditional distribution of D_2 given $D_1 = 1$ in Example 4
  (i) from the table of Example 4;
  (ii) using the fact that this conditional distribution is hypergeometric.

9. If U and V are the random variables defined in Problem 1.7.14 find the conditional distribution of V given that (i) $U = 0$, (ii) $U = 1$, (iii) $U = 2$, (iv) $U = 3$, (v) $U = 4$.

10. If U and W are the random variables defined in Problem 1.7.14 find the conditional distribution of W given that (i) $U = 0$, (ii) $U = 1$, (iii) $U = 2$, (iv) $U = 3$, (v) $U = 4$.

11. In Example 1,

    (i) find the conditional expectation of the random variable F, given each of the possible values of $G(G = 0, G = 1, G = 2)$;

    (ii) show that

$$P(G = 0)E(F|G = 0) + P(G = 1)E(F|G = 1) + P(G = 2)E(F|G = 2) = E(F).$$

12. (i) Find the conditional expectation of the random variable G of Example 3, given each of the possible values of B.

    (ii) In analogy to Problem 11(ii), show that $E(G)$ can be computed by adding the conditional expectations of part (i) each multiplied by the probability of the corresponding value of B.

13. Under the assumptions of Problem 1.7.14, find the conditional expectation of

    (i) V given each of the possible values of U, $(U = 0, 1, 2, 3, 4)$;

    (ii) D given each of the possible values of U.

14. For any two random variables Z and W, let $\varphi(w) = E(Z|W = w)$. Then $E\varphi(W) = E(Z)$. Prove this result for the case that Z can take on three values z_1, z_2, z_3, and W two values w_1, w_2.

15. Check the conditional distribution of G given $F = 3$ given in the text by enumerating all possible simple events (pairs of values of the two dice) and constructing a conditional model (given the event $F = 3$) using (4.1.1).

16. Prove that if Z and W are independent, then the conditional distribution of W given that Z has any particular value z agrees with the marginal distribution of W.

17. Let the joint distribution of Z, W, and V be given by

$$\tfrac{1}{4} = P(Z = 0 \text{ and } W = 0 \text{ and } V = 0) = P(Z = 0 \text{ and } W = 1 \text{ and } V = 1)$$
$$= P(Z = 1 \text{ and } W = 0 \text{ and } V = 1) = P(Z = 1 \text{ and } W = 1 \text{ and } V = 0).$$

    (i) Find the joint marginal distribution of Z and W and show that Z and W are independent.

    (ii) Show that Z and V are independent and that W and V are independent.

    (iii) Find the conditional distribution of V given $Z = 0$ and $W = 0$. Determine whether this is equal to the marginal distribution of V.

7.2 COVARIANCE AND CORRELATION

When developing a formula for the variance of the sum of two random variables Z and W in Section 5.7, we were led to define the covariance of Z and W by

$$(1) \qquad \text{Cov}(Z, W) = E[(Z - \zeta)(W - \eta)]$$

where $\zeta = E(Z)$ and $\eta = E(W)$. We shall now discuss and illustrate the meaning of covariance, derive certain laws of covariance, and use these to obtain certain results for variance. In particular, these will provide proofs of formulas (5.7.15) and (6.2.5).

As mentioned in Section 5.7, the covariance of two random variables is a measure of their tendency to vary in the same way. Suppose in fact that the joint distribution of Z and W is such that large values of Z and W usually occur together, and that small values tend to occur together. This would be the case for example if Z and W were the grades on the midterm and final examinations of a randomly selected student, or if they were the ages of husband and wife of a randomly selected couple. When this happens, the quantities $Z - \zeta$ and $W - \eta$ will usually both be positive or both negative, so that the product

$$(2) \qquad\qquad (Z - \zeta)(W - \eta)$$

will usually be positive, and its expectation (1) will therefore typically also be positive (but see Problem 7).

On the other hand, suppose that large values of Z tend to go together with small values of W and vice versa (for example, if Z is the number of cigarettes smoked between the ages of 20 and 25 and W is the age at time of death, or if Z denotes the number of hours spent watching TV and W the number of books read). Then if one of the factors of (2) is positive, the other factor, and hence the product, will usually be negative. The covariance (1) will then typically also be negative.

EXAMPLE 1 *Two pennies* Suppose that two fair pennies are tossed. Let Z denote the number of heads on the first penny and W the total number of heads on the two pennies. Here large values of Z and W, and small values of Z and W, tend to occur together, so that we expect the covariance to be positive. The joint distribution of Z and W is shown in the following table. The covariance of Z and W is therefore

$$\text{Cov}(Z, W) = \tfrac{1}{4}\cdot\tfrac{1}{2} + \tfrac{1}{4}\cdot 0 + \tfrac{1}{4}\cdot 0 + \tfrac{1}{4}\cdot\tfrac{1}{2} = \tfrac{1}{4},$$

which is positive as we expected.

Event	Probability	z	w	$z - \zeta$	$w - \eta$	$(z - \zeta)(w - \eta)$
HH	$\frac{1}{4}$	1	2	$\frac{1}{2}$	1	$\frac{1}{2}$
HT	$\frac{1}{4}$	1	1	$\frac{1}{2}$	0	0
TH	$\frac{1}{4}$	0	1	$-\frac{1}{2}$	0	0
TT	$\frac{1}{4}$	0	0	$-\frac{1}{2}$	-1	$\frac{1}{2}$

If instead we consider the covariance of Z with the total number of tails on the two tosses, we would expect a negative result (Problem 1).

EXAMPLE 2. *Double lottery.* As a second example let us obtain an expression for the covariance of the numbers Y and Z appearing on the randomly selected ticket of the double lottery discussed in Example 1.2, which we shall denote by

(3) $$\lambda = \text{Cov}(Y, Z).$$

Since each of the N tickets has probability $1/N$ of being drawn, and since $E(Y) = \bar{v}$ and $E(Z) = \bar{w}$, we see from (1) and the definition of expectation that

(4) $$\lambda = [(v_1 - \bar{v})(w_1 - \bar{w}) + \ldots + (v_N - \bar{v})(w_N - \bar{w})]/N.$$

The quantity on the right side of (4) is known as the "population covariance." Equations (3) and (4) are analogous to (5.6.3) and (5.6.4) concerning the population variance.

Covariance obeys a number of simple laws, somewhat analogous to the laws of variance. Two of these were developed in Section 5.7:

(5) $$\text{Cov}(Z, W) = E(ZW) - E(Z)E(W)$$

and

(6) $$\text{Cov}(Z, W) = 0 \qquad \text{if } Z \text{ and } W \text{ are defined on different factors of a product model.}$$

From the definition it is clear that covariance is commutative; that is, that

(7) $$\text{Cov}(Z, W) = \text{Cov}(W, Z).$$

It is also easily seen from (1) that the covariance of any random variable with a constant is zero,

(8) $$\text{Cov}(Z, c) = 0.$$

As further consequences of (1) we find that the covariance of Z and W is unchanged if any constants are added to Z and W,

(9) $$\text{Cov}(Z + a, W + b) = \text{Cov}(Z, W);$$

while multiplication by constants gives

(10) $$\text{Cov}(aZ, bW) = ab\,\text{Cov}(Z, W).$$

Finally, we have an *addition law* for covariance:

(11) $$\text{Cov}(Z + W, V) = \text{Cov}(Z, V) + \text{Cov}(W, V).$$

The proofs of these laws are left to the reader (Problem 8).

As mentioned in Section 5.7, equation (6) holds quite generally for any independent random variables Z and W. Thus, independent random variables always have zero covariance. The converse, however, is not true. Two random variables may be completely dependent, in the sense that the value of one is determined by the value of the other, but still have covariance equal to zero. For example, let Z take on the values $-1, 0, 1$ with probabilities $\frac{1}{4}, \frac{1}{2}, \frac{1}{4}$ and let $W = Z^2$. Here W is very strongly dependent on Z, being in fact known as soon as Z is known. On the other hand, since $E(Z) = 0$, it follows from (5) that

$$\mathrm{Cov}(Z, W) = E(ZW) = \tfrac{1}{4}(-1) + \tfrac{1}{2}(0) + \tfrac{1}{4}(1) = 0.$$

Equations (9) and (10) show that covariance is unchanged by addition of arbitrary constants to the two variables, but not if the variables are multiplied by constants. A measure of the tendency of the two variables to vary together, which is unchanged both by addition and by multiplication with *positive* constants, is the *correlation coefficient* defined by

$$(12) \qquad \rho(Z, W) = \frac{\mathrm{Cov}(Z, W)}{\sqrt{\mathrm{Var}(Z) \cdot \mathrm{Var}(W)}}.$$

(The correlation coefficient is undefined if either $\mathrm{Var}(Z)$ or $\mathrm{Var}(W)$ is zero.) Since the denominator of (12) is always positive, the correlation coefficient always has the same sign as the covariance. It follows that typically ρ is positive if Z and W tend to vary in the same direction, and is negative if they tend to vary in opposite directions. It is seen from (6) and (12) that

$$(13) \qquad \rho(Z, W) = 0 \qquad \text{if } Z \text{ and } W \text{ are defined on}$$
different factors of a product model

and indeed (13) holds also if Z and W are independent. Two random variables whose correlation coefficient is zero are said to be *uncorrelated*.

It is important to realize that the correlation coefficient is not, as is frequently believed, a measure of how strongly the variables Z and W depend on each other, but only of the extent to which they vary together or in opposite direction. As shown above, two variables may be very strongly dependent without varying either together or in opposite directions, and the correlation coefficient in such a situation may be zero.

We must still show that $\rho(Z, W)$ is unchanged if arbitrary constants are added to Z and W, or if Z and W are multiplied by positive constants. The first statement is immediate since $\mathrm{Cov}(Z, W)$, $\mathrm{Var}(Z)$, and $\mathrm{Var}(W)$ are all unchanged if Z and W are replaced by $Z + a$ and $W + b$. To see the second, we compute

$$\rho(aZ, bW) = \frac{\mathrm{Cov}(aZ, bW)}{\sqrt{\mathrm{Var}(aZ) \cdot \mathrm{Var}(bW)}} = \frac{ab\,\mathrm{Cov}(Z, W)}{\sqrt{a^2\,\mathrm{Var}(Z) \cdot b^2\,\mathrm{Var}(W)}} = \rho(Z, W).$$

The correlation coefficient has the further property that it always lies between -1 and $+1$ so that for any two random variables Z, W

$$(14) \qquad -1 \leqq \rho(Z, W) \leqq 1.$$

For a proof of this statement see Problem 15.

As an application of the laws of covariance, we shall now extend the addition law for variance (5.7.10),

(15) $\text{Var}(Z + W) = \text{Var}(Z) + \text{Var}(W) + 2 \, \text{Cov}(Z, W)$

to obtain a formula for the variance of three random variables, say Z, W, and V. If we think of $Z + W + V$ as the sum of $Z + W$ and V, then (15) shows that

$$\text{Var}(Z + W + V) = \text{Var}[(Z + W) + V] = \text{Var}(Z + W) + \text{Var}(V) + 2 \, \text{Cov}(Z + W, V).$$

Now apply (15) again and (11) to find

(16) $\begin{aligned} \text{Var}(Z + W + V) &= \text{Var}(Z) + \text{Var}(W) + \text{Var}(V) \\ &\quad + 2[\text{Cov}(Z, W) + \text{Cov}(Z, V) + \text{Cov}(W, V)]. \end{aligned}$

The argument leading to (16) may be extended to more than three summands (Problem 16). One then obtains the *general addition law of variance:*

$$\text{Var}(Z_1 + \ldots + Z_s) = \text{Var}(Z_1) + \ldots + \text{Var}(Z_s)$$

(17) $+\ 2$ [the sum of the covariances of the $\binom{s}{2}$ pairs of the random

variables $Z_1, \ldots, Z_s$].

In the particular case when the random variables $Z_1, \ldots, Z_s$ are defined on different factors of a product model (or indeed when they are independent), each of the covariances is zero and (17) implies (5.7.15).

As a further application of (17) we shall now derive the correction factor $(N - s)/(N - 1)$ for the variance of a hypergeometric random variable, encountered in Section 6.2. Consider first a somewhat more general problem.

EXAMPLE 3. *Variance in the lottery model.* Recalling the lottery model of Examples 5.3.5, 5.5.2, and 5.6.2, suppose that s tickets are drawn without replacement from a box containing N tickets labeled with values $v_1, \ldots, v_N$. Let the numbers appearing on the s successively drawn tickets be $Y_1, \ldots, Y_s$. These are jointly distributed random variables, and by the equivalence law of ordered sampling, each of the Y's has the distribution of the random variable Y of Example 5.6.2. Thus in particular

$$\text{Var}(Y_1) = \ldots = \text{Var}(Y_s) = \tau^2,$$

where τ^2 is defined in (5.6.3). Similarly, each of the $\binom{s}{2} = \frac{1}{2}s(s - 1)$ pairs of the Y's has the same joint distribution, and hence the same covariance, which we shall for the moment denote by γ. From (17), it then follows that

(18) $\text{Var}(Y_1 + \ldots + Y_s) = s\tau^2 + s(s - 1)\gamma.$

There remains the problem of determining the value of γ, which can be solved by means of a trick. Since formula (18) is valid for any value of $s \leq N$, let us consider the special case $s = N$, which gives

$$\text{Var}(Y_1 + \ldots + Y_N) = N\tau^2 + N(N - 1)\gamma.$$

But $Y_1 + \ldots + Y_N$ is the sum of the v values on all tickets and hence is equal to $v_1 + \ldots + v_N$, which is a constant, and which therefore by (5.7.1) has variance zero. Therefore

$$0 = N\tau^2 + N(N - 1)\gamma \quad \text{or} \quad \gamma = -\tau^2/(N - 1).$$

Substitution in (18) gives

$$\text{Var}(Y_1 + \ldots + Y_s) = s\tau^2 - s(s - 1)\tau^2/(N - 1)$$

and hence

(19)
$$\text{Var}(Y_1 + \ldots + Y_s) = \frac{N - s}{N - 1} s\tau^2.$$

We recognize the factor $(N - s)/(N - 1)$ as the correction factor for sampling without replacement discussed in Section 6.2. In fact, if the s tickets were drawn with replacement, the s drawings would be unrelated, so that a product model would be appropriate. In this model each of the random variables $Y_1, \ldots, Y_s$ would have variance τ^2, and by (5.7.15) we would have $\text{Var}(Y_1 + \ldots + Y_s) = s\tau^2$, which differs from (19) by just the factor in question.

Since the covariance γ of any two Y's is $-\tau^2/(N - 1)$ and the variance of each Y is τ^2, it is seen as a by-product of the above argument that the correlation coefficient of any two Y's is

(20)
$$\rho = -1/(N - 1),$$

provided only that all tickets do not bear the same number. It is remarkable that this value is independent of the values $v_1, \ldots, v_N$ marked on the tickets.

EXAMPLE 4. *Variance of the hypergeometric distribution.* The hypergeometric distribution may be viewed as the special case of the lottery model when r of the tickets have v values one and the remaining $N - r$ tickets have v value zero. In this special case, $\bar{v} = r/N$ and

$$v_1^2 + \ldots + v_N^2 = r.$$

Hence, by (5.6.5),

$$\tau^2 = r(N - r)/N^2.$$

Since the random variable D of Section 6.2 is equal to $Y_1 + \ldots + Y_s$, we see from (19) that

(21) $$\mathrm{Var}(D) = \frac{N - s}{N - 1} \cdot s \frac{r}{N} \left(1 - \frac{r}{N} \right),$$

which agrees with (6.2.5).

PROBLEMS

1. If U denotes the total number of tails on the two tosses of Example 1, find $\mathrm{Cov}(U, Z)$.

2. A sample of size $s \geq 2$ is drawn without replacement from a population consisting of r special and $N - r$ ordinary items. If I_1 indicates that a special item is obtained on the first draw and I_2 that a special item is obtained on the second draw, find $\mathrm{Cov}(I_1, I_2)$ using formula (5).

3. Apply (5) to show that the population covariance λ of Example 2 can be written as

(22) $$\lambda = \frac{1}{N} (v_1 w_1 + \ldots + v_N w_N) - \bar{v} \cdot \bar{w}.$$

4. Find the covariance of the random variables F and G of Example 1.1.

5. Find the covariance of the random variables Z and W of Problem 1.4.

6. Find the covariance of the random variables U and V of Problem 1.7.14.

7. Consider the four-point bivariate distribution, which assigns probability $\frac{1}{2}(1 - p)$ to each of the points $(1, 1)$ and $(-1, -1)$ in the (z, w)-plane, and probability $\frac{1}{2}p$ to each of the points $(10, -10)$ and $(-10, 10)$. Show that p may be chosen so that (2) is positive with high probability but (1) is negative.

8. Write out formal proofs of laws (9), (10), and (11).

9. Find the correlation coefficient of the pair of random variables whose covariance was found in (i) Example 1, (ii) Problem 1, (iii) Problem 2, (iv) Problem 5.

10. Let Z take on the values $-a, 0, a$ with probabilities $p/2, 1 - p, p/2$ respectively and let $W = Z^2$. Find the correlation coefficient of Z and W.

11. Let X, Y, Z be independent random variables with zero expectations and variances $\mathrm{Var}(X) = \sigma^2$, $\mathrm{Var}(Y) = \mathrm{Var}(Z) = \tau^2$.
  (i) Determine the correlation coefficient of $U = X + Y$ and $V = X + Z$. What is the value of this correlation coefficient when $\sigma = \tau$?
  (ii) Show that the correlation coefficient is unchanged if the expectations of X, Y, Z are $E(X) = \lambda, E(Y) = \mu, E(Z) = \nu$ instead of being zero.

12. For the distribution of Problem 7, obtain the correlation coefficient ρ of Z and W as a function of p.

13. What happens to the correlation coefficient ρ of two random variables Z and W
  (i) if Z is multiplied by a negative constant?
  (ii) if both Z and W are multiplied by negative constants?

14. Show that
$$\text{Var}(aZ + bW) = a^2\,\text{Var}(Z) + 2ab\,\text{Cov}(Z, W) + b^2\,\text{Var}(W).$$

15. For all random variables Z, W show that
  (i) for any constant a
$$\text{Var}(Z) + 2a\,\text{Cov}(Z, W) + a^2\,\text{Var}(W) \geqq 0;$$

  (ii) $[\text{Cov}(Z, W)]^2 \leqq \text{Var}(Z)\,\text{Var}(W);$
  (iii) $[\rho(Z, W)]^2 \leqq 1$, and hence (14).

[Hint: (i) Use the fact that $\text{Var}(Z + aW)$ cannot be negative; (ii) apply the result of (i) with $a = -\text{Cov}(Z, W)/\text{Var}(W)$; (iii) use (ii).]

16. Prove (17) for the case $s = 4$. [Hint: Obtain this result from (16) by the method used to obtain (16) from (15).]

17. Why does formula (20) require the assumption that all tickets do not bear the same number?

18. Suppose that in the double lottery of Example 2, s tickets are drawn without replacement. Let (Y_1, Z_1) denote the v and w values on the first ticket, (Y_2, Z_2) the values on the second ticket, etc.
  (i) If $\mu = \text{Cov}(Y_1, Z_2)$, show that μ is in fact the covariance of any Y and Z with different subscripts, and that
$$(23) \qquad\qquad \mu = -\lambda/(N - 1).$$

  (ii) If $\bar{Y} = (Y_1 + \ldots + Y_s)/s$ and $\bar{Z} = (Z_1 + \ldots + Z_s)/s$, use (i) and (3) to show that
$$(24) \qquad\qquad \text{Cov}(\bar{Y}, \bar{Z}) = \frac{N - s}{N - 1} \cdot \frac{\lambda}{s}.$$

  (iii) Find the correlation coefficient of $(\bar{Y}, \bar{Z})$.
[Hint for (i): Use (3), (7), and (11) to express $\text{Cov}(Y_1 + \ldots + Y_s, Z_1 + \ldots + Z_s)$ in terms of λ and μ, and put $s = N$ in the resulting expression as in Example 3.]

19. Obtain the results corresponding to the three parts of the preceding problem under the assumption that the s tickets are drawn with replacement.

20. Show that the variance of the random variable M of Section 6.8 (the number of matchings) is $\text{Var}(M) = 1$. [Hint: Use the representation $M = I_1 + \ldots + I_N$ of Section 6.8, the addition law (17), and the equivalence law of ordered sampling, to find
$$\text{Var}(M) = N\,\text{Var}(I_1) + N(N - 1)\,\text{Cov}(I_1, I_2).$$

Complete the solution by finding $\text{Var}\,I_1$ and $\text{Cov}(I_1, I_2).$]

7.3 THE MULTINOMIAL DISTRIBUTION

As we mentioned at the end of Section 3.3, the concept of binomial trials extends in a natural way to unrelated trials with more than two possible outcomes. Suppose that on each trial of a sequence there are k possible outcomes, just one of which will occur. In addition, suppose that the probability of the first outcome is the same, say p_1, on each of the trials; the probability of the second outcome is the same, say p_2, on each of the trials; etc. We must of course have $p_1 + p_2 + \ldots + p_k = 1$. These assumptions amount to representing each of the trials by a model with k simple events to which are assigned the probabilities $p_1, p_2, \ldots, p_k$. If in addition we are willing to assume that the trials are unrelated, a model for the whole experiment may be formed as the product of the models for the separate trials. This product will be called the *multinomial trials* model.

In deciding whether to use a multinomial trials model for a sequence of trials, two questions must be considered, just as in the binomial case:

(a) For each of the outcomes, is the chance of its occurrence the same on all trials?

(b) Are the trials unrelated; that is, are the chances on each trial unaffected by what happens on the other trials?

We shall now consider several examples, for which a multinomial trials model may be appropriate.

EXAMPLE *1*. *The loaded die.* Recall model (1.3.1) for a single throw of a loaded die. In the present notation, if the outcomes of interest are "one point showing," . . . , "six points showing," then $p_1 = .21$, $p_2 = \ldots = p_5 = .17$, $p_6 = .11$. For several throws of this die, the multinomial trials model may be used with $k = 6$.

EXAMPLE *2*. *Random digits.* On each trial with a random digit generator (Example 1.2.2) there are ten possible outcomes, the digits $0, 1, \ldots, 9$. If the generator is working properly, the multinomial trials model may be used with $k = 10$ and $p_1 = \ldots = p_{10} = .1$.

EXAMPLE *3*. *Quality control.* Recall Example 6.1.3, but suppose that each item coming off the production line is classified as having major defects, minor defects but no major ones, or no defects. If the process is in a state of control, we may use the multinomial model with $k = 3$. The model will be realistic provided p_1, p_2, p_3 are close to the long-run frequencies with which the three types of items are produced.

EXAMPLE *4*. *Election forecasting.* Suppose that a polling organization samples the electorate to find how many voters favor Candidate A, how many favor Candidate B, and how many are undecided. Each interview with a voter may be regarded as a trial having three possible outcomes, with probabilities equal to the fractions p_1, p_2, p_3 of the electorate who favor A, favor B, or are undecided. The trials are of course related (unless the sampling is done with replacement), but if the sampling fraction is small, the degree of relationship is negligible, and a multinomial model with $k = 3$ may be used (see the discussion in Example 3.3.3).

As in the binomial case, the interest in a sequence of multinomial trials is often centered on the numbers of occurrences of each possible outcome. In a sequence of n multinomial trials, let $B_1, B_2, \ldots, B_k$ be the numbers of occurrences of the k possible outcomes, where of course $B_1 + B_2 + \ldots + B_k = n$. (Thus, in Example 4, n would be the number of voters interviewed, and B_1, B_2, B_3 would be the numbers of them who favor Candidate A, who favor Candidate B, and who are undecided.) The random variables $B_1, B_2, \ldots, B_k$ are known as *multinomial* random variables. Since the k random variables have the fixed sum n, it is enough to consider any $k - 1$ of them, for example $B_1, B_2, \ldots, B_{k-1}$; their values determine the value of the remaining one.

In a multinomial model, the marginal distribution of any one of the B's, for example B_1, is binomial. This may be seen by labeling the first outcome "success" and lumping the other $k - 1$ outcomes together as "failure." Since the trials are unrelated, and since on each trial the probability of "success" is p_1, it follows that B_1 has the binomial distribution (n, p_1). It follows that

$$
(1) \qquad
\begin{aligned}
E(B_1) &= np_1, & \ldots, & & E(B_k) &= np_k \\
\text{Var}(B_1) &= np_1(1 - p_1), & \ldots, & & \text{Var}(B_k) &= np_k(1 - p_k).
\end{aligned}
$$

Furthermore, it is easy to verify (Problems 4 and 5) that

$$
(2) \qquad \text{Cov}(B_1, B_2) = -np_1 p_2
$$

and similarly for the other pairs.

The joint distribution of any $k - 1$ of the random variables $(B_1, B_2, \ldots, B_k)$ is known as a *multinomial* distribution, or more specifically as a *k-nomial* distribution. In the special case $k = 2$, there are only two possible results, and one is dealing with a sequence of binomial trials; then either of the random variables B_1 or B_2 has a 2-nomial, or binomial, distribution, in agreement with our earlier terminology. When $k = 3$, the joint distribution of any pair of (B_1, B_2, B_3) is 3-nomial, or *trinomial*. To keep the discussion simple we shall treat only this case.

Let us denote by S_1, S_2, S_3 that a trial results in the first, second, or third of the three possible outcomes. (Thus $S_2 S_1 S_1 S_3$, for example, indicates that the first trial results in outcome 2, the second and third in outcome 1, and the fourth in outcome 3). Then the probability of getting b_1 results S_1, b_2 results S_2, and $b_3 = n - b_1 - b_2$ results S_3 in some specified order (such as for example $S_1 S_1 \ldots S_1 S_2 S_2 \ldots S_2 S_3 S_3 \ldots S_3$) is

$$
(3) \qquad p_1^{b_1} p_2^{b_2} p_3^{b_3}.
$$

(This corresponds to formula (3.3.1) in the binomial case.)

Let us denote by $\begin{pmatrix} n \\ b_1, b_2 \end{pmatrix}$ the number of ways to choose b_1 of the trials for S_1 and b_2 other trials for S_2. Then $P(B_1 = b_1 \text{ and } B_2 = b_2)$ is the sum of $\begin{pmatrix} n \\ b_1, b_2 \end{pmatrix}$ terms, each equal to (3), and hence

(4)
$$P(B_1 = b_1 \text{ and } B_2 = b_2) = \begin{pmatrix} n \\ b_1, b_2 \end{pmatrix} p_1^{b_1} p_2^{b_2} p_3^{b_3}.$$

The number $\begin{pmatrix} n \\ b_1, b_2 \end{pmatrix}$ can easily be expressed in terms of numbers of combinations. There are $\begin{pmatrix} n \\ b_1 \end{pmatrix}$ ways to choose the b_1 trials for S_1. For each such choice there remain $n - b_1$ trials, of which b_2 may be chosen in $\begin{pmatrix} n - b_1 \\ b_2 \end{pmatrix}$ ways for S_2. Therefore

(5)
$$\begin{pmatrix} n \\ b_1, b_2 \end{pmatrix} = \begin{pmatrix} n \\ b_1 \end{pmatrix} \begin{pmatrix} n - b_1 \\ b_2 \end{pmatrix}.$$

EXAMPLE 1. *The loaded die (continued).* Suppose that the die is loaded so that the face with one point (ace) appears .2 of the time and the opposite face (six) only .1 of the time. The other faces together then appear .7 of the time. The die is thrown $n = 10$ times and the numbers B_1 of aces and B_2 of sixes are observed. The joint (trinomial) distribution of B_1 and B_2 is shown in the table (where only those entries are given that are at least .0001 to four decimal accuracy).

b_2 \ b_1	0	1	2	3	4	5	6	7	Total
0	.0282	.0807	.1038	.0791	.0395	.0136	.0032	.0005	.3486
1	.0404	.1038	.1186	.0791	.0339	.0097	.0018	.0002	.3875
2	.0259	.0593	.0593	.0339	.0121	.0028	.0004		.1937
3	.0099	.0198	.0169	.0081	.0023	.0004			.0574
4	.0025	.0042	.0030	.0012	.0002				.0111
5	.0004	.0006	.0003	.0001					.0014
6	.0001								.0001
Total	.1074	.2684	.3019	.2015	.0880	.0265	.0054	.0007	

As an illustration, let us check one of the entries, say that corresponding to $b_1 = 2$ and $b_2 = 3$. According to (4),

$$P(B_1 = 2 \text{ and } B_2 = 3) = \begin{pmatrix} 10 \\ 2, 3 \end{pmatrix} (.2)^2 (.1)^3 (.7)^5$$

where

$$\begin{pmatrix} 10 \\ 2, 3 \end{pmatrix} = \begin{pmatrix} 10 \\ 2 \end{pmatrix} \begin{pmatrix} 8 \\ 3 \end{pmatrix} = 45 \cdot 56 = 2520$$

and on carrying through the calculation, the desired probability is seen to be .0169.

It is an interesting property of the multinomial distribution that all of the marginal and conditional distributions associated with it are again multinomial. For example, if $k = 5$, the *joint* marginal distribution of B_1 and B_2 is trinomial with n trials and probabilities p_1, p_2; while the conditional distribution of B_3, B_4 given $B_1 = b_1$ and $B_2 = b_2$ is trinomial with $n - b_1 - b_2$ trials and probabilities $p_3/(p_3 + p_4 + p_5)$ and $p_4(p_3 + p_4 + p_5)$ (Problem 11).

Just as the binomial distribution has been generalized to the multinomial distribution, so the hypergeometric distribution may be generalized to the *multiple hypergeometric* distribution, providing a model for a more precise analysis of experiments such as that of Example 4 where a sample is drawn without replacement from a population of items of more than two types (see Example 7.1.4 and Problems 12–16).

PROBLEMS

1. Check the entry .1186 in the table of Example 1.

2. Suppose that in a production line on the average 3% of the items have irreparable defects while 10% have reparable defects. In a day's production of 100 items, let B_1 and B_2 be the numbers of items with irreparable and reparable defects.

  (i) Use (1) and (2) to find $E(B_1)$, $E(B_2)$, $\mathrm{Var}(B_1)$, $\mathrm{Var}(B_2)$, $\mathrm{Cov}(B_1, B_2)$.

  (ii) Suppose that it costs \$10 to repair an item, and there is a \$50 loss when an item is irreparable. Let L be the loss due to defects in the day's production. Find $E(L)$, $\mathrm{Var}(L)$. [Hint: See Problems 7.2.14 and 6.5.19.]

  (iii) Use the normal approximation to find the probability that L exceeds \$350.

3. In a multinomial model, what is the distribution of $B_1 + B_2$? [Hint: Use an argument analogous to that showing the marginal distribution of B_1 to be binomial.]

4. Find $\mathrm{Var}(B_1 + B_2)$ from Problem 3, and use this result together with the addition law of variance to verify (2).

5. Verify (2) by representing B_1 and B_2 as sums of indicators.

6. In how many ways can ten councilmen be distributed among three committees so that just two councilmen are put on Committee A, three others on Committee B, and the remaining five on Committee C? [Hint: See derivation of (5).]

7. Generalize (5) to obtain a formula for the number $\binom{n}{b_1,\ b_2,\ b_3}$ of ways to choose b_1 trials for S_1, b_2 trials for S_2, b_3 trials for S_3 in a sequence of n quadrinomial trials.

8. In generalization of (4) and (5), let $k = 4$ and find the quadrinomial distribution of B_1, B_2, B_3.

9. Using the table of Example 1 and Table B,

(i) check the distribution of B_1 against the binomial distribution ($n = 10$, $p = .2$);

(ii) find the distribution of B_3 and check it against the binomial distribution ($n = 10$, $p = .7$).

10. Let B_1 and B_2 have the trinomial distribution (4). Using (5) and (4.2.1), show that the conditional distribution of B_2, given that $B_1 = b_1$, is binomial $(n - b_1, p_2/(p_2 + p_3))$. Check this result on the table of Example 1 for the case $b_1 = 4$.

11. Show that in a multinomial distribution with $k = 5$ and probabilities $p_1, p_2, \ldots, p_5$

(i) the joint marginal distribution of B_1 and B_2 is trinomial with n trials and probabilities p_1, p_2;

(ii) the conditional distribution of B_3, B_4 given $B_1 = b_1$, $B_2 = b_2$ is trinomial with $n - b_1 - b_2$ trials and probabilities $p_3/(p_3 + p_4 + p_5)$, $p_4/(p_3 + p_4 + p_5)$.

Problems 12–16 are concerned with the multiple hypergeometric distribution.

12. In a population of N items there are r_1, r_2, r_3 items of types 1, 2, 3 respectively ($r_1 + r_2 + r_3 = N$). If a random sample of s items is drawn and D_1, D_2, D_3 denote the numbers of items of the three types in the sample, show that

$$P(D_1 = d_1 \text{ and } D_2 = d_2) = \binom{r_1}{d_1}\binom{r_2}{d_2}\binom{r_3}{d_3} \Big/ \binom{N}{S}.$$

13. In the preceding problem

(i) what are the marginal distributions of D_1, D_2, D_3;

(ii) what is $E(D_1)$ and $\text{Var}(D_1)$?

14. In Problem 12, what is the conditional distribution of D_2 given $D_1 = d_1$?

15. In Problem 12, what is the distribution of $D_1 + D_2$?

16. Find $\text{Cov}(D_1, D_2)$

(i) by specializing formula (2.4);

(ii) in analogy with Problem 4, using Problem 15.

[Hint for (i): For items of type 1, set $v = 1$, $w = 0$; for items of type 2, set $v = 0$, $w = 1$; for items of type 3, set $v = w = 0$.]

PART II
STATISTICS

.

INTRODUCTION

In Part II we shall take up the subject of statistics. The meaning of this word has undergone a great change during the last two centuries. The word statistics, which is cognate with the word state, originally denoted the art and science of government: the first professors of statistics at German universities in the Eighteenth Century would today be called political scientists. As governmental decisions are to some extent based on data concerning the population, trade, agriculture, and so forth, the statistician naturally became interested in such data, and gradually statistics came to mean the collection of data about the state, and then the collection and handling of data generally. This is still a popular meaning of the word, but a further change is now taking place. There is no point in collecting data unless some use is made of it, and the statistician naturally became involved in the interpretation of the data. The modern statistician is a student of methods of drawing conclusions about a population, on the basis of data that ordinarily is collected from only a sample of the population.

To make clear the relation as well as the distinction between statistics and probability theory, let us consider an example. Suppose a customer is interested in the fraction of a boxcar-load of electric motors that conform to a specified quality standard. As it would be too expensive to take all of the motors apart, he chooses a random sample for examination. Using the notation of the preceding chapters, we suppose that there are N motors in the shipment of which r are below standard, that s of them are examined, and that D of those examined are below standard. In the model we developed for this problem in Section 6.2, D is a random variable having the hypergeometric distribution, and we calculated the probability that D will assume one of its possible values d to be

$$(1) \qquad P(D = d) = \binom{r}{d}\binom{N-r}{s-d} \bigg/ \binom{N}{s}.$$

This is a typical result in the theory of probability.

But is the result relevant to the problem facing the customer? Before we can compute (1), we must know r, and if r were known no inspection would be necessary. Also, (1) gives merely the chance that D will take on various values, while the customer will in practice observe the value that D actually has. The problem at hand then is not to find the distribution of D knowing r, but rather to draw a conclusion about r after observing D. This is a statistical problem: to draw a conclusion about a population on the basis of a sample. We may think of statistics and probability theory as working in opposite directions. Probability theory proceeds from a known population to derive distributions related to a sample from the population, while statistics proceeds from the observed sample to draw conclusions about the unknown features of the population.

The domain of statistics is far broader than sampling, however. In general, a statistical problem arises when we observe a random variable whose distribution is only partially known to us. The quantity D above is an example, since the distribution of D involves the unknown quantity r. As another illustration, suppose B is the number of males among n births. We may be willing to assume that B has the binomial distribution (n, p), but not know the probability p that a child is male.

In both of the examples cited, there is in the probability model a quantity (r, p) knowledge of whose value would completely specify the distribution. We shall refer to such a quantity as a *parameter*. We may define statistics as the branch of mathematics concerned with methods of drawing conclusions about parameters on the basis of the observed values of random variables whose distributions depend on the parameters. Statistics leans heavily on probability theory, to specify the distributions corresponding to various values of the parameters, but in a sense goes beyond probability theory to deal with the more complicated questions that arise when the underlying distributions are not completely specified.

Statistical problems fall into certain classes according to the nature of the conclusion that is to be drawn. Sometimes we wish merely to make the closest estimate that we can of an unknown parameter: this will confront us with a problem of *estimation*. If our estimate is a single numerical value, it is called a *point estimate;* in other problems we may try to specify a range within which the parameter is thought to lie—such a range is called an *interval estimate*. Various ideas of point estimation will be introduced and illustrated in Chapters 8–10.

Often we are confronted with the necessity of choosing between two clearly defined courses of action, the proper course depending on the value of an unknown parameter. For example, the customer of our illustration above must either purchase the consignment of electric motors or return them to the manufacturer. If he knew the value of r, he would presumably know which action to take. Again, an anthropologist who believes he may have discovered a primitive tribe with an unusually high proportion of male births, must decide whether to publish his finding or to wait until he has collected more data. Finally, the public health authorities may have to decide, on the basis of data collected from a clinical trial of a new vaccine, the distribution of which depends on the unknown degree of protection it affords, whether to recommend its use by the public. In each of these cases, it is necessary to choose between two actions on the basis of the observed value of a random variable whose distribution depends on a parameter, the value of which would indicate the correct action. The branch of statistics concerned with such problems, known as *hypothesis testing*, is taken up in Chapters 11–13.

For any particular statistical problem, the mathematical analysis will characteristically develop through several stages. At the beginning, when the problem is first recognized and formally described, interest centers on finding some plausible technique for dealing with it. The technique is usually advanced on a quite intuitive basis at first. For example, in the lot inspection problem described above it seems intuitively plausible to use the observed fraction D/s of defectives in the sample, as an estimate for the unknown fraction r/N of defectives in the entire lot.

A second stage of development is reached when the properties of the technique are studied, and its performance is measured according to some standard. Thus, the distribution of the estimate D/s can be worked out, and we can see how satisfactory the estimate is by calculating how frequently the estimate will be reasonably close to the quantity r/N being estimated.

A third stage is concerned with a comparison of several different techniques for the same problem on the basis of their performance, in order to select the one most suitable for the purpose at hand. It can for example be shown that the estimate D/s is, in terms of the criteria of performance for estimates we shall advance in Chapter 8, the best of all possible estimates for r/N.

A fourth stage, in many ways the most interesting, involves questions of *design of experiments*. How large should the sample size s be? Is it best to select the sample completely at random, or could more information be gained at less cost in some other way? To illustrate, suppose that the shipment of N motors is packed in three different crates, each the product of a different factory, and that there is reason to believe that the three factories differ in the quality of their output. In this case, it would be better to think of the entire lot of N motors as consisting of the three sub-lots or *strata*, and to draw independent smaller samples separately from each stratum. The possibilities of improving the sampling design by stratification will be taken up in Section 10.3.

In addition to estimation and hypothesis testing there are many other types of statistical problems—for example, the problem of choosing between three or more actions—but the ones we have mentioned will serve to introduce the subject. The field is at present in a rapid state of development, with exciting new ideas and results appearing nearly every year. Statistics is in close contact with mathematics and with nearly every field of science, and statistical ideas are often involved in new advances in many scientific disciplines.

CHAPTER 8

ESTIMATION

8.1 UNBIASED ESTIMATION

The term estimation is used in statistics in a way very similar to its use in everyday language. The contractor estimates the cost of building a house; the physician estimates a patient's length of stay in a hospital; the aircraft pilot estimates the time of arrival; the surveyor estimates the distance which he is about to measure; and the city planner estimates the population of the city.

In statistics the quantity to be estimated is one of the parameters of the probability model, or some quantity whose value depends on the parameters. The available information consists of the observed values of the random variables and certain known aspects of the experiment. The estimate is computed from these values, using the assumptions of the model.

EXAMPLE 1. *Binomial model.* Let B be a random variable having the binomial distribution corresponding to n trials with success probability p. Here n will usually be known but p is an unknown parameter. It may be desired to use the observed value of B, and the known number n, to estimate p. Thus we may wish for example to estimate the probability of a cure in a sequence of patients who are given a new treatment; the probability of a defective in a sequence of items produced by a certain manufacturing process; the probability of twins in a sequence of births; and so forth. The binomial model is of course appropriate in these examples only if the trials are unrelated and the chance for success is constant.

Since probability corresponds to frequency, it seems natural to use the observed frequency of success B/n as an estimate of the unknown success probability p. Thus, if a slot machine is observed to pay off 17 times in 1000 trials, we would estimate the pay-off probability to be .017. Similarly if there are 521 males in 1000 births of some species, we would estimate the probability of a male to be .521 in the species.

An estimate, since it is computed from random variables, will itself be a random variable. Thus, B/n is a random variable whose possible values are

$$0, \frac{1}{n}, \frac{2}{n}, \ldots, \frac{n-1}{n}, 1.$$

The distribution of the estimate will in general depend on the distribution of the random variables, and hence on the parameters of the model. The estimate B/n for example will be equal to b/n if and only if $B = b$, so that $P(B/n = b/n)$ is the same as $P(B = b) = \binom{n}{b} p^b (1 - p)^{n-b}$, which depends on p. Quite generally, an estimate T for a parameter θ will be a random variable, whose distribution depends on the value of θ.

When is an estimate a good one? The objective of estimation is, of course, to produce an estimate T which is close to the actual value of the quantity θ that is being estimated. Since T is a random variable, it is not usually possible to be sure that T is close to θ, but one may be able to find an estimate T having a high probability of being close to θ.

To insure that an estimate T will be close to θ with high probability, one would want to require as a first step that the distribution of T be centered at or near θ. The situation is illustrated in Figure 1, which shows the

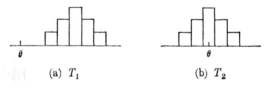

(a) T_1 (b) T_2

FIGURE 1.

distributions of two different estimates T_1 and T_2 of the same parameter θ. Clearly T_2 is a better estimate than T_1 in the sense that it is much more likely to fall close to θ. The reason is that the distribution of T_2 is centered at θ, while the distribution of T_1 is centered at a point some distance away from θ.

An analogy may help to make the discussion more concrete. Consider a rifleman who fires repeatedly at a target. His shots will not all fall at the same point, but will be distributed in a "pattern of fire." Other things being equal, he will want to have his pattern of fire centered on the bull's-eye, rather than to its right or to its left. Similarly, the statistician wants his estimate to have a distribution centered at the parameter; otherwise he is not even "shooting at the right target."

There are many different ways of defining the center of a distribution, but most commonly the center is taken to be the expected value (see Section 5.4). The proposed requirement, that the distribution of an estimate T should be centered on the parameter value, then becomes the requirement that the expected value $E(T)$ should equal the value θ of the parameter, whatever this value may be.

Definition. An estimate T of the parameter θ is *unbiased* if

(1) $E(T) = \theta$,   whatever θ may be.

If (1) does not hold, T is said to be *biased*, being biased toward large values if $E(T) > \theta$, and toward small values if $E(T) < \theta$. Since the expected value represents the average value in the long run, we may say that an unbiased estimate will be "right on the average."

EXAMPLE 1. *Binomial model (continued).* Let us check whether the estimate B/n proposed in Example 1 is unbiased. Since $E(B) = np$ (see (6.1.6)),

(2) $E(B/n) = p$.

This is true, whatever value p may happen to have. Thus, B/n is an unbiased estimate for p.

EXAMPLE 2. *Hypergeometric model.* Let D be the hypergeometric variable discussed in Section 6.2. The distribution of D depends on the parameter r (the number of special items in the population) as well as on the known quantities N (population size) and s (sample size). Let us consider how one might use the observed value of D to estimate the parameter r/N, the fraction of special items in the population. This may for example be the fraction of voters in a city favoring a proposed issue of school bonds; the proportion of defective items in a shipment; the proportion of television viewers watching a certain program; and so forth. The hypergeometric model will of course be appropriate in these examples only if the sample is a *random* sample from the population in question.

It seems intuitively reasonable that the fraction D/s of special items in the sample could be used to estimate the fraction r/N of special items in the population. Since $E(D) = sr/N$ (see (6.2.3)), it follows that

(3) $E(D/s) = r/N$,

whatever the value of r may be, so that D/s is an unbiased estimate for r/N. It is interesting to notice that this estimate does not require knowledge of the population size N.

To illustrate, suppose that a sample of $s = 50$ items drawn at random from a lot of N items is found to contain $D = 7$ defective items. Then $D/s = \frac{7}{50}$ is our estimate of the fraction of defective items in the lot.

EXAMPLE *3*. *Difference of two binomials.* Consider two unrelated sequences of n_1 and n_2 binomial trials with success probabilities p_1 and p_2 respectively. The numbers of successes in the two sequences are observed to be B_1 and B_2, and we wish to estimate the difference $p_2 - p_1$. We may want to determine, for example, how much the addition to a medication of a somewhat toxic ingredient increases the chance of a cure; how much an expensive refinement of a production process cuts down the probability of defectives; by how much a certain biochemical treatment will increase the chance that a calf in a dairy herd will be female. In each of these cases, a number of trials are performed under two conditions (with and without the toxic ingredient, etc.). If all the trials are unrelated, and if the probabilities are constant under each of the two conditions, a product of two binomial models will be appropriate; the numbers B_1 and B_2 of successes under the two conditions will be defined on different factors of the model, and each will have a binomial distribution.

The natural estimate for $p_2 - p_1$ is the difference of the estimates B_2/n_2 of p_2 and B_1/n_1 of p_1. Since both of these estimates are unbiased, it is clear from (2) and the laws of expectation that

$$\frac{B_2}{n_2} - \frac{B_1}{n_1}$$

is an unbiased estimate of $p_2 - p_1$.

This example indicates a general method for constructing unbiased estimates of the difference of two parameters. If T_2 is an unbiased estimate of θ_2 and T_1 is an unbiased estimate of θ_1, then $T_2 - T_1$ will be an unbiased estimate of $\theta_2 - \theta_1$.

The property of unbiasedness is desirable, but we do not wish to leave the impression that it is essential. A small bias may not be very important and, as is illustrated in Figure 2, an estimate that is slightly biased but tightly distributed may be preferable to one which is unbiased but whose distribution is widely dispersed.

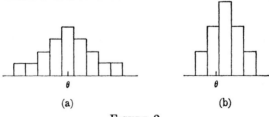

(a)   (b)

FIGURE 2.

It is clear that a reasonable choice between estimates (a) and (b) will require that one consider not only the center of the distribution of T, but also its dispersion. This consideration will be explored in the next section.

We conclude this section by illustrating another possible difficulty in applying the method of unbiased estimation: namely that an unbiased estimate may not exist.

EXAMPLE 1. *Binomial model (continued)*. In binomial trials one is sometimes interested in the ratio $p/(1 - p)$ rather than in p itself. (In studies of human populations, for example, the sex ratio is usually defined as the proportion of males to females.) It turns out that no unbiased estimate for $p/(1 - p)$ exists. To show this, let us consider the case $n = 2$ and suppose there did exist an unbiased estimate T. Denote by a, b, c the values taken on by T when $B = 0, 1, 2$ respectively. Then, if we denote $1 - p$ by q, unbiasedness would mean that

(4) $$E(T) = aq^2 + 2bpq + cp^2 = p/q, \text{ for all } p \text{ between 0 and 1}.$$

To see that no a, b, c exist for which this holds, consider what happens to the right-hand side of the displayed equation as p gets close to 1. The numerator will then be close to 1 and the denominator close to 0, so that the ratio p/q will be very large. In fact, by choosing p sufficiently close to 1, p/q can be made as large as we please. To make $p/q = 1000$ for example, let $p = 1000/1001$. Then $q = 1/1001$ and $p/q = 1000$. Similarly $p/q = 1,000,000$ if $p = 1,000,000/1,000,001$, and so on. To match this, the left-hand side would also have to become arbitrarily large as p gets close to 1. Since p and q are between 0 and 1, the quantities p, pq, and q will all be at most 1, and the left-hand side therefore cannot exceed the number $a + 2b + c$, no matter how close p gets to 1. Therefore, however large may be the fixed values assigned to a, b, and c, there will exist p sufficiently close to 1 so that the right side of (4) will exceed the left side. The displayed equation therefore cannot hold for all p, and no unbiased estimate exists for p/q. A similar argument holds for other values of n.

The nonexistence of an unbiased estimate does not mean that no reasonable estimate of p/q can exist. In fact, when n is large, B/n should be near p and $(n - B)/n$ near q, so that their ratio $[B/n] \div [(n - B)/n] = B/(n - B)$ should be near p/q. Thus at least for large n, $B/(n - B)$ is a reasonable estimate of p/q.

PROBLEMS

1. In n binomial trials, find an unbiased estimate for
   (i) the failure probability $q = 1 - p$;
   (ii) the difference $q - p = 1 - 2p$.

2. Find the probability that the estimate B/n of a binomial probability p does not differ from p by more than .1 if $p = \frac{1}{2}$ and (i) $n = 5$, (ii) $n = 10$, (iii) $n = 20$.

3. Solve the three parts of the preceding problem when $p = .4$. [Hint: Use the normal approximation for part (iii).]

4. Let T be a statistic with expectation $E(T) = 2\theta - 1$. Find an unbiased estimate of θ.

5. Solve the preceding problem when (i) $E(T) = \frac{1}{3}\theta + 2$; (ii) $E(T) = 2 - 3\theta$.

6. Under the assumptions of Example 2, find an unbiased estimate
   (i) for the number r of special items in the population;
   (ii) for the number of ordinary items in the population.

7. If T_1 and T_2 are unbiased estimates for θ_1 and θ_2 respectively, find unbiased estimates for

(i) $\theta_1 + \theta_2$

(ii) $(\theta_1 + \theta_2)/2$

(iii) $(\theta_1 + 2\theta_2)/3$

(iv) $a_1\theta_1 + a_2\theta_2$.

8. If T_1 and T_2 are unbiased estimates for θ,

(i) show that $(T_1 + 2T_2)/3$ is also an unbiased estimate for θ;

(ii) find other combinations of T_1 and T_2 that are unbiased estimates of θ.

9. Let T_1, T_2 be statistics with expectations $E(T_1) = \theta_1 + \theta_2$, $E(T_2) = \theta_1 - \theta_2$. Find unbiased estimates of (i) θ_1; (ii) θ_2.

10. Solve both parts of the preceding problem when

(i) $E(T_1) = \theta_1 + \theta_2$, $E(T_2) = 2\theta_1 + 3\theta_2$

(ii) $E(T_1) = \theta_1 + 2\theta_2$; $E(T_2) = 2\theta_1 - 3\theta_2$.

11. If $T_1, \ldots, T_n$ are unbiased estimates of θ,

(i) show that $(T_1 + \ldots + T_n)/n$ is also an unbiased estimate of θ;

(ii) show that $(c_1T_1 + \ldots + c_nT_n)/(c_1 + \ldots + c_n)$ is an unbiased estimate of θ for any constants $c_1, \ldots, c_n$ with $c_1 + \ldots + c_n \neq 0$.

12. If T_1, T_2 are independent and if both are unbiased estimates of θ, find an unbiased estimate of (i) θ^2; (ii) $\theta(1 - \theta)$.

13. The voters in a city are listed in five precincts with $N_1, \ldots, N_5$ voters respectively. A random sample of voters is taken from each precinct, the sample sizes being $s_1, \ldots, s_5$ and the numbers of voters in the five samples (say $D_1, \ldots, D_5$) who favor school bonds are observed. Use $D_1, \ldots, D_5$ to find an unbiased estimate of the total number r of voters in the city who favor the bonds. [Hint: If $r_1, \ldots, r_5$ denote the numbers of voters in the five precincts that favor the bonds, then $r = r_1 + \ldots + r_5.$]

14. (i) If B has the binomial distribution (n, p) and if p is known but n unknown, find an unbiased estimate for n.

(ii) If you are told that in a number of tosses with a fair coin the coin fell heads ten times, what is your estimate of the total number of tosses?

15. If B has the binomial distribution (n, p), of what quantity is the random variable B^2 an unbiased estimate? [Hint: For any variable Z, it follows from (5.6.2) that $E(Z^2) = \text{Var}(Z) + [E(Z)]^2.$]

16. If D has the hypergeometric distribution (6.2.1), of what quantity is the random variable D^2 an estimate?

17. Use the result of the preceding problem to find an unbiased estimate of (i) r^2; (ii) $\text{Var}(D)$.

18. Use the result of the preceding problem to find an unbiased estimate of (i) p^2, (ii) $\text{Var}(B)$.

19. A lot of $N = 1000$ items contains $r = 200$ special items. A random sample of unknown size s contains $D = 18$ special items. Find an unbiased estimate of s.

20. We wish to estimate $1/(1 + p)$ in Example 1. Suggest a reasonable estimate. Is it unbiased in the case $n = 1$?

21. To estimate the number N of fish in a lake, a biologist catches $r = 100$ fish, tags them, and then releases them. A few days later, a sample of fish is obtained from the lake and it is found that 3% of the fish have tags. Suggest a reasonable estimate for N. [Hint: Equate the fraction of tagged fish in the lake to 3%.]

22. If X has the uniform distribution (6.8.1), find an estimate of the form $aX + b$ which is unbiased for estimating N. [Note: This problem is based on the first half of Section 6.8.]

8.2 THE ACCURACY OF AN UNBIASED ESTIMATE

The condition of unbiasedness was introduced in the preceding section as a first step toward obtaining a good estimate, that is, one whose distribution is concentrated near the parameter θ being estimated. Unbiasedness alone is not enough however, since a distribution can be centered on θ without being highly concentrated there. This is illustrated in Figure 1, which shows the histograms of two estimates T_1 and T_2. Both are unbiased but the distribution of T_2 is more concentrated than that of T_1, and hence T_2 will usually be closer to θ than T_1. An estimate, even if unbiased, will be satisfactory only if its distribution is not too widely dispersed.

In terms of the analogy of target shooting, it is not enough that the pattern of fire be centered at the bull's-eye, the pattern should also be tightly concentrated. If both of these conditions hold, most of the shots will land close to the bull's-eye. Similarly, an estimate which is unbiased and has a tightly concentrated distribution will usually give values that are close to the true value being estimated.

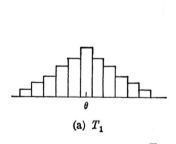

(a) T_1 (b) T_2

FIGURE 1.

Just as there are many ways of specifying the center of a distribution, there are many different measures of its degree of dispersion. We introduced in Section 5.6 the variance as the most commonly used such measure. By definition, the variance of T is the expected squared deviation of T from its expected value $E(T)$. If T is an unbiased estimate of θ, then $E(T) = \theta$. Therefore

(1) $\mathrm{Var}(T) = E(T - \theta)^2,$ provided T is unbiased.

The variance of an unbiased estimate T is thus the expected squared deviation of T from the parameter value θ being estimated, and we shall adopt this as the measure of accuracy of an unbiased estimate.

An estimate becomes much more useful if it is accompanied by an indication of its accuracy. After all, any number might be regarded as an estimate, perhaps a very inaccurate one, of any parameter. The user of an estimate needs to know how much reliance he may place on it. The most common way to provide this information is to give, along with an estimate, its variance or (equivalently) its standard deviation. As we shall see, in most cases the standard deviation cannot be given exactly but can be itself estimated. A common style for presenting an estimate and its (estimated) standard deviation is in the form

(estimate) ± (standard deviation)

For example, if in a scientific paper one reads "The weight gain is 12.6 ± 1.8 kilograms," this would mean that the estimated weight gain is 12.6 kilograms, and that this estimate has the standard deviation 1.8 kilograms, or at least 1.8 is an estimate for the standard deviation. At the end of this section, we shall give a more specific justification of the use of the standard deviation as a measure of accuracy.

Let us now illustrate the computation of variances (and hence of standard deviations) by considering the unbiased estimates proposed in the examples of Section 1.

EXAMPLE 1. *Binomial model.* In Example 1.1 we found B/n to be an unbiased estimate for p in the binomial model. What is the standard deviation of this estimate? Recalling (6.1.7) that $\mathrm{Var}(B) = np(1 - p)$, and (5.7.3), we see that

(2) $\mathrm{Var}(B/n) = \mathrm{Var}(B)/n^2 = p(1 - p)/n$

and hence

(3) $\mathrm{SD}(B/n) = \sqrt{p(1 - p)}/\sqrt{n}.$

Notice that the standard deviation of the estimate B/n for p will become smaller as n is made larger. In fact, $\mathrm{SD}(B/n)$ can be made as small as one pleases by taking n large enough. That is, the frequency of success will estimate the probability p of success with high accuracy if the number of trials is sufficiently large. This corresponds (within the model) to our basic concept of probability as long-run frequency.

EXAMPLE 2. *Hypergeometric model.* In a similar way, formula (6.2.5) enables us to compute the variance of the unbiased estimate D/s for the parameter r/N of Example 1.2. Since $\mathrm{Var}(D/s) = \mathrm{Var}(D)/s^2$, it follows that

(4) $$\operatorname{Var}\left(\frac{D}{s}\right) = \frac{N-s}{N-1} \cdot \frac{r}{N}\left(1 - \frac{r}{N}\right) \cdot \frac{1}{s}$$

and

(5) $$\operatorname{SD}\left(\frac{D}{s}\right) = \sqrt{\frac{N-s}{N-1} \cdot \frac{r}{N}\left(1 - \frac{r}{N}\right) \cdot \frac{1}{s}}.$$

As the sample size s is increased, these quantities become smaller. In the extreme case $s = N$, they are in fact equal to zero. This is reasonable since the estimate is perfect (that is, always coincides with the true value of the parameter) when the entire population is taken into the sample.

EXAMPLE 3. *Difference of two binomials.* For estimating the difference $p_2 - p_1$, we proposed in Example 1.3 the estimate $(B_2/n_2) - (B_1/n_1)$. Since B_1 and B_2 are defined on different factors of a product model, it follows from (2) and (5.7.14) that

(6) $$\operatorname{Var}\left(\frac{B_2}{n_2} - \frac{B_1}{n_1}\right) = \frac{p_2(1 - p_2)}{n_2} + \frac{p_1(1 - p_1)}{n_1}.$$

Again, the variance may be made as small as one pleases by taking n_1 and n_2 sufficiently large.

We have pointed out earlier (Sections 5.4 and 5.6) that expectation and variance are not always satisfactory measures of the center and dispersion of a distribution. However, for those distributions for which the normal approximation of Chapter 6 is adequate, the standard deviation of an unbiased estimate T of θ can be directly related to the accuracy of the estimate in terms of the probability that T will be close to θ.

Suppose that T is an unbiased estimate of θ such that the histogram of the standardized variable

$$T^* = (T - \theta)/\operatorname{SD}(T)$$

is well approximated by the normal curve. The probability that T be within a given distance d of θ is then

(7) $$\Pi = P(-d < T - \theta < d),$$

which can also be written as

$$\Pi = P\left(-\frac{d}{\operatorname{SD}(T)} < T^* < \frac{d}{\operatorname{SD}(T)}\right).$$

Since our normal approximation applies to T^*, this probability is approximately equal to

$$\Phi\left(\frac{d}{\operatorname{SD}(T)}\right) - \Phi\left(\frac{-d}{\operatorname{SD}(T)}\right) = 2\Phi\left(\frac{d}{\operatorname{SD}(T)}\right) - 1,$$

so that equation (7) becomes (approximately)

(8) $$\Pi = 2\Phi\left(\frac{d}{\operatorname{SD}(T)}\right) - 1.$$

This equation relates the probability Π that an unbiased estimate T will be within a distance d of θ, to the ratio $d/\text{SD}(T)$. Reading the right-hand side of (8) from a normal table, we obtain the following values,

TABLE 1

$d/\text{SD}(T)$	1.0	1.5	2.0	2.5	3.0
Π	.683	.866	.954	.988	.997

In order to have for example Π equal to .95, we must have $d/\text{SD}(T) = 2$ and hence require the standard deviation $\text{SD}(T)$ to be $d/2$.

As an illustration of this method, suppose we would like to have an 80 percent chance of T being within .25 units of θ. How small must $\text{SD}(T)$ be? It follows from equation (8) that $\Pi = .8$ if $2\Phi(d/\text{SD}(T)) = 1.8$ and hence if $\Phi(d/\text{SD}(T)) = .9$. From the auxiliary entries of Table E we read $\Phi(1.282) = .9$. We should therefore have $d/\text{SD}(T) = 1.282$ and hence

$$\text{SD}(T) = \frac{.25}{1.282} = .195.$$

It is worth noting that if for a certain $\text{SD}(T)$ the probability is Π of T lying within a distance d of θ, then for any smaller value of $\text{SD}(T)$, the probability of T lying within a distance d of θ is even higher than Π. To see this, consider the right-hand side of (8). If $\text{SD}(T)$ is decreased, $d/\text{SD}(T)$ is increased; then $\Phi(d/\text{SD}(T))$ is also increased, and so therefore is Π. The smaller $\text{SD}(T)$ is, the larger is therefore the probability that T will differ from θ by less than any given amount d. This agrees with our idea that the accuracy of the estimate T improves as $\text{SD}(T)$ gets smaller.

PROBLEMS

1. Find the variance of the unbiased estimates of (i) Problem 1.1(i); (ii) Problem 1.1(ii); (iii) Problem 1.6(i); (iv) Problem 1.6(ii).

2. Find the variance of the unbiased estimate of (i) Problem 1.4; (ii) Problem 1.5(i); (iii) Problem 1.5(ii). [Express the answers in terms of $\text{Var}(T)$.]

3. Assuming T_1 and T_2 to be independent, find the variance of each of the four unbiased estimates of Problem 1.7 in terms of $\text{Var}(T_1)$ and $\text{Var}(T_2)$.

4. Under the assumptions of Problem 1.8 determine which of the unbiased estimates $(T_1 + T_2)/2$ and $(T_1 + 2T_2)/3$ of θ you would prefer if T_1 and T_2 are independent and (i) $\text{Var}(T_1) = \text{Var}(T_2)$, (ii) $\text{Var}(T_1) = 3\,\text{Var}(T_2)$.

5. Assuming the statistics T_1 and T_2 of Problems 1.9 and 1.10 to be independent, find the variances of the unbiased estimates of θ_1 and θ_2 in (i) Problem 1.9; (ii) Problem 1.10(i); (iii) Problem 1.10(ii).

6. Let T_1, T_2 be unbiased estimates of θ, independent, and have variances $\operatorname{Var}(T_1) = \sigma^2$, $\operatorname{Var}(T_2) = \tau^2$.

(i) Show that $(T_1 + T_2)/2$ is a better unbiased estimate than T_1 if $\tau^2 = \sigma^2$.

(ii) Find values of σ^2 and τ^2 for which T_1 is better than $(T_1 + T_2)/2$.

7. If T_1, T_2, T_3 are unbiased estimates of θ, are independent, and all have the same variance, determine which of the three unbiased estimates of θ

$$(T_1 + 2T_2 + T_3)/4, \qquad (2T_1 + T_2 + 2T_3)/5, \qquad (T_1 + T_2 + T_3)/3$$

you would prefer.

8. Solve the preceding problem if

(i) $\operatorname{Var}(T_2) = \sigma^2$, $\operatorname{Var}(T_1) = \operatorname{Var}(T_3) = \tfrac{1}{4}\sigma^2$

(ii) $\operatorname{Var}(T_2) = \sigma^2$, $\operatorname{Var}(T_1) = \operatorname{Var}(T_3) = 4\sigma^2$.

9. A merchant receives shipments of 100 fuses from each of two manufacturers, and wishes to estimate the total number of defectives among the 200 fuses. He considers two procedures, each requiring him to inspect 40 fuses:

   (a) to take a random sample of 40 fuses from the 200 and to observe the number D of defectives;

   (b) to take unrelated samples of 20 fuses from each shipment and to observe the numbers D_1 and D_2 of defectives.

Suppose that the actual numbers of defectives in the two shipments are r_1 and r_2.

   (i) Construct unbiased estimates for $r_1 + r_2$ based on methods (a) and (b).

   (ii) Compare the variances of these two estimates in case $r_1 = 7$ and $r_2 = 18$.

10. In Example 3, determine which of three pairs of numbers of trials

$$\text{(a) } n_1 = 500, \ n_2 = 500; \quad \text{(b) } n_1 = 400, \ n_2 = 600;$$
$$\text{(c) } n_1 = 300, \ n_2 = 700$$

produces the most accurate estimate of $p_2 - p_1$ if (i) $p_1 = .5$, $p_2 = .5$; (ii) $p_1 = .2$, $p_2 = .5$.

11. In order to estimate the difference between the proportions of voters favoring a candidate in two cities having N_1 and N_2 voters respectively, a random sample of s_1 voters is taken from the first city and an unrelated random sample of s_2 voters from the second city. If D_1 and D_2 denote the number of voters in the two samples favoring the candidate, find an unbiased estimate of the difference in question and give its variance.

12. Let T be an unbiased estimate of θ whose distribution can adequately be approximated by the normal curve. For each of the values $\operatorname{SD}(T) = .5, 1, 2$, find the probability that T lies within d units of θ when (i) $d = .5$, (ii) $d = 1$, (iii) $d = 2$, and present the nine answers in a table.

13. Under the assumptions of the preceding problem, suppose that $\operatorname{SD}(T) = .8$. Within what distance of θ can we be (i) 80%, (ii) 90%, (iii) 95% sure that T will lie?

14. Under the assumptions of Problem 12, how small must $\operatorname{SD}(T)$ be if we require a 90% chance of T being within .41 units of θ? (Use the approximation $\Phi(1.64) \doteq .95$.)

15. Supplement Table 1 by finding the entires in the second row corresponding to the following entries in the first row: (i) 1.25, (ii) 1.75, (iii) 2.25, (iv) 2.75.

16. Find the entires for the first line of Table 1 that would correspond to the following entries in the second line: (i) .7, (ii) .75, (iii) .8, (iv) .9, (v) .95.

17. What is the probability that the estimate B/n of p given in Example 1 falls between .4 and .6 (inclusive) when $p = .5$ and $n = 25$? Find the exact probability and compare it with the normal approximation.

18. If $n = 64$, what is the probability that the estimate B/n of p given in Example 1 falls between
   (i) .4 and .6 when $p = .5$;
   (ii) .3 and .5 when $p = .4$;
   (iii) .2 and .4 when $p = .3$?

19. What is the probability that the estimate of $p_2 - p_1$ given in Example 3 falls between $-.1$ and .1 when (a) $p_2 = p_1 = .3$, (b) $p_2 = p_1 = .4$, (c) $p_2 = p_1 = .5$ and (i) $n_1 = n_2 = 40$; (ii) $n_1 = 30$, $n_2 = 50$; (iii) $n_1 = 20$, $n_2 = 60$?

20. Is it possible for an estimate of θ to have high probability of being close to θ, even if its variance is large? [Hint: Suppose a random variable X takes on the value θ with probability .99 and the value $\theta + 1000$ with probability .01. What is the variance of X, what the probability of X being close to θ?]

21. A statistician wants his estimate to have the maximum possible probability of falling within two units of the parameter θ. Estimate T_1 is unbiased and has standard deviation 1. Estimate T_2 is biased but has standard deviation $\frac{1}{2}$. How large must the bias of T_2 be before the statistician would prefer T_1? Assume that both T_1 and T_2 are approximately normally distributed. [Hint: Note that $E(T_1) = \theta$ and $E(T_2) = \theta + b$, where b is the bias of T_2. Use the normal approximation to find $P(|T_1 - \theta| \leq 2)$ and $P(|T_2 - \theta| \leq 2)$.]

8.3 SAMPLE SIZE

In the planning of almost any random experiment the question arises: "How many observations will be needed?" To answer this question, one must of course first be clear about the objectives of the experiment. In estimation problems, the objective relates to the degree of accuracy that is required for the estimate. In the preceding section, we have proposed the variance (or standard deviation) as a measure of accuracy. Alternatively, one might be interested in controlling the probability with which the estimate falls within a given distance of the parameter being estimated. As we have seen, for estimates to which the normal approximation is applicable, this can again be achieved by controlling the variance of the estimate.

Let us now translate the specification of the variance of an estimate into a specification of sample size. We have seen in the examples of Section 2 that in each case the variance becomes small as the number of observations becomes large, and can be made arbitrarily small by taking sufficiently many observations. Unfortunately, the same formulas show that the variances depend not only on the sample sizes but also on the parameters of the model, which of course are unknown. Does this mean that sample size cannot be rationally determined? Usually, one has some idea in advance of the experiment as to the possible or likely values of the unknown quantities, and this permits at least a rough calculation of the appropriate sample size, as is illustrated by the following examples.

EXAMPLE 1. *Binomial model.* Suppose a biologist wishes to estimate the frequency of twin births in a species. Let p denote the probability that a birth will produce twins, and suppose that he wishes his estimate of p to have standard deviation .01 (the estimate will then by Table 2.1 have probability about .95 of falling within .02 of the correct value).

He observes B twin births in a sequence of n births, and (as suggested in Example 1.1) uses B/n to estimate p. The standard deviation of B/n is, by (2.3), $\sqrt{p(1-p)}/\sqrt{n}$. Setting this equal to the desired value .01 gives

$$n = \left(\frac{\sqrt{p(1-p)}}{.01}\right)^2 = 10{,}000\, p(1-p),$$

which depends on p. If it is known from earlier studies that p is near .1, the biologist should plan to observe about $n = 10{,}000 \times .1 \times .9 = 900$ births.

To emphasize the dependence of n on p, let us change the problem to that of estimating the frequency of males, which we shall assume to be near .5. Then, to have an estimate with standard deviation .01 would require $n = 10{,}000 \times .5 \times .5 = 2500$ observations. The number n of observations required to estimate binomial p with standard deviation Δ is in general

$$(1) \hspace{3cm} n = p(1-p)/\Delta^2,$$

which is proportional to $p(1-p)$. In Figure 1 we show $p(1-p)$ plotted against p. It is seen that $p(1-p)$ reaches a maximum of $\frac{1}{4}$ at $p = \frac{1}{2}$

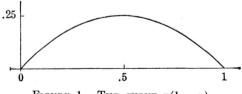

.25

0   .5   1

FIGURE 1. THE CURVE $p(1-p)$

(Problem 5). This implies that if one uses $n = 1/(4\Delta^2)$, which is the appropriate value of n in case $p = .5$, one will have enough observations to give an estimate for p with standard deviation Δ or less. Furthermore, the curve of Figure 1 is flat near its maximum; hence, if the value of p is near .5, the proportion of observations taken unnecessarily as the result of using $n = 1/(4\Delta^2)$ will be small. For example, if $p = .4$ or $.6$, $1/(4\Delta^2)$ is only 4% larger than would be required (Problem 6).

The situation is quite different when estimating a probability p known to be near 0 or near 1, because then the appropriate value of n depends heavily on the true value of p. For example, if one believes that $p = .02$, and wishes to have an estimate with standard deviation $\Delta = .001$, one might plan to take $n = .02 \times .98/(.001)^2 = 19,600$. If in fact $p = .04$, the proper value would be $n = .04 \times .96/(.001)^2 = 38,400$, nearly twice as large. In such a situation it is often reasonable to use a completely different design. For example, one might conduct a pilot study, the result of which would give a preliminary estimate for p which could be used to determine the number of observations in the main study. This is an example of what is known as "two-stage" design.

Inspection of formula (1) shows that the number n of observations required increases very rapidly when the desired standard deviation Δ is decreased. The reason is that $SD(B/n)$ is inversely proportional to the square root of n, and hence decreases very slowly as n increases. This is known as the *square root law*. It means that one must take four times as many observations to obtain twice as much accuracy, when accuracy is measured by standard deviation. If one wishes to cut the standard deviation of the estimate to $\frac{1}{10}$ its value, one must take 100 times as many observations, etc.

EXAMPLE 2. *Hypergeometric model.* Planning the sample size s for the hypergeometric model is rather similar to the corresponding problem in the binomial case. The unbiased estimate D/s for the fraction r/N of special items in a population has variance (see (2.4))

$$\text{Var}\left(\frac{D}{s}\right) = \frac{N - s}{N - 1} \cdot \frac{r}{N}\left(1 - \frac{r}{N}\right) \cdot \frac{1}{s}.$$

To illustrate the determination of sample size in this case, suppose that a polling organization wishes to estimate the proportion of voters who favor candidate A in a forthcoming election for Mayor. There are only two candidates, and the election is expected to be close so that it is reasonable to assume the proportion r/N to be close to $\frac{1}{2}$. The number of voters is $N = 10,000$, and the standard deviation of the estimate is not to exceed .01 and hence the variance is not to exceed .0001. (This will give a probability $\Pi = .95$ that the estimate D/s will be within .02 of r/N.) The desired accuracy will be achieved if the number s of voters to be interviewed satisfies the equation

$$.0001 = \frac{10,000 - s}{9999} \cdot \frac{1}{4s},$$

or $s = 10,000/4.9996 = 2000$.

Let us now change the problem by supposing that the election is for Governor (and is again close) and that the number of voters is $N = 1,000,000$. If the estimate D/s is to achieve the same precision, the sample size s must satisfy the equation

$$.0001 = \frac{1,000,000 - s}{999,999} \cdot \frac{1}{4s}$$

which gives $s = 1,000,000/4000.9996 = 2494$.

Surprisingly, only 25% more interviews are required to forecast an election with 100 times as many voters. It seems very strange at first that the sample size depends so little on the size of the population. The result may appear less paradoxical if one realizes that the sample size would not depend on the size N of the population at all if the sampling were done with replacement. In that case, the voters drawn would be unrelated, each having probability r/N of favoring candidate A. Therefore, D would have the binomial distribution ($n = s$, $p = r/N$), so that $\mathrm{Var}(D/s)$ would be $\frac{1}{s} \cdot \frac{r}{N}\left(1 - \frac{r}{N}\right)$. In two populations having the same fraction r/N favoring A, the precision of the estimate D/s would therefore be the same regardless of the sizes of the two populations. The only way population size influences $\mathrm{Var}(D/s)$ is through the correction factor

$$\frac{N - s}{N - 1} = \frac{1 - (s/N)}{1 - (1/N)}$$

and this will be very nearly 1 (and therefore essentially constant) as soon as N is large enough to make s/N very small. Even when forecasting a presidential election involving 50,000,000 voters, a sample of a few thousand would suffice, if it were feasible to draw a simple random sample from the voting population and if there were no complicating factors such as the Electoral College. (Of course in practice many of those interviewed will refuse to state how they intend to vote, or will give false information, or change their minds, or not vote, so that no sample size will be large enough to permit really accurate election forecasts.)

PROBLEMS

1. The biologist of Example 1 wishes to estimate p within .05 with probability .8. Calculate the needed sample size under the assumption that (i) $p = .1$, (ii) $p = .2$, (iii) $p = .3$.

2. In the preceding problem, what is the approximate probability that the estimate will be within .05 of the true value if p is in fact (i) .2 in part (i); (ii) .3 in part (ii); (iii) .4 in part (iii)?

3. Solve the three parts of Problems 1 and 2 under the assumption that the biologist wishes to estimate p within .05 with probability .9.

4. A doctor wishes to estimate the fraction of cases that will be cured by a new treatment.

  (i) To how many cases does he have to apply the treatment if he wishes his estimate to have standard deviation .05, and if he believes that the treatment will cure about $\frac{3}{4}$ of all cases?

  (ii) With the experiment of part (i), what approximately is the probability that the estimate will exceed .8 when actually the treatment cures only 60% of all cases?

5. (i) Show that $p(1 - p) = \frac{1}{4} - (p - \frac{1}{2})^2$.

  (ii) Use (i) and the fact that a square is always ≥ 0 to prove that $p(1 - p)$ reaches its maximum value $\frac{1}{4}$ at $p = \frac{1}{2}$.

  (iii) Use (i) to show that $p(1 - p)$ increases as p increases from 0 to $\frac{1}{2}$ and decreases as p increases from $\frac{1}{2}$ to 1.

6. (i) In Example 1, show that $1/(4\Delta^2)$ observations exceed by only 4% the number that is necessary to achieve standard deviation Δ if $p = .4$ or $p = .6$.

  (ii) By what percentage do $1/(4\Delta^2)$ observations exceed the number necessary to achieve standard deviation Δ if $p = .3$ or $p = .7$?

7. In Example 2, compute the required size s of the poll if the population size N is (i) 1000, (ii) 5000, (iii) 20,000, (iv) 50,000, (v) 50,000,000. Exhibit the results of this problem together with those of Example 2 in a table which shows how the size of the poll increases with the size of the population.

8. Solve the five parts of the preceding problem if it is only required that the standard deviation of the estimate not exceed .02 (rather than .01).

9. Suppose in Example 2 that the race is between three candidates all of which have about equal chances of getting the plurality. If r denotes as before the number voting for candidate A, it is then reasonable to assume r/N to be close to $\frac{1}{3}$. If the standard deviation of the estimate D/s is not to exceed .01, find the sample size required if (i) $N = 1,000$, (ii) $N = 4,000$, (iii) $N = 10,000$, (iv) $N = 100,000$, (v) $N = 10,000,000$.

10. In the preceding problem, let r' denote the number of persons voting for candidate B. Find an unbiased estimate of the difference $r' - r$ in the numbers of votes for candidates B and A.

11. A customer who receives a shipment of 8000 parts wishes to estimate the fraction that will withstand a given stress. He believes this fraction to be near .8. How many parts should be tested if the chance that the estimate will be off by not more than .1 is to be (i) 70%, (ii) 75%, (iii) 90%, (iv) 95%?

12. Solve the four parts of the preceding problem if the customer believes the fraction to be near (a) .7, (b) .9.

13. Use Problem 5(i) to show that the variance of D/s in Example 2 never exceeds
$$\frac{N - s}{N - 1} \cdot \frac{1}{4s}.$$

14. In Example 2.3 suppose that we believe both p_1 and p_2 to be close to $\frac{1}{2}$ and that we plan an experiment with $n_1 = n_2$. How many trials are required altogether if the standard deviation of the estimate is to be .01? Compare this with the number of trials required to estimate a single binomial probability p with the same standard deviation.

15. Suppose in Example 2.3 that we believe p_1 to be approximately $\frac{1}{2}$ and p_2 approximately $\frac{1}{4}$. It is desired to have $\mathrm{Var}(B_2/n_2 - B_1/n_1) \leqq \frac{1}{5}$. This result may be achieved with various combinations of n_1 and n_2. By trial and error find the combination using the smallest total number of trials.

16. Suppose that p_1 and p_2 are the success probabilities in n_1 and n_2 binomial trials respectively and that it is desired to estimate $p_2 - p_1$ within .02 with probability .9. If $n_1 = n_2 = n$, calculate the needed sample size n under the assumption that (i) $p_1 = p_2 = .5$; (ii) $p_1 = p_2 = .25$; (iii) $p_1 = p_2 = .1$.

17. Solve the preceding problem under the assumption that (i) $p_1 = .4$, $p_2 = .6$; (ii) $p_1 = .3$, $p_2 = .5$; (iii) $p_1 = .2$, $p_2 = .4$.

CHAPTER 9

ESTIMATION IN MEASUREMENT
AND SAMPLING MODELS

9.1 A MODEL FOR SAMPLING

We have already discussed (Section 2.1) the idea of obtaining information about a population by examining a sample drawn at random from the population. In Sections 2.1 and 6.2 we considered examples in which the quantity of interest was the proportion of the population having a given property or attribute, for example the proportion of defective items in a lot, or of persons in some group belonging to a specified political party. We shall now generalize the discussion. Suppose there is attached to each item in the population a value of some variable quantity capable of taking on a spectrum of values, and that we are interested in knowing something about these values. A few examples will indicate in how many different contexts such situations arise.

EXAMPLE 1. *Wheat acreage.* An agricultural agency needs to know the total acreage planted to wheat in a geographical district. There is a list of the farms in the district. Each farm has a certain acreage planted to wheat. The agency takes a random sample of the farms, and sends a surveyor to each farm in the sample to measure the acreage in wheat on that farm.

EXAMPLE 2. *Shaft diameter.* An automatic lathe turns out shafts that are machined to a specified diameter. The lathe is not perfect, and each shaft has an actual diameter that departs more or less from the desired value. The inspector takes a sample from a lot and measures the actual diameters of the sampled shafts.

EXAMPLE 3. *Medical expenses.* A labor union is considering starting a health insurance scheme for its members. It wants to know how much the members spend on medical care. A sample of members are asked to submit a careful record of their medical expenses for one year.

For dealing with the problem of "sampling by variables" we shall use the model already developed in Chapter 5 in the context of a lottery. (The reader should at this point review Examples 5.3.5, 5.5.2, and 5.6.2.) In conformance with our earlier notation, let N denote the number of items in the population (e.g., the number of tickets in the box, the number of farms, the number of shafts, the number of union members). The quantity of interest (prize, wheat acreage, diameter, medical expenses) will be denoted by v; the value of this quantity for the first item in the population is v_1, for the second item is v_2, and so forth. The average value of v in the population is (Example 5.3.5)

$$(1) \qquad \bar{v} = (v_1 + \ldots + v_N)/N$$

while the mean square deviation of these values is (see (5.6.3))

$$(2) \qquad \tau^2 = [(v_1 - \bar{v})^2 + \ldots + (v_N - \bar{v})^2]/N.$$

An (ordered) random sample of size s is drawn from the population. Let Y_1 denote the v value of the first item drawn, Y_2 the v value of the second item drawn, and so forth up to Y_s. By the equivalence law of ordered sampling, each of the N items has the same chance of being selected on any given draw. It follows that the random variables $Y_1, \ldots, Y_s$ all have the same distribution. Their common expectation and variance is given by (see (5.3.2))

$$(3) \qquad E(Y_1) = \ldots = E(Y_s) = \bar{v}$$

and (see (5.6.4))

$$(4) \qquad \text{Var}(Y_1) = \ldots = \text{Var}(Y_s) = \tau^2.$$

The purpose of sampling is to gain information about the v values of the population. In particular, we shall be interested in estimating the average v value $\bar{v}$. It seems intuitively natural to use the mean

$$(5) \qquad \overline{Y} = (Y_1 + Y_2 + \ldots + Y_s)/s$$

of the values observed in the sample as an estimate of the mean $\bar{v}$ of the values in the population.

Before discussing the expectation and variance of $\overline{Y}$, let us first consider an alternative sampling scheme. Suppose that s items are drawn *with replacement*, i.e., each item drawn is replaced before the next item is drawn. As was pointed out in Section 6.2, when sampling with replacement it is reasonable to assume that the s draws are unrelated, and to use a product model. In this model, each factor coincides with the model of Example 5.3.5 for the random drawing of a single item, so that the v values $Y_1, Y_2, \ldots, Y_s$ appearing on the successive draws each have the distribution of the random variable Y of Example 5.3.5.   Therefore formulas (3) and (4) are valid whether the drawing is with or without replacement.

Since each of the Y's has $\bar{v}$ as its expected value, it follows from the laws of expectation that

$$(6) \hspace{4cm} E(\overline{Y}) = \bar{v},$$

so that $\overline{Y}$ is an unbiased estimate of $\bar{v}$.   This is true whether the sampling is done with or without replacement.

The variance of the estimate $\overline{Y}$ will however depend on which method of sampling is used.   If one samples with replacement, the random variables $Y_1, Y_2, \ldots, Y_s$ are defined on different factors of a product model, and since each has variance τ^2, it follows from (4), the addition law for variance (5.7.15), and (5.7.3) that

$$(7) \hspace{3cm} \mathrm{Var}(\overline{Y}) = \tau^2/s \hspace{2cm} \text{(with replacement)}.$$

On the other hand, if the sampling is without replacement, the random variables $Y_1, \ldots, Y_s$ are dependent.   In this case, the formula for the variance must be modified by the factor $(N - s)/(N - 1)$ which corrects for dependence, and which we have already encountered in Section 6.2. Thus, for sampling without replacement,

$$(8) \hspace{3cm} \mathrm{Var}(\overline{Y}) = \frac{N - s}{N - 1} \cdot \frac{\tau^2}{s} \hspace{1.5cm} \text{(without replacement)}.$$

(This formula follows from (7.2.19).)

In planning a sampling investigation the question will always arise: how large a sample should be taken?   As pointed out in Section 8.3, this depends on the degree of accuracy required for the estimate.   Since $\mathrm{Var}(\overline{Y})$ depends on τ^2, one must also have at least some idea of the value of τ^2 before a rational choice of s is possible.   Information about τ^2 may be available from studies made earlier on the same or on similar populations. Once the desired variance of the estimate is specified, and a value for τ^2 is obtained, the solution of equation (8) for s is exactly similar to that in the hypergeometric Example 8.3.2.   (In fact, the hypergeometric model is a special case of the sampling model, as was discussed in Example 7.2.4.)

EXAMPLE 4.   *Pre-school children.*   To aid its planning, a school board wishes to know the number of pre-school children in its district.   A sample of s is taken from the N households in the district, and the numbers $Y_1, \ldots, Y_s$ of pre-school children in these s households is determined. The board would like to estimate the average number $\bar{v}$ of pre-school children, where $v_1, \ldots, v_N$ are the numbers in the N households, with a standard deviation of .1.

In a study already made in a neighboring city it was found that 54%

of the households had no pre-school children, while 21% had 1, 16% had 2, 7% had 3, and 2% had 4. For that population,

$$\bar{v} = 0(.54) + 1(.21) + 2(.16) + 3(.07) + 4(.02) = .82$$

and (using (5.6.5))

$$\tau^2 = [0(.54) + 1(.21) + 4(.16) + 9(.07) + 16(.02)] - (.82)^2 = 1.13.$$

The board believes that its district is sufficiently similar to that of the other city that it may use the value $\tau^2 = 1.13$ in planning its study. There are $N = 1200$ households in its district. Thus s is determined by the equation

$$(.1)^2 = \frac{1200 - s}{1199} \cdot \frac{1.13}{s},$$

or $s = 103$.

It is of course possible that in the board's district the numbers of pre-school children are somewhat more variable than in the neighboring city. In this case, the estimate based on 103 observations will be somewhat less precise than desired. It will however still be an unbiased estimate of the average number of pre-school children among the households in the district.

In many cases a sufficiently accurate prior value for τ^2 may not be available. It may then be desirable to conduct a pilot study whose purpose is to determine a reasonable size for the main sample. From the data gained in such a pilot study, the value of τ^2 may be estimated by the methods developed in Section 5.

PROBLEMS

1. Of the $N = 1000$ families in a town, 140 own no car, 675 own one car, 158 own two cars, and 27 own three cars. Calculate $\bar{v}$ and τ^2 where the v value of each family is the number of cars it owns.

2. For the $N = 63$ law firms of Problem 2.1.17, calculate $\bar{v}$ and τ^2 where the v value of each firm is the number of lawyers who belong to it.

3. How will $\bar{v}$ and τ^2 be changed if every v value is
(i) increased by a constant amount a;
(ii) multiplied by a constant amount b?
(Note the analogy with the laws of expectation and variance.)

4. A population of $N = 10$ items has the following v values: 876, 871, 875, 872, 871, 875, 871, 876, 878, 879. Compute $\bar{v}$ and τ^2 for this population. [Hint: Use the idea of Problem 3(i); that is, work first with the "coded" values 876 − 870 = 6, 871-870 = 1, etc.]

5. Solve the preceding problem if the population has $N = 12$ items with the following v values: 1012.1, 1013.4, 1012.6, 1012.1, 1012.8, 1013.7, 1013.1, 1011.9, 1018.3, 1012.0, 1012.4.

6. If you believe that the "outlying" value 1018.3 represents an error (perhaps a copying mistake?) you might compute $\bar{v}$ and τ^2 for the "trimmed" set of the eleven values remaining after the outlier has been removed. Carry this out and compare the resulting values of $\bar{v}$ and τ^2 with those of the preceding problem.

7. Solve Problem 4 if $N = 8$ and the v values are: $\frac{3}{12}, \frac{5}{12}, \frac{3}{12}, \frac{8}{12}, \frac{1}{12}, \frac{4}{12}, \frac{7}{12}, \frac{11}{12}$. [Hint: Use the idea of Problem 3(ii).]

8. A sample of size $s = 2$ is drawn without replacement from the $N = 5$ integers 1, 2, 3, 4, 5. Let Y_1 and Y_2 denote the first and second integers drawn,
   (i) obtain the distribution of $\bar{Y} = (Y_1 + Y_2)/2$;
   (ii) calculate $E(\bar{Y})$ and $\text{Var}(\bar{Y})$ from this distribution and compare them with the values given by (6) and (8).

9. It is planned to draw a sample of $s = 50$ items from a population of $N = 100$ items, with replacement. If instead the sampling were done without replacement, how many fewer observations would be required to produce an estimate of $\bar{v}$ with the same variance?

10. What is $\text{Var}(\bar{Y})$ for sampling with replacement when $s = N$? Explain this result intuitively.

11. Suggest an unbiased estimate for the population total $v_1 + \ldots + v_N$ based on $Y_1, \ldots, Y_s$. Give formulas for the variance of your estimate when sampling (i) with (ii) without replacement.

12. A population of size N is divided into strata of sizes $N', N'', \ldots (N = N' + N'' + \ldots)$. Samples of sizes $s', s'', \ldots$ are drawn from these strata, and the sample averages $\bar{Y}', \bar{Y}'', \ldots$ are to be used to estimate the over-all population average $\bar{v}$.
   (i) Express $\bar{v}$ in terms of the strata averages $\bar{v}', \bar{v}'', \ldots$.
   (ii) Give an unbiased estimate of $\bar{v}$.
   (iii) State the variance of your estimate.

13. The v value of r of the N items in a population is equal to v', and that of the remaining $N - r$ items is equal to v''.
   (i) In a sample of size s, where s is smaller than both r and $N - r$, what are the possible values of $\bar{Y}$?
   (ii) What is the probability that $\bar{Y} = v'$?

14. Two years after the complete count was made that is reported in Problem 1, a survey is planned to estimate the average number of cars per family in the town. How large a sample should be taken if the standard deviation of the estimate is to be about .05? [Hint: Assume that N and τ^2 have not changed materially during the two years.]

15. In Example 4, how large a sample is required if $\bar{v}$ is to be estimated with standard deviation (i) .05; (ii) .02?

9.2   A MODEL FOR MEASUREMENTS

In the first section of the book we gave, as an example of a random experiment, the measurement of the distance between two points. When a distance is measured repeatedly, the value that is obtained will usually not be exactly the same on the various trials. The differences are attributed to *measurement error*, and much work in science and technology is related to attempts to control and reduce errors of measurement. Not only the measurement of distance but the measurement of all sorts of quantities is afflicted with error.

EXAMPLE 1. *Velocity of light.* A variety of techniques has been devised in the attempt to reduce the error of measuring the velocity of light in a vacuum, but even the most refined methods will yield varying values on repeated determinations.

EXAMPLE 2. *Intelligence.* Psychologists have devised tests whose scores are intended to measure the intelligence quotient of individuals. When the IQ of an individual is measured several times, the values obtained will in general be somewhat different.

EXAMPLE 3. *Bioassay.* The potency of a drug is often measured by finding out how much of it is required to elicit a given response in experimental animals. This process of the bioassay of the drug will not yield quite the same value when it is repeated.

In each of these examples, and in countless others, if the measurement is repeated many times under carefully controlled conditions, the results will portray the features of a random experiment. While the value obtained will differ from trial to trial, the frequency with which a given value is obtained will tend to be stable in a long sequence of trials. This empirical fact suggests that probability models might be useful for the analysis of measurement error. We shall in this section develop a simple model, and in this and later sections show how statistical methods can be applied to gain some knowledge concerning the parameters of the model.

Let us denote the different values that may be obtained when a particular measurement is made, by $u_1, u_2, \ldots$. On any one trial, exactly one of these values will occur. These values may serve as the simple events of our model. If the measurement were to be repeated many times under stable conditions, the value u_1 would occur with a certain frequency $f(u_1)$. Corresponding to this frequency we shall in the model assign a probability p_1 to the simple event u_1. Similarly to u_2 will be assigned the probability p_2, etc. The probabilities $p_1, p_2, \ldots$ must of course be nonnegative, and must add up to 1.

Now consider any one trial. On this trial, the measured value will depend on the result of the random experiment, and will therefore be a random quantity. Let us represent this quantity by the random variable X. A possible value of X is u_1, and in fact $P(X = u_1) = p_1$. Similarly, $P(X = u_2) = p_2$, etc. We shall be particularly interested in $E(X)$, the expected value of the measured quantity. For brevity we shall denote $E(X)$ by the single letter μ. According to (5.4.2),

(1) $$\mu = E(X) = u_1 p_1 + u_2 p_2 + \ldots.$$

The variance of X, which we shall denote by σ^2, is according to (5.6.1) given by

(2) $$\sigma^2 = \mathrm{Var}(X) = (u_1 - \mu)^2 p_1 + (u_2 - \mu)^2 p_2 + \ldots.$$

We shall refer to the model we have just developed as the *measurement model*.

Let us consider next a more complicated experiment, in which two quantities are to be measured. For example, a physicist may wish to measure the velocity of light in a vacuum and in air, or the distance between two mountain peaks may be measured by two different teams of surveyors. For each of the two measurements separately, a measurement model of the kind described above would be appropriate. Let X and Y be the random variables representing the first and second measurements. The two measurements may be regarded as two parts of the experiment, and if they are made in such a way that their results are unrelated, we may use the product of the two measurement models as a model for the experiment as a whole. In such a product model, the random variables X and Y are independent (see Section 5.2).

These considerations extend at once to three or more measurements, represented by random variables $X, Y, Z, \ldots$. For each measurement separately, a measurement model may be used. If, in addition, the several measurements are unrelated, the product of these separate models will serve as a model for the whole experiment, within which $X, Y, Z, \ldots$ are independent random variables. We shall refer to this product model as a *model for several measurements*.

In the statistical analysis of such experiments, it is often necessary to consider linear combinations of the measurements, of the form $aX + bY + cZ + \ldots$ where $a, b, c \ldots$ are constants. It follows immediately from (5.5.5) and (5.5.3) that the expectation of such a linear combination is given by

(3) $$E(aX + bY + cZ + \ldots) = aE(X) + bE(Y) + cE(Z) + \ldots.$$

Furthermore, since $aX, bY, cZ, \ldots$ relate to separate factors of a product model, it follows from the laws (5.7.15) and (5.7.3) of variance that

(4) $$\mathrm{Var}(aX + bY + cZ + \ldots)$$
$$= a^2 \mathrm{Var}(X) + b^2 \mathrm{Var}(Y) + c^2 \mathrm{Var}(Z) + \ldots.$$

An important special case of the model for several measurements arises when several, say n, measurements are made on the same quantity. In this case it is convenient to represent the n successive measurements by $X_1, X_2, \ldots, X_n$. Since the measurements are of the same quantity, it is natural to use the same factor model for each. The resulting product model will be called the *model for repeated measurements*. The common expectation of the X's, defined by (1), will be denoted by μ:

$$\mu = E(X_1) = E(X_2) = \ldots = E(X_n);$$

and the common variance, defined by (2), will be denoted by σ^2:

$$\sigma^2 = \mathrm{Var}(X_1) = \mathrm{Var}(X_2) = \ldots = \mathrm{Var}(X_n).$$

In this model, μ and σ^2 are parameters whose values are usually unknown. The purpose of taking the measurements is to obtain information concerning these unknown parameters. In the present section we shall consider the problem of estimating μ, which one may think of as the quantity being measured. (The estimation of σ^2, the precision of the measuring process, will be taken up in Section 5.) For the repeated measurements model, formulas (3) and (4) simplify to

(5) $$E(a_1 X_1 + \ldots + a_n X_n) = (a_1 + \ldots + a_n)\mu$$

and

(6) $$\mathrm{Var}(a_1 X_1 + \ldots + a_n X_n) = (a_1^2 + \ldots + a_n^2)\sigma^2.$$

For estimating the quantity μ, it is common practice to use the arithmetic mean $\overline{X}$ of the measurements,

$$\overline{X} = \frac{X_1 + \ldots + X_n}{n} = \frac{1}{n} X_1 + \ldots + \frac{1}{n} X_n.$$

Application of (5) with $a_1 = \ldots = a_n = 1/n$ gives

(7) $$E(\overline{X}) = \mu.$$

Thus, $\overline{X}$ is an unbiased estimate of the common expected value μ of the measurements. It follows similarly from (6) that

(8) $$\mathrm{Var}(\overline{X}) = \sigma^2/n,$$

and hence that

(9) $$\mathrm{SD}(\overline{X}) = \sigma/\sqrt{n}.$$

Thus the standard deviation of the arithmetic mean of n measurements equals the standard deviation σ of a single measurement, divided by $\sqrt{n}$. The estimate $\overline{X}$ for μ can therefore be made as precise as desired by taking a sufficiently large number of measurements.

There is another way of examining the behavior of the estimate $\overline{X}$ when the number n of repeated measurements is very large. Recall the Chebyshev inequality, developed in Section 6.9. Formula (6.9.5) with $\overline{X}$ in place of $\overline{Z}$ (and μ in place of ζ) states that

$$P(|\overline{X} - \mu| \geq c) \leq \sigma^2/nc^2,$$

or equivalently

$$P(\mu - c < \overline{X} < \mu + c) \geq 1 - \frac{\sigma^2}{\mu c^2}.$$

This inequality holds for any positive number c, and we shall now suppose that c is very small, so that $(\mu - c, \mu + c)$ is a very small interval. Although σ^2/c^2 is then very large, we can—by taking n sufficiently large—make σ^2/nc^2 as small as we please, and hence

$$P(\mu - c < \overline{X} < \mu + c)$$

as close to 1 as we please. That is, for any fixed interval $(\mu - c, \mu + c)$, no matter how small, we can make it as nearly certain as we wish that $\overline{X}$ will fall in this interval, by taking sufficiently many measurements. A method of estimation (such as estimating μ by the average $\overline{X}$), which has the property that the estimate is practically certain to fall very close to the parameter μ being estimated, provided there are sufficiently many observations, is called *consistent*.

When planning to make a number of measurements the question usually arises: how many should be made? As in the binomial and hypergeometric cases, the number of observations required will depend on the desired degree of precision. It also depends on σ, the standard deviation of an individual measurement. Unfortunately, this quantity is not usually known. Just as with the analogous problem of determining the size of a sample (Section 1) where τ plays the same role as σ plays here, one must either take a value of σ from previous work with measurements of similar precision, or carry out a pilot study to estimate σ by the method to be discussed in Section 5. Once a value of σ is obtained, the required number n of measurements is from (8)

$$n = \sigma^2/\mathrm{Var}(\overline{X}),$$

where $\mathrm{Var}(\overline{X})$ is the desired variance of the estimate.

EXAMPLE 4. *Weighing with a chemical balance.* An object weighing about 3 grams is to be weighed with a standard deviation of 0.03 gram. From previous experience, the balance is known to provide weights for such objects with a standard deviation of approximately $\sigma = 0.1$ gram. Therefore we will need to make approximately $(.1)^2/(.03)^2$, or $n = 11$ determinations.

The model for several measurements and its special case, the model for repeated measurements, are product models. Accordingly they can be realistic only if the several measurements are unrelated (see Section 3.2). In particular, if the measurements are made in such a way that the results found on some of the measurements exert an influence on the values obtained on others, unrelatedness would not hold. The danger of such an influence is greatest in the case of repeated measurements by the same observer of the same quantity. The observer remembers what he found previously, and this knowledge may influence his later findings, especially if the measurement involves a large degree of subjective judgement. Before adopting the model for repeated measurements, one should carefully consider whether the measurements are unrelated, or nearly so.

Just as the hypergeometric model can be viewed as a special case of the sampling model by assigning to each item in the population a v value that is 1 or 0, so the binomial model can be viewed as a special case of the model for repeated measurements by assigning to each trial a u value that is 1 or 0. Depending on whether a trial is a success or failure, we assign to it the "measurement" 1 or 0. The "measurements" $X_1, \ldots, X_n$ are then just the indicator variables that indicate success on the n trials. The total number

$$B = X_1 + \ldots + X_n$$

of successes has the binomial distribution corresponding to n independent trials each with success probability $p = P(X = 1)$. The expectation μ of the X's in this case is p and the estimate $\bar{X}$ of $\mu = p$ reduces to B/n, the estimate proposed for p in Section 8.1.

The model for repeated measurements has important applications also to the problem of sampling. As we remarked in the preceding section, if the sampling is done with replacement, the variables $Y_1, \ldots, Y_s$ (representing the v values of successive items) not only all have the same distribution but in addition are independent. The model for sampling with replacement is therefore formally identical with the measurement model, with the distinct values among $v_1, v_2, \ldots$ playing the role of $u_1, u_2, \ldots$, with $\bar{v} = \mu$, etc. Suppose now that sampling is without replacement but that the *sampling fraction* s/N is small. Since the sample is then only a small fraction of the population, a replaced item is unlikely to be drawn again, so that the distinction between sampling with and without replacement becomes unimportant. (Formally, the Y's are dependent but the dependence is so slight that it may be disregarded.) In this case, the model for sampling with replacement, and hence the measurement model, frequently serves as a good approximation.

While the estimate $\bar{X}$ for μ may be made as precise as desired by taking n large enough, there is an important practical difficulty that the experimenter must keep in mind. The quantity near which $\bar{X}$ will probably be is μ, the expected value of any single measurement. Unfortunately, μ may not be the quantity that one wants to measure, in which case the measurements are said to be *biased*. Suppose, for example, that several IQ tests are given to a candidate in order to obtain a very accurate determination of his intelligence. No increase in the number of tests will improve the determination if he has persuaded a more intelligent friend to take the tests in his place. Similarly, it is useless to multiply thermometer readings to obtain a very accurate determination of a temperature if the thermometer being used is not calibrated correctly. Averaging a large number of biased measurements will produce a result $\bar{X}$ that is very *precise* (in the sense that its variability is small) but not very *accurate* (in the sense that $\bar{X}$ will not be close to the quantity of interest). To take another simple illustration, suppose we measure the length of a ship one hundred times using a meter stick very carefully, and average the results. Our final value will be very precise but inaccurate if by mistake we have used a stick that is one yard instead of one meter long.

The purpose of making repeated measurements on a quantity μ is of course to obtain a more precise estimate for μ than is possible with a single measurement. As we see from (9), the standard deviation of $\overline{X}$, which measures its precision, can be made as small as desired by increasing the number n of measurements sufficiently. Unfortunately, just as in the binomial model (Section 8.3), the standard deviation goes down only slowly as n is increased; the "square root law" holds also in the measurement model. To increase precision tenfold, 100 times as many measurements are required.

PROBLEMS

1. Cut a piece of paper into 20 strips of equal width. On the upper edge of each strip, make a pencil mark at the point which appears to you (by eye) to be the midpoint. After all 20 strips have been bisected visually, determine the midpoint of each strip using a ruler. In all 20 cases measure the error of the midpoint obtained visually, and estimate the expected value of your error, that is, of the bias of the visual method. (Keep the observations for later reference; see Problem 5.1.)

2. Make 10 measurements of the width of a room using a foot rule. Use these measurements to estimate the width of the room. (Keep the observations for later reference; see Problem 5.1.)

3. Explain the relevance of the measurement model when the numbers of points are observed on repeated throws with a die. What are the possible values u_1, $u_2, \ldots$ of this "measurement"? What is the meaning of μ? How would you estimate μ?

4. Is the measurement model appropriate for a set of n digits produced by a random digit generator?

5. The following are 29 measurements of a physical quantity:

6.59	6.62	6.88	6.08	6.26	6.65	6.36	6.29	6.58	6.65
6.55	6.63	6.39	6.44	6.34	6.79	6.20	6.37	6.39	6.43
6.63	6.34	6.46	6.30	6.75	6.68	6.85	6.57	6.49	

Obtain an estimate of this quantity. [Hint: See Problem 1.4.]

6. Repeated measurements $X_1, \ldots, X_n$ are to be taken of a physical quantity μ, and it is known that the variance of each measurement is approximately equal to 4. How many observations are required if we wish the standard deviation of the estimate $\overline{X}$ of μ to be (i) .4; (ii) .1?

7. Solve the two parts of the preceding problem if the variance of each measurement is equal to (i) 2; (ii) 16.

8. Let X_1, X_2 be two unrelated measurements of μ, with $E(X_1) = E(X_2) = \mu$.
  (i) Show that for any constant a, the statistic $T = aX_1 + (1 - a)X_2$ is an unbiased estimate for μ.
  (ii) Assuming that $\mathrm{Var}(X_1) = \mathrm{Var}(X_2) = \sigma^2$, find the variance of T for the values $a = \frac{1}{4}, \frac{1}{3}, \frac{1}{2}, \frac{2}{3}, \frac{3}{4}$. Which value of a would you prefer?

(iii) Assuming instead that $\text{Var}(X_1) = \frac{1}{2} \text{Var}(X_2)$, find the variance of T for the same values of a as in part (ii). Which value of a would you prefer now?

9. Three objects A, B, and C have true weights α, β, and γ. They are weighed in pairs: A and B together have weight X; B and C together have weight Y; C and A together have weight Z. We assume that the weighing process is free of bias, so that $E(X) = \alpha + \beta$, and so on.

(i) Find unbiased estimates of α, β, and γ.

(ii) Find the variances of the unbiased estimates of part (i) and compare them with the variances of direct single weighings of each object, which also provide estimates of α, β, and γ in three weighings. (Assume that any weighing has variance σ^2, and that the weighings are unrelated.)

10. Three objects A, B, and C have true weights α, β, and γ. In each of three weighings, two of the objects are put on one side of the scale and the third on the other side. If X is the weight that has to be added to C to balance the scale when A and B are on the other side, $E(X) = \alpha + \beta - \gamma$. Similarly $E(Y) = \alpha + \gamma - \beta$ and $E(Z) = \beta + \gamma - \alpha$.

(i) Find unbiased estimates of α, β, and γ.

(ii) Find the variances of the unbiased estimates of part (i) and compare them with the variances of the estimates of the preceding problem. (Assume as before that any weighing has variance σ^2, and that the weighings are unrelated.)

11. Four objects A, B, C, and D have true weights α, β, γ, and δ. Suggest three different methods of using four weighings which provide unbiased estimates of α, β, γ, and δ, and compare the variances of the estimates for the three different methods. (Hint: Methods are suggested by Problems 9 and 10.)

12. A chemical balance produces weighings with standard deviation 24 units. How many repeated measurements must be averaged to produce an estimate with standard deviation equal to (i) 2 units; (ii) 1 unit?

13. From a large population, n persons are selected at random and are asked whether they prefer product A or B, or whether they are indifferent. Let p, q, r denote the probabilities that a person prefers A, is indifferent, or prefers B. Each person is assigned a score: $+1$ for preferring A, -1 for preferring B, 0 for indifference. In a model for repeated measurements these scores are denoted by $X_1, \ldots, X_n$.

(i) Express μ in terms of p, q, r.

(ii) Express σ^2 in terms of p, q, r.

(iii) Express the estimate $\bar{X}$ in terms of the numbers of persons in the three categories.

14. Measurements now cost \$1 each. For \$3 we could obtain measurements with standard deviation only half as large as the present measurements. Would this be a good idea?

15. Suppose, in the model for repeated measurements, we happen to know that $\mu = 7$, while σ is unknown. Suggest an estimate for σ^2. [Hint: What is the value of $E(X_1 - 7)^2$? of $E(X_2 - 7)^2$?]

9.3 COMPARING TWO QUANTITIES OR POPULATIONS

One frequently wishes to determine the difference between two quantities, for example the difference in the velocity of light in two media, or the difference in potency of two drugs.   Similarly, one may wish to know the difference between the average value of a certain quantity, such as income or intelligence, in two different populations.   In such cases, where one is comparing two quantities or two populations, the experimental data will consist of observations made under two different circumstances.   We shall now use the ideas of the preceding sections to develop models for analyzing such comparative data.   These models together with those of the next section form the basis for the statistical analysis of comparative experiments.

EXAMPLE 1.   *Comparison of two measured quantities.*   Let us first consider experiments for comparing two quantities.   To provide a basis for the comparison, suppose that a number of measurements are made on each quantity, say n measurements on the first and k measurements on the second.   If the $n + k$ measurements are unrelated, we may use the model for several measurements of Section 2; denoting by $X_1, X_2, \ldots, X_n$ the random variables which represent the n measurements on the first quantity, and by $X_1', X_2', \ldots, X_k'$ the random variables which represent the k measurements on the second quantity.   All $n + k$ random variables are independent.   Since $X_1, X_2, \ldots, X_n$ represent measurements on the same (first) quantity, it is natural to assume that they all have the same distribution, and hence in particular the same expectation, say μ, and the same variance, say σ^2.   Similarly $X_1', X_2', \ldots, X_k'$ all have the same distribution, and hence the same expectation, say μ', and the same variance, say σ'^2.

In accordance with the discussion of the repeated measurement model of Section 2, one may use $\overline{X} = (X_1 + X_2 + \ldots + X_n)/n$ to estimate μ, and $\overline{X}' = (X_1' + X_2' + \ldots + X_k')/k$ to estimate μ'.   Then it is natural to use $\overline{X}' - \overline{X}$ as the estimate for $\mu' - \mu$.   Since

$$\overline{X}' - \overline{X} = \frac{X_1'}{k} + \ldots + \frac{X_k'}{k} - \frac{X_1}{n} - \ldots - \frac{X_n}{n}$$

we can apply (2.5) to find that

(1) $$E(\overline{X}' - \overline{X}) = \mu' - \mu,$$

so that $\overline{X}' - \overline{X}$ is an unbiased estimate for $\mu' - \mu$; and apply (2.4) to find

(2) $$\mathrm{Var}(\overline{X}' - \overline{X}) = \sigma'^2/k + \sigma^2/n.$$

In designing an experiment to compare two measurements, two sample sizes, n and k, must be selected.   The problem of making this choice will be discussed in Section 10.2.

The above model for comparison of two measured quantities may be thought of as the product of a model for n repeated measurements on the first quantity, and of a model for k repeated measurements on the second quantity. With this point of view, (2.7) gives $E(\overline{X}) = \mu$ and $E(\overline{X}') = \mu'$, from which (1) follows. Furthermore, (2.8) gives $\mathrm{Var}(\overline{X}) = \sigma^2/n$ and $\mathrm{Var}(\overline{X}') = \sigma'^2/k$, from which (2) follows using the fact that $\overline{X}$ and $\overline{X}'$ are independent. This approach gives alternative proofs of (1) and (2).

EXAMPLE 2. *Comparison of populations.* Now let us turn to the problem of determining the difference between the average values of a quantity in two different populations. A sample is drawn from each population, and the value of the quantity is determined for each item in each sample. Suppose for example that one wishes to know the difference in the average amount of engine wear in two fleets of trucks. A sample is taken from each fleet, and the wear determined on the trucks in the two samples. Again, in a study to compare caries rates of children in two cities with different amounts of fluorides in the water supply, samples of children are drawn from the first-grade children in each city, and dentists determine the number of caries for each child in each sample. For each sample separately, a model for sampling (as discussed in Section 1) is appropriate; and, if the samples are so drawn that they are unrelated, the product of the two models may be used for the whole experiment.

Let the first population consist of N individuals, having values $v_1, \ldots, v_N$ for the quantity of interest, and the second population of K individuals, having values $v'_1, \ldots, v'_K$. In accordance with our earlier notation we shall denote the average of the v values by $\bar{v}$ and that of the v' values by $\bar{v}'$. Similarly we shall denote by τ^2 and τ'^2 respectively the population variance of the v values and of the v' values. Let us represent the observed values in the sample of s from the first population by the random variables $Y_1, \ldots, Y_s$, and those in the sample of t from the second population by $Y'_1, \ldots, Y'_t$. We know from (1.6) that $\overline{Y}$ is an unbiased estimate for $\bar{v}$ and $\overline{Y}'$ is an unbiased estimate for $\bar{v}'$. Therefore

$$(3) \qquad\qquad \overline{Y}' - \overline{Y}$$

is an unbiased estimate for $\bar{v}' - \bar{v}$, the difference in the mean values of the quantity with two populations. Since $\overline{Y}$ and $\overline{Y}'$ are independent, formula (1.8) gives

$$(4) \qquad \mathrm{Var}(\overline{Y}' - \overline{Y}) = \frac{N - s}{N - 1} \cdot \frac{\tau^2}{s} + \frac{K - t}{K - 1} \cdot \frac{\tau'^2}{t}.$$

PROBLEMS

1. Use (2.5) and (2.6) to write out a formal proof of (1) and (2).

2. Suppose in Example 1 that $\sigma'^2 = 4\sigma^2$.

(i) For each possible combination of n and k for which the total number $n + k$ of measurements is equal to nine, find the variance (2), and plot these variances as a function of n for $n = 1, 2, \ldots, 8$.

(ii) Among the combinations of part (i), find that which makes the variance (2) as small as possible.

3. In Example 1 suppose that $n = k$. Find the number n of measurements required if the standard deviation of the estimate $\bar{X}' - \bar{X}$ is to be .1 when (i) $\sigma'^2 = \sigma^2 = 1$; (ii) $\sigma^2 = 1, \sigma'^2 = 4$.

4. The following are ten measurements of μ' and eight measurements of μ:

$$\mu': 17.3, 17.1, 18.2, 17.5, 15.8, 16.9, 17.0, 17.5, 17.8, 17.1$$

$$\mu: 3.2, 3.2, 3.9, 3.3, 2.7, 3.4, 4.0, 2.9.$$

Obtain an estimate of $\mu' - \mu$.

5. Samples of 100 families are taken in each of two towns. The number of cars owned by each of the 200 families is observed and the information summarized in the following table.

	No cars	One car	Two cars	Three cars	Total
Sample 1	23	58	14	5	100
Sample 2	17	51	26	6	100

(i) Estimate the difference in the average number of cars owned by families in the first town and in the second town.

(ii) Estimate the difference in the (relative) frequency with which a family owns at least one car in the first town and in the second town.

6. In Example 2 devise an unbiased estimate of

(i) $(v_1 + \ldots + v_N) - (v_1' + \ldots + v_K')$

(ii) $(v_1 + \ldots + v_N) + (v_1' + \ldots + v_K')$

(iii) $\bar{v} + \bar{v}'$.

7. Find the variances of the estimates proposed for the three parts of the preceding problem.

8. Suppose in Example 2 that $N = K = 101$ and that $\tau'^2 = 2\tau^2$. Find by trial and error the combination of sample sizes s and t which make the variance (4) as small as possible, subject to the condition that the total sample size $s + t$ is equal to six.

9. In Example 2, assume $N = K = 101$ and suppose that $s = t$. Find the sample size s required if $\tau = 1, \tau' = 1.5$ and if the standard deviation of the estimate $\bar{Y}' - \bar{Y}$ is to be (i) .1; (ii) .05.

10. Solve part (i) of the preceding problem if the sample sizes, instead of being equal, are to be in the relationship $t/s = \frac{3}{2}$.

11. Show that the method of estimating $\mu' - \mu$ by $\bar{X}' - \bar{X}$ proposed in Example 1 is consistent as k and u get large.

9.4 ESTIMATING THE EFFECT OF A TREATMENT

In scientific and technical work, one is often interested in determining the magnitude of the effect a treatment will have when it is applied to subjects of a specified kind. In the present section we shall develop a model that may be used in designing and analyzing experiments for estimating the average effect of a treatment in a population of subjects. We begin by giving examples of such situations.

EXAMPLE *1. Vaccine boosters.* The passage of time will usually attenuate the protection afforded by a vaccine, as is reflected by a slow decrease in the level of antibody in the blood. It may then be desirable to administer a "booster," intended to raise the antibody level again and thus to restore the protection. A question of interest about such a booster is: how much does it raise the level?

EXAMPLE *2. Engine-oil additives.* Petroleum engineers devise many chemicals of which it is hoped that they will prolong the life of automobile engines when added to the engine oil. For example, consider an additive which is intended to increase the number of miles a car may be driven before it begins to require oil between changes. For such an additive, it is of importance to know how many miles of additional driving it will afford.

EXAMPLE *3. Rainmaking.* There are firms which offer, for a fee, to "seed" the clouds of a storm in order to increase the amount of precipitation. The question then arises: by how much will seeding increase the precipitation?

In each of these examples we may distinguish a *treatment* (booster, additive, cloud-seeding) that may be applied to *subjects* (persons, engines, storms) in an attempt to change a *response* (antibody level, miles before oil must be added, amount of rainfall). Let us now specify just what is meant by the "effect" of such a treatment.

Consider one particular subject, and suppose that his response would be equal to w if he were given the treatment, while it would be equal to v if he were not given the treatment. Then the difference $w - v$ is the *additional* response elicited by the treatment, above what it would have been without the treatment. We shall define this difference to be the *effect* of the treatment *on that subject*, and denote it by Δ; that is,

$$\Delta = w - v.$$

For example, if a particular storm would produce 2.4 inches of rain without seeding, and $w = 2.7$ inches with seeding, then the effect of seeding on that storm is $\Delta = 2.7 - 2.4 = .3$ inch.

It is not to be expected that a treatment has exactly the same effect on all subjects, and what is usually of primary interest is its *average* effect on subjects of a specified kind. Consider a population of N subjects of the kind in question, and suppose that the first subject would give responses w_1 and v_1 with and without treatment, so that the effect of the treatment on the first subject is $\Delta_1 = w_1 - v_1$. Similarly, for the second subject the effect is $\Delta_2 = w_2 - v_2$, and so on up to $\Delta_N = w_N - v_N$. Then the average effect of the treatment *in this population* is $\bar{\Delta} = (\Delta_1 + \ldots + \Delta_N)/N$.

While the effects $\Delta_1, \ldots, \Delta_N$ of the treatment on the different subjects in the population are never exactly equal, the estimation of $\bar{\Delta}$ is of interest mainly in those cases when they are not too different. Indeed, if the treatment effects are drastically different, this fact would usually be of greater importance than the value of $\bar{\Delta}$ itself. For example, if a drug will speed up the recovery of half the patients by two months, while slowing down the recovery of the other half by two months, $\bar{\Delta}$ is 0; but the important question would be to identify the patients whom the drug will benefit. Fortunately, in many areas of experimentation the treatment effects are rather similar for the different subjects, so that most of them are close to $\bar{\Delta}$. The ideal case would arise in the *constant effect model*, in which the effects $\Delta_1, \ldots, \Delta_N$ are all equal so that

(1) $$\Delta_1 = \ldots = \Delta_N = \bar{\Delta}.$$

Experiments for estimating $\bar{\Delta}$ fall into two main types, according to whether it is, or is not, feasible to observe both the v value and the w value of the same subject.

Type (a). In Example 1, it is possible to measure a subject's antibody level, give him the booster, and then (after an appropriate interval) measure the level again. We may then interpret the first measurement as the subject's level v without treatment, and the second measurement as the subject's level w with treatment, and by subtraction find the magnitude $\Delta = w - v$ of the treatment effect for this subject.

Type (b). In Examples 2 and 3 it is difficult to see how both v and w could be determined for the same subject. For example, if a storm is seeded, one can observe how much rain w it produces when treated, but it is then not possible to observe how much rain v that same storm would have produced if it had not been seeded.

We shall now propose experimental designs suitable for these two types of experiments.

DESIGN A: *Both responses observable on the same subject.* In this case, an estimate for $\bar{\Delta}$ may be obtained as follows. Draw a random sample of, say, s subjects from the population of N subjects. For each subject in the sample, observe both v and w values, and by taking their difference find the treatment effect for that subject. Let these effects be denoted by

$D_1, \ldots, D_s$ for the s subjects. Then the model of Section 1 applies, with $\Delta_1, \ldots, \Delta_N$ playing the role of $v_1, \ldots, v_N$, and $D_1, \ldots, D_s$ playing the role of $Y_1, \ldots, Y_s$. By (1.6), $\overline{D} = (D_1 + \ldots + D_s)/s$ is an unbiased estimate of $\overline{\Delta}$, and the variance of the estimate is by (1.8)

$$(2) \qquad \qquad \mathrm{Var}(\overline{D}) = \frac{N - s}{N - 1} \cdot \frac{\delta^2}{s}$$

where δ^2 is the population variance of $\Delta_1, \ldots, \Delta_N$, i.e.,

$$(3) \qquad \qquad \delta^2 = [(\Delta_1 - \overline{\Delta})^2 + \ldots + (\Delta_N - \overline{\Delta})^2]/N.$$

Now let us turn to experiments of type (b), in which it is not possible to observe both the v value and the w value on the same subject. In this case, one can never discover the Δ value of any subject; yet, fortunately, it is still possible to estimate the average Δ value in the population. By averaging the equations $\Delta_1 = w_1 - v_1, \ldots, \Delta_N = w_N - v_N$, we find that

$$(4) \qquad \qquad \overline{\Delta} = \overline{w} - \overline{v};$$

that is, the average treatment effect in the population is the difference of the average responses with and without treatment. The following design makes it possible to estimate $\overline{w}$ and $\overline{v}$, and hence $\overline{\Delta}$.

DESIGN B: *Only one response observed on a given subject.* Draw two distinct samples, of sizes s and t, from the population. On the first sample of s subjects observe the v values, which may be denoted by $Y_1, Y_2, \ldots, Y_s$. The model of Section 1 applies, so that $\overline{Y}$ is an unbiased estimate of $\overline{v}$. On the second sample of t subjects observe the w values, which may be denoted by $Z_1, Z_2, \ldots, Z_t$. Again using the model of Section 1, $\overline{Z}$ is seen to be an unbiased estimate of $\overline{w}$. It follows that $\overline{Z} - \overline{Y}$ is an unbiased estimate of $\overline{w} - \overline{v} = \overline{\Delta}$. Thus, the difference between the average response of the t treated subjects and the average response of the s untreated subjects provides an unbiased estimate of the average effect of the treatment in the population from which the two samples come. (The validity of the above argument rests on the fact that the second sample behaves like a random sample drawn from the entire population. This follows from the generalized equivalence law of ordered sampling (Section 2.3), applied to the total sample of $s + t$ observations.)

The calculation of the exact variance of the estimate $\overline{Z} - \overline{Y}$ is somewhat technical (see Problem 6), but there is an easy approximation in case both of the sampling fractions s/N and t/N are small. In this case, it would make little difference whether we sampled with or without replacement. For sampling with replacement, a product model is appropriate for the $s + t$ draws and hence we may use for the random variables $Y_1, \ldots, Y_s$, $Z_1, \ldots, Z_t$ essentially the same special case of the model for several measurements already employed in Example 3.1. If we again denote the population variance of $v_1, \ldots, v_N$ by τ^2, and similarly denote the population variance of $w_1, \ldots, w_N$ by ω^2, it follows from (3.2) that

(5) $$\mathrm{Var}(\overline{Z} - \overline{Y}) = \frac{\tau^2}{s} + \frac{\omega^2}{t} \quad \text{(approximation for } s/N, t/N \text{ small).}$$

In applications of (5), the requirement that s/N and t/N be small often presents a difficulty. An experimenter is seldom so fortunate as to be able to draw random samples from a number N of suitable subjects so large that the sizes s and t of the two experimental groups constitute only small fractions of N. Rather, he will assemble (perhaps with difficulty) as many suitable subjects as he needs, so that often all N subjects are used for the experiment. In this case, $s + t = N$ and s/N and t/N are not small. Fortunately, there is another situation in which (5) holds. In the constant effect model (1), we have (see Problem 3) $\tau^2 = \omega^2$, and unless N is very small (see Problem 7)

(6) $$\mathrm{Var}(\overline{Z} - \overline{Y}) = \frac{\tau^2}{s} + \frac{\tau^2}{t}, \quad \text{(approximation for constant effect model, } N \text{ not small).}$$

Thus, our simple approximation for the variance of the estimate of $\overline{\Delta}$ holds if either the sampling fractions are small, or the effect is constant.

Since there often is available much past experience with responses in the absence of treatment, so that the value of $\bar{v}$ may be taken as known, it is a common practice to dispense with the first sample of design B, using the known value of $\bar{v}$ rather than its estimate $\overline{Y}$. This is, however, not good practice, since the circumstances under which the past information about $\bar{v}$ was acquired may differ materially from those under which the current observations are made. It is seldom safe to dispense with current observations $Y_1, \ldots, Y_s$ without treatment. These are known as the *control* observations, since they provide the basis against which the treated responses may be gauged. (In this language, in design A we may say that "each subject is its own control.") For further discussion of the need for controls see Section 12.1.

Design B can of course be used even in those cases in which design A is applicable, and the question then arises, which design to use. This choice involves two main issues.

(i) When it is applicable, design A is usually much more efficient than design B, for the following reasons. We have remarked that the estimation of $\overline{\Delta}$ is of interest principally in those cases where $\Delta_1, \ldots, \Delta_N$ are rather similar to each other, so that their population variance δ^2 is small. It then follows from (2) that $\mathrm{Var}(\overline{D})$ is small even with a sample whose size s is quite modest. In the ideal case of the constant effect model (1), it is easy to see from (3) that $\delta^2 = 0$, so that $\overline{D}$ provides a perfect estimate even with a sample of size 1. While the ideal is never attained in practice, it is usually the case that Δ is less variable than either v or w. Then δ^2 is smaller than either τ^2 or ω^2, and hence $\text{Var}(\overline{D})$ is smaller than $\text{Var}(\overline{Z} - \overline{Y})$ as given by (5), when equal numbers of observations are used in the two designs.

(ii) Design A, however, may involve a risk of bias that is absent from design B. When both v and w values are observed on the same subject, it is often difficult to avoid the possibility that the two observations are taken under different conditions. Thus, in Example 1, it may take some time for the booster shot to achieve its full effect, so that the measurement w of antibody level with treatment is made at a considerably later time than the measurement v of antibody level without treatment. The fact that the levels are measured at widely separated times tends to reduce their comparability. In other cases, the mere act of making the first measurement may influence the value of the second measurement. For example, a psychologist wishes to investigate the possibility that a stimulating drug improves performance on an IQ test. He administers the test to a subject to measure his response v without treatment, then gives the drug and re-administers the test to measure his response w with treatment. Clearly, the fact that the subject has taken the test before might tend to improve his performance the second time, apart from any specific influence of the drug.

For clarity of exposition, we have written of comparing treated subjects with untreated controls. Actually, the interest may reside in comparing two treatments with each other. In such cases, the machinery we have developed can be applied, with one of the treatments playing the role of "no treatment." Thus, $v_1, \ldots, v_N$ become the responses to the first treatment while $w_1, \ldots, w_N$ are the responses to the second treatment, and $\overline{\Delta}$ is the average superiority of the second treatment over the first.

While the question is not really within the technical domain of statistics, it should be pointed out that a crucial matter to be settled in planning a treatment-response experiment is the choice of the response variable to be observed. Two issues govern this choice: the response should be easily and accurately observable; but, more important, it must be closely related to the actual effect of interest. Thus, in an experiment to compare two methods of teaching the calculus, it is easy to observe a student's score on a standardized examination, but much more difficult to determine the depth of understanding which the student has acquired. If the examination score is chosen as the response variable, there is a real danger that the experiment will favor a superficial instructional method that prepares the students for the specific examination. Again, in Example 1, the response variable was taken to be the antibody level in the person's blood. The actual effect of interest is of course the degree of protection against the disease, which is very difficult to observe, but which according to theory should be closely related to antibody level.   If this theory is wrong, the experimental conclusions could be most misleading.

PROBLEMS

1. Suppose the untreated and treated responses in a population of size $N = 5$ are
$$v_1 = 15,\ v_2 = 9,\ v_3 = 27,\ v_4 = 21,\ v_5 = 6$$
$$w_1 = 22,\ w_2 = 8,\ w_3 = 34,\ w_4 = 24,\ w_5 = 12.$$
   (i) If design B is used with $s = 3$ and $t = 2$, obtain the distribution of $\bar{Z} - \bar{Y}$.
   (ii) If design A is used, obtain the distribution of $\bar{D}$ for $s = 1$.
   (iii) Graph the histograms for the distributions of (i) and (ii) on a common scale and decide which design seems best. [Hint: (i) For each of the $\binom{5}{2} = 10$ possible choices of Z_1 and Z_2 list the values of $\bar{Z}$, $\bar{Y}$, and $\bar{Z} - \bar{Y}$.]

2. Compute τ^2, ω^2, λ, and δ^2 for the population of Problem 1, where λ is the population covariance (7.2.22).

3. (i) Show that if (1) holds, then
(7)
$$\tau^2 = \omega^2 = \lambda.$$
   (ii) Show conversely that (7) implies (1).   [Hint: (ii) Note that (7) implies $\delta^2 = 0$.]

4. Under the assumptions of Problem 1(i) compute $\operatorname{Var}(\bar{Z} - \bar{Y})$
   (i) from the distribution of $\bar{Z} - \bar{Y}$ obtained in Problem 1(i),
   (ii) from formula (8) below.

5. Compute $\operatorname{Var}(\bar{D})$ under the assumptions of Problem 1(ii) and compare this variance with $\operatorname{Var}(\bar{Z} - \bar{Y})$ of the preceding problem.

6. (This problem depends on Section 7.2.)   Let $\bar{Y}$ and $\bar{Z}$ be the estimates of $\bar{v}$ and $\bar{w}$ of design B.
   (i) Show that $\operatorname{Cov}(\bar{Y}, \bar{Z}) = -\lambda/(N - 1)$ where λ is the population covariance (7.2.4).
   (ii) Show that
(8)
$$\operatorname{Var}(\bar{Z} - \bar{Y}) = \frac{N - s}{N - 1}\frac{\tau^2}{s} + \frac{N - t}{N - 1}\frac{\omega^2}{t} + \frac{2\lambda}{N - 1}.$$
   (iii) Explain why you would expect (8) to be close to (5) if s and t are small but N is large.
   [Hint: (i) It is a consequence of (7.2.23) that the covariance of the v value of any randomly drawn item with the w value of any other randomly drawn item is $-\lambda/(N - 1)$.]

7. Use (3) to show that in the constant effect model, (8) reduces to
(9)
$$\frac{N}{N - 1}\left(\frac{\tau^2}{s} + \frac{\tau^2}{t}\right)$$

9.5 ESTIMATION OF VARIANCE

We have pointed out, in Section 8.2, that it is customary when reporting an estimate to give its standard deviation, so that the user of the report has an idea of the accuracy and reliability of the estimate. The estimate is then frequently presented as

estimate ± standard deviation of estimate.

Thus, a chemist may write: "The material was found to contain 7.32 ± .16 grams of calcium." This means that his estimate for the amount of calcium is 7.32 grams, and that the standard deviation of the estimate is .16 gram. Similarly, an agronomist may report: "The fertilizer increased the yield at the rate of 120 ± 30 pounds per acre," or the report of a public opinion poll may say: "Candidate A has the support of 62 ± 3% of the voters." Note that this is not a guarantee that the support falls between $62 - 3 = 59\%$ and $62 + 3 = 65\%$; the figure following the ± is the standard deviation rather than the extreme limit of error, and it is quite possible that an estimate will be off by more than one standard deviation. In fact, if the distribution of the estimate may be approximated by the normal curve, there is something like a 32% chance that this will occur (see Table 8.2.1).

The variance of an unbiased estimate was worked out in a number of examples in the preceding sections. In each case, the reader may observe that the formula for the variance of the estimate involves unknown quantities. Thus, in Example 8.1.1 we suggested that B/n be used to estimate p in the binomial model. The variance of the estimate is $\mathrm{Var}(B/n) = p(1 - p)/n$, which involves the unknown quantity p. Again, the estimate $\overline{X}$ for μ in the model for repeated measurements (Section 2) has variance σ^2/n, and σ^2 is usually not precisely known. How then is it possible to report the variance of the estimate? What is done in practice is to report not the exact value of the variance or standard deviation but an estimate of that value. Naturally it would be nicer if the exact value could be given, but this is impossible, and practical needs are usually served by giving the estimated value. When the chemist reports 7.32 ± .16 grams of calcium, he is really saying "I estimate that there are 7.32 grams. This estimate has a standard deviation of about .16 gram; that is, the standard deviation of the estimate 7.32 is itself estimated to be .16." In this section we shall take up the problem of estimating the variance and standard deviation of an estimate.

EXAMPLE 1. *Repeated measurements.* On the basis of unrelated measurements, $X_1, X_2, \ldots, X_n$ with common expectation μ and common variance

$$\sigma^2 = E(X_1 - \mu)^2 = E(X_2 - \mu)^2 = \ldots = E(X_n - \mu)^2,$$

we wish to estimate σ^2. If μ were known, we could compute the n quantities

$$(X_1 - \mu)^2, (X_2 - \mu)^2, \ldots, (X_n - \mu)^2$$

and, by the interpretation of expectation as "long-run average," would expect the arithmetic mean of these n quantities to be close to their common expected value σ^2. This would suggest

$$(1) \qquad \frac{1}{n}[(X_1 - \mu)^2 + (X_2 - \mu)^2 + \ldots + (X_n - \mu)^2]$$

as a reasonable estimate for σ^2. Actually, μ is unknown, but it can be estimated by $\overline{X}$. Replacing in (1) the unknown quantity μ by its estimate $\overline{X}$ gives

$$(2) \qquad \frac{1}{n}[(X_1 - \overline{X})^2 + (X_2 - \overline{X})^2 + \ldots + (X_n - \overline{X})^2].$$

Expression (2) should be approximately equal to (1), and hence may serve as an estimate for σ^2. Using the identity (5.7.19) we see that the estimate (2) is equal to

$$(3) \qquad \frac{1}{n}[X_1^2 + X_2^2 + \ldots + X_n^2] - \overline{X}^2.$$

Since each of the equivalent quantities (2) and (3) provides an estimate of σ^2, the square root of either quantity may be used to estimate σ. Further, since $\mathrm{Var}(\overline{X}) = \sigma^2/n$, division of either (2) or (3) by n provides an estimate for $\mathrm{Var}(\overline{X})$, whose square root will then be an estimate for $\mathrm{SD}(\overline{X}) = \sqrt{\mathrm{Var}(\overline{X})}$.

To illustrate, suppose that the distance between two points has been measured 20 times, yielding the observations (in feet)

8715	8714	8733	8725	8714
8718	8722	8729	8712	8725
8714	8734	8722	8721	8729
8706	8712	8708	8719	8728.

The average of the twenty measurements is $\overline{X} = 8720$. Subtracting this from each measurement gives the 20 differences

-5	-6	13	5	-6
-2	2	9	-8	5
-6	14	2	1	9
-14	-8	-12	-1	8.

These differences will of course add up to zero. The sum of the squares of the differences is $(-5)^2 + (-2)^2 + \ldots + (8)^2 = 25 + 4 + \ldots + 64 = 1256$. This is divided by $n = 20$ to give the estimate 62.8 for the variance σ^2 of the measurements, or the estimate $\sqrt{62.8} = 7.92$ for σ.

Further, $62.8/20 = 3.14$ is an estimate for $\mathrm{Var}(\overline{X})$, and its square root $\sqrt{3.14} = 1.8$ is an estimate for $\mathrm{SD}(\overline{X})$. The final report would read: "The distance is estimated to be 8720 ± 1.8 feet."

FXAMPLE 2. *Sampling.* The estimate $\overline{Y}$ of $\bar{v}$ was seen in (1.7) and (1.8) to have variance proportional to τ^2. To estimate this variance we need first to estimate τ^2. An argument similar to that of Example 1 suggests that τ^2 be estimated by

(4) $$\frac{1}{s}[(Y_1 - \overline{Y})^2 + (Y_2 - \overline{Y})^2 + \ldots + (Y_s - \overline{Y})^2].$$

This quantity multiplied by $(N - s)/(N - 1)s$ will then be a reasonable estimate of (1.8),

$$\frac{N - s}{N - 1} \cdot \frac{\tau^2}{s} = \mathrm{Var}(\overline{Y}).$$

EXAMPLE 3. *Binomial model.* As was pointed out in Section 2, the binomial model is a special case of the measurement model, with $p(1 - p)$ playing the role of σ^2 and B/n the role of $\overline{X}$. The estimate (3) can be simplified in this case, by using the fact that an indicator is always equal to its square (see Problem 5.4.11). Since in the binomial case $X_1, \ldots, X_n$ indicate success on the binomial trials, $X_1^2 + \ldots + X_n^2 = X_1 + \ldots + X_n = B$, and (3) becomes

$$\frac{B}{n} - \left(\frac{B}{n}\right)^2 = \frac{B}{n}\left(1 - \frac{B}{n}\right)$$

which is an estimate for $p(1 - p)$. It is interesting to note that this estimate is obtained by replacing p in the quantity $p(1 - p)$ by its estimate B/n.

In Sections 1 and 2 we pointed out that knowledge of τ^2 or σ^2 is needed for the calculation of sample size when planning an investigation in which the models for sampling or repeated measurements are used. Sometimes a pilot study will be undertaken for the purpose of estimating one of these parameters so that the required size of the main study may be determined. These estimates can then be obtained from the observations in the pilot study using formulas (4) or (2).

As with other estimates, we should also inquire into the accuracy of the estimates (2) for σ^2 and (4) for τ^2. This is, however, a more difficult problem, about which we shall only say that, under realistic assumptions about the distribution of the measurements X and the population values v, a rather large sample is required to produce satisfactorily precise estimates.

Are the estimates (2) and (4) unbiased? To answer this question it is necessary to determine their expected values.   As we shall show below, (2) has the expected value

(5)
$$\frac{n-1}{n}\sigma^2$$

while (4) can be shown to have the expected value

(6)
$$\frac{N}{N-1}\cdot\frac{s-1}{s}\tau^2.$$

The estimates are therefore slightly biased, but the bias is negligible if n and s are not too small, and can if desired be corrected (Problems 5 and 8).

To illustrate the calculations required to obtain (5) and (6), consider the estimate (2). Applying the identity (5.7.18) with $X_1, \ldots, X_n$ in place of $v_1, \ldots, v_N$ and μ in place of a, we find that the estimate (2) is equal to

(7)
$$\frac{1}{n}[(X_1 - \mu)^2 + \ldots + (X_n - \mu)^2] - (\bar{X} - \mu)^2.$$

Now $E(X_1 - \mu)^2 = \ldots = E(X_n - \mu)^2 = \sigma^2$, and $E(\bar{X} - \mu)^2 = \mathrm{Var}(\bar{X}) = \sigma^2/n$. The expected value of (7), and hence of the estimate (2), is therefore

$$\frac{1}{n}[\sigma^2 + \ldots + \sigma^2] - \frac{\sigma^2}{n} = \sigma^2 - \frac{\sigma^2}{n} = \frac{n-1}{n}\sigma^2.$$

PROBLEMS

1.   Compute the estimate (2) of the variance σ^2 of the measurements of (i) Problem 2.1, (ii) Problem 2.2, (iii) Problem 2.5.

2.   Let $X_1, \ldots, X_n$ be n measurements of μ, each having variance σ^2; and let $X'_1, \ldots, X'_k$ be k measurements of μ' each having variance σ'^2. If all measurements are unrelated, suggest an estimate of the variance $\mathrm{Var}(\bar{X}' - \bar{X})$.

3.   How is the estimate (2) for σ^2 altered
   (i) if every measurement is increased by the constant amount a?
   (ii) if every measurement is multiplied by the constant amount b?

4.   A pilot sample of $s = 10$ from a population of $N = 500$ gives the following v values:

$$17 \quad 25 \quad 8 \quad 19 \quad 21 \quad 17 \quad 18 \quad 12 \quad 3 \quad 10.$$

   (i) Calculate the estimate (4) of the population variance τ^2.
   (ii) Assuming that this estimate is correct, determine the sample size required if it is desired to estimate the population mean $\bar{v}$ with standard deviation $\frac{1}{2}$.

5.   Show that the estimate (2) becomes unbiased if the factor $1/n$ is replaced by $1/(n-1)$.

6.   Give an unbiased estimate of $p(1 - p)$ by specializing the result of Problem 5 to the binomial case.   [Hint: Use the method of Example 3.]

7. Check the result of Problem 6 by comparing it with part (ii) of Problem 8.1.18.

8. Use (6) to find an unbiased estimate of τ^2.

9. (i) Using the fact that (2) and (3) are equal, and letting s and $Y_1, \ldots, Y_s$ play the roles of n and $X_1, \ldots, X_n$, show that (4) is equal to

(8)
$$\frac{1}{s}[Y_1^2 + \ldots + Y_s^2] - \bar{Y}^2.$$

   (ii) Using (5.6.2), (1.6), and (1.8), show that the expected value of (8) is given by (6).

10. (i) In the sampling model of Section 1, suppose that r of the v's are equal to 1 and $N - r$ equal to 0. Show that $\bar{v} = r/N$ and $\tau^2 = \frac{r}{N}\left(1 - \frac{r}{N}\right)$.

   (The second of these equations is proved in Example 7.2.4.)

   (ii) The number D of items in the sample having v value 1 is given by $D = s\bar{Y}$ and has a hypergeometric distribution.

   (iii) The estimate (4) of $\tau^2 = \frac{r}{N}\left(1 - \frac{r}{N}\right)$ reduces to $\frac{D}{s}\left(1 - \frac{D}{s}\right)$.

[Hint: (iii) The argument parallels that of Example 3.]

CHAPTER 10

OPTIMUM METHODS OF ESTIMATION

10.1 CHOICE OF AN ESTIMATE

For each of the problems considered in Chapters 8 and 9 an intuitively appealing estimate was available for the parameter being estimated: the frequency of success as the estimate of the probability of success; the average of the measurements as the estimate of the quantity being measured; and so forth. Each of these estimates was seen to be unbiased, and its variance was calculated. This corresponded to the first two stages of the development of statistical methodology outlined in the introduction to Part II: an inference method is devised on intuitive grounds for dealing with a statistical problem, and the performance properties of the method are investigated.

In more complex problems there is frequently a choice between a number of reasonable estimates. The problem of selecting the best of these, which constitutes the third of the four stages, will be illustrated in the present section. In the following section we shall consider the perhaps even more important fourth stage, which is concerned with the proper design of the experiment; that is, the question of what to observe and how to conduct the experiment.

EXAMPLE 1. *Several measurements of the same quantity.* Suppose that there are available several measurements $X, Y, Z, \ldots$ for the same quantity μ. How should these measurements be combined to produce an estimate for μ?

If the way in which the measurements were made justifies treating them as unrelated, it is natural to use the model for several measurements introduced in Section 9.2. If in addition we are willing to assume that each measurement is free of bias, we may specialize that model by assuming

$$(1) \qquad E(X) = \mu, \quad E(Y) = \mu, \quad E(Z) = \mu, \ldots.$$

The measurements $X, Y, Z, \ldots$ could in principle be combined in many different ways. One type of combination that is frequently employed is the *weighted average*

(2) $aX + bY + cZ + \ldots$ where $a + b + c + \ldots = 1$.

Here we say that measurement X is given weight a, measurement Y is given weight b, etc. The weights are required to add up to one.

There are of course many different weighted averages. Under assumption (1), it is easily seen from (9.2.3) that all of them are unbiased estimates for μ:

$$E(aX + bY + cZ + \ldots) = aE(X) + bE(Y) + cE(Z) + \ldots$$
$$= (a + b + c + \ldots)\mu = \mu.$$

Since all weighted averages are unbiased, the considerations of Section 8.2 suggest that the choice among them may be made on the basis of variance. If we write for brevity

(3) $\mathrm{Var}(X) = \sigma^2, \quad \mathrm{Var}(Y) = \tau^2, \quad \mathrm{Var}(Z) = \omega^2, \ldots$

it follows from (9.2.4) that

(4) $\mathrm{Var}(aX + bY + cZ + \ldots) = a^2\sigma^2 + b^2\tau^2 + c^2\omega^2 + \ldots.$

Which of the weighted averages (2) has the smallest variance? This is a typical problem of the third stage of the development of statistics—that of selecting the "best" of several possible procedures—and as we shall see below, it turns out to have a very simple and elegant solution.

Before discussing the general solution, let us consider the simplest special case, in which all the measurements have the same variance,

(5) $\sigma^2 = \tau^2 = \omega^2 = \ldots.$

Since in this case the individual measurements are all equally good, it would seem natural to give all of them the same weight. The weights must add up to 1; so if there are (say) n measurements, each would receive weight $1/n$. In this special case, the weighted average becomes the ordinary arithmetic mean.

If however the measurements are not equally precise, intuition suggests that they should not receive equal weight but that more weight should be given to the more precise measurements than to the less precise ones. To make this consideration specific, suppose there are two measurements X and Y, and that $\sigma^2 = 1$ while $\tau^2 = 4$. If we insisted on giving equal weight to X and Y, using the estimate $\frac{1}{2}X + \frac{1}{2}Y = \frac{1}{2}(X + Y)$, we would have variance

$$\mathrm{Var}(\tfrac{1}{2}X + \tfrac{1}{2}Y) = \tfrac{1}{4}\sigma^2 + \tfrac{1}{4}\tau^2 = \tfrac{1}{4}(1 + 4) = \tfrac{5}{4}.$$

In this example, the arithmetic mean of the two measurements is a 25% poorer estimate (in the sense of having 25% larger variance) than the single measurement X alone! There exist however weighted averages of the two measurements which are better than X. For example, $\frac{4}{5}X + \frac{1}{5}Y$ has variance

$$\text{Var}(\tfrac{4}{5}X + \tfrac{1}{5}Y) = \tfrac{16}{25}\sigma^2 + \tfrac{1}{25}\tau^2 = \tfrac{4}{5},$$

which is 20% smaller than the variance of X alone.

We are now ready to state the solution to the problem of finding the best or *optimum* weights. As mentioned above, it is intuitively clear that the weight should be the smaller, the larger is the variance of the measurement. It turns out that in fact *the optimum weights for the several measurements are inversely proportional to the variances of those measurements.* This "rule of inverse proportionality," combined with the requirement that the weights add up to one, yields the weights

(6)
$$a = \frac{1/\sigma^2}{1/\sigma^2 + 1/\tau^2 + 1/\omega^2 + \ldots}, \quad b = \frac{1/\tau^2}{1/\sigma^2 + 1/\tau^2 + 1/\omega^2 + \ldots},$$

$$c = \frac{1/\omega^2}{1/\sigma^2 + 1/\tau^2 + 1/\omega^2 + \ldots}, \quad \ldots$$

It is easily checked that these weights do add up to one, and that they are proportional to the reciprocals of the variances, as was required.

We shall postpone the proof of the optimality of the weights (6) to Section 4, but will now illustrate their meaning and use.

EXAMPLE 2. *Repeated measurements.* Recall the model for repeated measurements of Section 9.2, which is a special case of Example 1. Since in this case the variances of the n measurements are all equal, (6) implies that the weights should all be equal, leading to the estimate $\overline{X}$. This estimate, presented on intuitive grounds in Section 9.2, is therefore shown by (6) to be the best weighted average of the repeated measurements.

EXAMPLE 3. *Two laboratories.* Suppose that a physical constant μ has been measured by two laboratories, and that the first laboratory reports the value X while the second reports Y. Past experience suggests that the first laboratory is three times as precise as the second, in the sense that $\tau^2 = 3\sigma^2$. Then the optimum weights for combining X and Y are

$$a = \frac{1/\sigma^2}{1/\sigma^2 + 1/\tau^2} = \frac{\tau^2}{\sigma^2 + \tau^2} = \frac{3\sigma^2}{\sigma^2 + 3\sigma^2} = \frac{3}{4}$$

$$b = \frac{1/\tau^2}{1/\sigma^2 + 1/\tau^2} = \frac{\sigma^2}{\sigma^2 + \tau^2} = \frac{\sigma^2}{\sigma^2 + 3\sigma^2} = \frac{1}{4}$$

so that the "best" weighted average is $\frac{3}{4}X + \frac{1}{4}Y$. Its variance is $\frac{9}{16}\sigma^2 + \frac{1}{16}\tau^2 = \frac{3}{4}\sigma^2$.

Note that we have not assumed knowledge of the actual values of the variances σ^2 and τ^2, but only that they are in the ratio $1:3$. (In general, the optimum weights (6) require only knowledge of the ratios of the variances of the several measurements.) But even the ratio will not in practice be exactly known. Fortunately, the variance of a weighted average is only very slightly increased if the weights are not quite the optimum ones, so that if the assumed ratio of the variances of the measurements is not quite correct, the weights (6) will still be nearly optimal.

As an illustration, suppose that we believe that $\tau^2 = 4\sigma^2$ when in reality $\tau^2 = 3\sigma^2$. We shall then use the weighted average $\frac{4}{5}X + \frac{1}{5}Y$ instead of using the best combination $\frac{3}{4}X + \frac{1}{4}Y$, and the variance of the estimate will be $\frac{16}{25}\sigma^2 + \frac{1}{25}\tau^2 = \frac{19}{25}\sigma^2 = .76\,\sigma^2$ instead of the smallest possible variance $\frac{3}{4}\sigma^2 = .75\,\sigma^2$ which we could achieve by using $a = \frac{3}{4}$, $b = \frac{1}{4}$. The increase is negligible.

EXAMPLE 4. *Sampling.* In Section 9.1, we proposed on intuitive grounds that the arithmetic mean $\overline{Y}$ of the observations in a sample might be used as an estimate of the population mean $\bar{v}$, and showed that $\overline{Y}$ is an unbiased estimate for $\bar{v}$. The ideas of the present section suggest that one might consider more general estimates, and in particular a weighted average

$$a_1 Y_1 + \ldots + a_s Y_s, \quad \text{where} \quad a_1 + \ldots + a_s = 1.$$

It is easy to check from (9.1.3) that all such weighted averages are unbiased estimates for $\bar{v}$. Which has the smallest variance? The symmetry of the problem, and analogy with Example 2, suggests that $\overline{Y}$ is the optimum weighted average, and this is indeed the case. In fact, the argument of Example 2 would apply here if the sampling were done with replacement.

To prove that $\overline{Y}$ is the weighted average having minimum variance is less simple when the sampling is done without replacement, since then we cannot treat $Y_1, \ldots, Y_s$ as independent random variables. We shall give the proof only for the case $s = 2$.

It was seen earlier ((9.1.4) and Example 7.2.3) that

$$\text{Var}(Y_1) = \text{Var}(Y_2) = \tau^2 \quad \text{and} \quad \text{Cov}(Y_1, Y_2) = -\tau^2/(N-1).$$

To show that $\frac{1}{2}Y_1 + \frac{1}{2}Y_2$ has smaller variance than any other weighted average, consider an arbitrary weighted average and denote by $\frac{1}{2} + a$ the weight it attaches to Y_1. Since the weights must add to 1, it then gives weight $1 - (\frac{1}{2} + a) = \frac{1}{2} - a$ to Y_2, so that it can be written as

$$(\tfrac{1}{2} + a)Y_1 + (\tfrac{1}{2} - a)Y_2.$$

The variance of this estimate is

$$(\tfrac{1}{2} + a)^2 \text{Var}(Y_1) + (\tfrac{1}{2} - a)^2 \text{Var}(Y_2) + 2(\tfrac{1}{2} + a)(\tfrac{1}{2} - a) \text{Cov}(Y_1, Y_2).$$

Multiplying out the coefficients, and substituting the variances and covariance, this becomes

$$(\tfrac{1}{4} + a + a^2)\tau^2 + (\tfrac{1}{4} - a + a^2)\tau^2 - 2(\tfrac{1}{4} - a^2)\tau^2/(N-1).$$

By combining the terms in a^2 and those not involving a at all (the terms in a cancel out), this reduces to

$$\left[\frac{1}{2}\left(1 - \frac{1}{N-1}\right) + 2a^2\left(1 + \frac{1}{N-1}\right)\right]\tau^2.$$

Since a^2 is positive for any a different from 0, this quantity is least when $a = 0$, as was to be proved.

While the inverse variance rule gives the best weighted average of independent estimates, there are estimation problems in which no weighted average does a satisfactory job, and some entirely different type of estimate is preferable. We can best make the point by means of an artificial example.

EXAMPLE 5. *Outliers.* Suppose that in the preceding example all of the v values are the same except two of them, one of which is far above, and the other equally far below, the others. To be specific, put

$$v_1 = \bar{v} + \Delta, \quad v_2 = \bar{v} - \Delta$$
$$v_3 = \bar{v}, \quad v_4 = \bar{v}, \ldots, v_N = \bar{v}$$

where Δ is some large number. We do not know the values of $\bar{v}$ or Δ, and wish to estimate $\bar{v}$ from the v values $Y_1, \ldots, Y_s$ of a sample. This is an extreme version of the fairly common practical situation that most of the v values are clustered near some point, while there are a few *outliers* or *extreme values* that lie far from this point. In our artificial example, v_1 and v_2 are the outliers.

The arithmetic mean of the N v values is of course $\bar{v}$, and the variance τ^2 of a single observation is

$$\tau^2 = \frac{1}{N}\left[(v_1 - \bar{v})^2 + (v_2 - \bar{v})^2 + \ldots + (v_N - \bar{v})^2\right] = \frac{1}{N}(\Delta^2 + \Delta^2) = \frac{2\Delta^2}{N}.$$

If Δ is sufficiently large, τ^2 may be enormous, and since $\text{Var}(\bar{Y}) = \frac{N-s}{N-1} \cdot \frac{1}{s}\tau^2$, $\text{Var}(\bar{Y})$ may also be very large, except in the case $(s = N)$ when the entire population is taken into the sample.

Thus, the best weighted average may be a very poor estimate. And yet, provided s is at least equal to 3, there exists for this problem an unbiased estimate whose variance is zero! Since there is only one outlier in each direction, if outliers appear in a sample of three or more they may be identified and eliminated, leaving only sample values equal to $\bar{v}$, so that the value of $\bar{v}$ can always be exactly determined from a sample of only three items.

Although the example is admittedly highly artificial, it points the way to reasonable methods of handling outliers. If in the sample most values are clustered together, while a few are far away in one direction or the other, it is natural to regard the latter as outliers and ignore them when estimating the central value of the population. A variety of methods has been developed to put this idea into practice. For example, one may arrange the s sample values in increasing order, eliminate the smallest and largest, and then average the $s - 2$ that remain. Or one may eliminate the two smallest and two largest, etc. The most extreme version of this procedure calls for using only the middle one of the s sample values, which is known as the sample *median*. Except under rather artificial assumptions, as in our example above, all of these estimates will be biased estimates for $\bar{v}$, and may be heavily biased. However, they may be very reasonable estimates for the central v value if this is defined in some other way. For example, the sample median is often a reasonable estimate for the (population) median mentioned at the end of Section 5.4.

We have discussed outliers in the context of sampling, but the same issues arise with the model for repeated measurements. If the X's have a common distribution most of which is concentrated near μ, but with a small probability of observing an X value far from μ, some other estimate (such as the sample median) may be a more accurate estimate for μ than is $\overline{X}$. The situation arises when the measurement is subject to being occasionally very far off. These extreme errors are called *gross errors*. They typically arise when from time to time the measuring process is subject to erratic behavior that will produce a measurement drastically different from those routinely obtained. Even a rather small risk of gross errors can then sway the balance in favor of the median.

PROBLEMS

1. One laboratory gives an estimate $\overline{X}$ for μ that is based on averaging m measurements $X_1, \ldots, X_m$; another gives the estimate $\overline{Y}$ based on averaging n measurements $Y_1, \ldots, Y_n$. If the measurements are equally accurate, what weighted average of $\overline{X}$ and $\overline{Y}$ should be used if (i) $m = 3$, $n = 5$; (ii) $m = 5$, $n = 10$; (iii) $m = 3$, $n = 9$?

2. Solve the three parts of the preceding problem if the variance of the X's is σ^2 and that of the Y's is $\tau^2 = 2\sigma^2$.

3. Three laboratories give estimates $\overline{X}$, $\overline{Y}$ and $\overline{Z}$ for μ that are based on averaging observations $X_1, \ldots, X_m$; $Y_1, \ldots, Y_n$; and $Z_1, \ldots, Z_k$ respectively. If the measurements are equally accurate, what weighted average of $\overline{X}$, $\overline{Y}$ and $\overline{Z}$ should be used if (i) $m = 2$, $n = 5$, $k = 10$; (ii) $m = n = 3$, $k = 9$.

4. Solve the two parts of the preceding problem if the variance of the X's is σ^2 that of the Y's is $2\sigma^2$, and that of the Z's is $\frac{1}{2}\sigma^2$.

5. Measurements $X_1, \ldots, X_4$ of μ with variance σ^2 are produced by one method, measurements $Y_1, \ldots, Y_6$ of μ with variance $2\sigma^2$ by another method. What weighted average of the ten measurements should be used for estimating μ?

6. Show that the variance of the optimum estimate, that is, the estimate with weights given by (6), is equal to

$$\frac{1}{(1/\sigma^2) + (1/\tau^2) + (1/\omega^2) + \ldots}.$$

7. Prove directly that in Example 3 with $\tau^2 = 3\sigma^2$ the best weights are $a = \frac{3}{4}$, $b = 1 - a = \frac{1}{4}$, as follows. The variance of $aX + (1 - a)Y$ is $[a^2 + 3(1 - a)^2]\sigma^2$. To minimize this variance, write $a = \frac{3}{4} + \Delta$ and show that $a^2 + 3(1 - a)^2 = \frac{3}{4} + 4\Delta^2$. This quantity takes on its minimum value when $\Delta = 0$ and hence when $a = \frac{3}{4}$.

8. Use the method of the preceding problem to prove that when $\tau^2 = 4\sigma^2$ the best weights are $a = \frac{4}{5}$ and $b = 1 - a = \frac{1}{5}$.

9. In Example 3 with $\tau^2 = k\sigma^2$, how large must the factor k be before the estimate $\frac{1}{2}(X + Y)$ is a worse estimate of μ than X alone?

10. In Problem 1, is it possible for Y to be a better estimate of μ than $(X_1 + \ldots + X_m + Y_1 + \ldots + Y_n)/(m + n)$?

11. In Example 3 with $\tau^2 = k\sigma^2$, how large must k be so that the estimate $\frac{1}{2}(X + Y)$ has twice the variance of the best estimate (6)? [Hint: Use the result of Problem 4.]

12. Two unbiased estimates X and Y for μ have $\mathrm{Var}(X) = 1$, $\mathrm{Var}(Y) = 2$, and $\mathrm{Cov}(X, Y) = 1$. By trial and error find approximately the best weights.

13. (i) In Example 3, we assumed that the separate estimates X and Y are independent. Suppose they are dependent but that their covariance is zero. Would $a = \frac{3}{4}$ still give the best weighted average?

    (ii) More generally, what happens to rule (6) if the assumption of independence is replaced by the assumption that each pair of measurements has zero covariance?

14. In Example 4, suppose that $N = 5$, $v_1 = 0$, $v_2 = -1$, $v_3 = 1$, $v_4 = -\Delta$, $v_5 = \Delta$, and $s = 3$.

    (i) Use (9.1.8) to find $\mathrm{Var}(\overline{Y})$. (Note that your result will involve Δ.)

    (ii) Find the distribution of the median $\widetilde{Y}$ for each of the following values of Δ: $\Delta = 0, .5, 1, 1.5$.

    (iii) For each of the Δ values of part (ii) find $\mathrm{Var}(\widetilde{Y})$.

    (iv) For every value of $\Delta \geqq 0$ determine whether $\overline{Y}$ or $\widetilde{Y}$ is the better estimate.

10.2 EXPERIMENTAL DESIGN

An important aspect of statistical theory is to devise experiments that will yield the maximum amount of useful information for a given cost. A properly designed experiment of modest size will often be more informative than a considerably larger experiment badly laid out. The ramifications of the theory of experimental design are complex, but even at the elementary level of this book the savings made possible by good design can be exemplified.

ÉXAMPLE 1. *Allocation in estimating difference.* Recall Example 9.3.1 where we studied the problem of estimating the difference $\mu' - \mu$ of two quantities each of which may be estimated by averaging repeated unrelated measurements.

As before, let $X_1, \ldots, X_n$ represent the n measurements of the first and $X'_1, \ldots, X'_k$ those of the second quantity. For these $n + k$ random variables we assumed the model for several measurements introduced in Section 9.2, with the additional assumption that $X_1, \ldots, X_n$ have common expectation μ and variance σ^2, and that $X'_1, \ldots, X'_k$ have common expectation μ' and variance σ'^2. Then $\overline{X}' - \overline{X}$ was shown to be an unbiased estimate of the difference $\mu' - \mu$ and its variance was found to be

$$(1) \qquad \mathrm{Var}(\overline{X}' - \overline{X}) = \mathrm{Var}(\overline{X}') + \mathrm{Var}(\overline{X}) = \frac{\sigma'^2}{k} + \frac{\sigma^2}{n}$$

The total number of measurements is $n + k$, which we may suppose to be fixed at $n + k = m$ by the available budget or experimental time. The question now arises as to how one should allocate the total number m of measurements between those made on μ and those made on μ' so that $\mathrm{Var}(\overline{X}' - \overline{X})$ may be as small as possible. The best values are said to constitute the *optimum allocation.*

If measurements of the two quantities μ and μ' are equally precise (i.e., if $\sigma = \sigma'$) then it seems plausible to use *equal allocation*, and to make the same number of measurements on each quantity (i.e., take $n = k$). On the other hand, if the measurements are of different precision, it might be desirable to devote more observations to the quantity which is more difficult to measure. In fact, it turns out that the optimum allocation divides the measurements between the two quantities just in the ratio of their standard deviations,

$$(2) \qquad \frac{k}{n} = \frac{\sigma'}{\sigma}.$$

This elegant formula will be proved in Section 4, but let us now illustrate its use. Suppose that a total of $n + k = 40$ measurements are to be made, and that $\sigma' = 3\sigma$. Then (2) gives $k = 3n$ or $n = 10$, $k = 30$. With these values, (1) shows that the estimate $\overline{X}' - \overline{X}$ has variance $(3\sigma)^2/30 + \sigma^2/10 = \frac{2}{5}\sigma^2$.

To see the meaning of optimum allocation more clearly, let us suppose that in the above problem one were instead to use equal allocation, i.e.,

$n = k = 20$. Then the estimate $\overline{X}' - \overline{X}$ would have variance $(3\sigma)^2/20 + \sigma^2/20 = \frac{1}{2}\sigma^2$. This variance is 25% larger than that achieved with the same total number of observations allocated optimally. To put it another way, with equal allocation one would need 50 observations rather than 40 to achieve the variance $\frac{2}{5}\sigma^2$. (See Problem 2.)

EXAMPLE 2. *Allocation in estimating treatment effect.* Very similar considerations apply to the allocation of a total sample when estimating the effect of a treatment. Recalling design B of Section 9.4, let $Y_1, \ldots, Y_s$ be the responses of a sample of s subjects serving as controls, and $Z_1, \ldots, Z_t$ be the responses of a sample of t different subjects receiving the treatment. Then $\overline{Z} - \overline{Y}$ estimates the difference $\overline{w} - \overline{v}$ attributable to the superiority of the treatment. If the sampling fractions s/N and t/N are small, the variance of the estimate $\overline{Z} - \overline{Y}$ is by (9.4.5) approximately

$$(3) \qquad\qquad \frac{\tau^2}{s} + \frac{\omega^2}{t}$$

where ω^2 and τ^2 are the population variances for the population of N responses when the treatment is and is not applied respectively. If the total number of subjects $s + t$ is fixed, how should it be divided between the two samples? Except for notation, this problem is identical with that of Example 1, and the optimum allocation by (2) is such that $s/t = \tau/\omega$.

The application of this formula requires knowledge of the ratio τ/ω. In most problems, the ratio is near 1, so that equal allocation $s = t$ will often be nearly the best. In particular, if the constant effect model is appropriate (Section 9.4), then $\tau^2 = \omega^2$ (Problem 9.4.3), and equal allocation is optimum.

EXAMPLE 3. *Estimating the effects of two treatments.* Suppose there are two treatments both of which are of interest, whose effects are to be estimated. For example, an agronomist may be interested in the effect on yield of each of two fertilizers, or of a fertilizer and also of planting date. Or a surgeon may wish to know the effect on time of recovery from an operation of a special diet and of certain exercises. Or an educator may be interested in the effect of television instruction and a new kind of textbook on the amount students learn in a calculus course. Let us denote the two treatments by A and B, and assume that they have the constant effects α and β. That is, if a subject has response v when not treated, we assume that if given treatment A his response will be $v + \alpha$, while given treatment B his response will be $v + \beta$. How should an experiment be designed to provide good estimates of α and β?

In each of the four designs to be considered below, we shall give unbiased estimates for α and β; the designs can then be compared in terms of the variances of these estimates. The assumption of constant treatment effect made above implies (Problem 9.4.3) that the variance of response will be the same, say τ^2, regardless of treatment, and that the difference of mean response in two treatment groups will have the variance given by (9.4.6).

Design 1. One might think of simply conducting two separate experiments, one for estimating each of the effects. To illustrate, suppose that a total of 200 subjects is drawn from a population so large that all observations can be treated as unrelated. We could devote 100 of these to each effect. Let us consider first the estimation of α. As in Example 2 it may be shown that if we consider only the problem of estimating α, the 100 subjects should be equally allocated, 50 to the group to receive treatment A and 50 to the control group. Let us denote the mean response of the treated subjects by $\overline{Y}_A$, and the mean response of the control subjects by $\overline{Y}_C$. Then $\overline{Y}_A$ has expectation $\bar{v} + \alpha$ while $\overline{Y}_C$ has expectation $\bar{v}$, so that $\overline{Y}_A - \overline{Y}_C$ is an unbiased estimate for α. By (9.4.6) with $s = t = 50$ this estimate has variance approximately $.04\,\tau^2$. (The correct variance is $.0404\,\tau^2$; see Problem 9.4.7.) The other experiment would be similar and produce an equally good estimate for β.

Design 2. A little thought will show how design 1 can be greatly improved. Each of the two parts of the experiment has a control group whose purpose is to estimate the average untreated response $\bar{v}$. Why do this twice? The estimate $\overline{Y}_C$ for $\bar{v}$ from the first part could also be used in the second part, saving the 50 control observations of that part. Thus, instead of using 200 subjects one could use only 150, divided into two treated groups of 50 each (giving mean responses $\overline{Y}_A$ and $\overline{Y}_B$) and a control group of 50 (with mean response $\overline{Y}_C$). Then $\overline{Y}_A - \overline{Y}_C$ and $\overline{Y}_B - \overline{Y}_C$ are unbiased estimates for α and β respectively, and each estimate has the same variance $.04\,\tau^2$ as with design 1. We have saved one-fourth of the observations without loss of accuracy.

Design 3. Is it best to allocate the 150 subjects of design 2 into three equal groups? When estimating a single treatment effect equal allocation was correct, but now things are different. Since $\overline{Y}_C$ is used twice, it is perhaps more important to have it accurate than to have $\overline{Y}_A$ and $\overline{Y}_B$ accurate. A little trial and error will show that the allocation of 60 subjects to the control group and 43 to each treatment group will (by 9.4.6) give estimates of variance $(\frac{1}{60} + \frac{1}{43})\tau^2 = .0399\,\tau^2$. This estimate, very slightly better than that of design 2, requires only $60 + 43 + 43 = 146$ subjects instead of 150. The reduction is small, but in some experiments a saving of even 4 subjects may be worthwhile. (If more than two effects are to be estimated, still larger savings are possible by using a control group somewhat bigger than the treatment groups; see Problem 6.)

Design 4. Under an additional assumption, a more dramatic saving is possible. Suppose that the treatments A and B are such that both may be applied at the same time, and that their operations on subjects do not interfere with each other. One might then be willing to make the assumption of *additive effects:*

(4)   a subject with untreated response v would have response
        $v + \alpha + \beta$ if both treatments A and B are applied.

We shall examine the reasonableness of this assumption below, but first let us see what its consequences are for the design.

The idea suggests itself of using four groups, the three of design 2 plus a fourth group of subjects receiving both treatments. Let us denote the mean response in this group by $\overline{Y}_{AB}$. For illustration suppose that each group is assigned 25 subjects, or 100 subjects in all. Then $\overline{Y}_C$, $\overline{Y}_A$, $\overline{Y}_B$, and $\overline{Y}_{AB}$ have expected values $E(\overline{Y}_C) = \bar{v}$, $E(\overline{Y}_A) = \bar{v} + \alpha$, $E(\overline{Y}_B) = \bar{v} + \beta$, $E(\overline{Y}_{AB}) = \bar{v} + \alpha + \beta$. Thus $\overline{Y}_A - \overline{Y}_C$ and $\overline{Y}_{AB} - \overline{Y}_B$ are both unbiased estimates for α, and so therefore is their average

(5) $\frac{1}{2}[(\overline{Y}_{AB} + \overline{Y}_A) - (\overline{Y}_B + \overline{Y}_C)].$

By Problem 13, the variance of this estimate is approximately $.04\,\tau^2$. Similarly

(6) $\frac{1}{2}[(\overline{Y}_{AB} + \overline{Y}_B) - (\overline{Y}_A + \overline{Y}_C)]$

is an unbiased estimate for β with variance $.04\,\tau^2$. These estimates are just as good as those of the earlier designs, but require only a total of 100 observations, instead of the 146 required by design 3!

This is the simplest example of a *factorial* design. In effect, we are now using all 100 observations to estimate α, and at the same time are using all 100 observations to estimate β, and are getting estimates just as precise as we would get if 100 observations had been made for doing just one of the two jobs. The basic idea of factorial experimentation, introduced mainly by R. A. Fisher, is capable of great elaboration to deal with a wide variety of experimental problems.

Let us now return to examine the meaning of the assumption (4), that the effects are additive. Is it reasonable to suppose that the increased response elicited when two treatments are applied simultaneously is just the sum of the increases that they would separately produce? For example, if one fertilizer will increase yield by three pounds and another will increase yield by two pounds, may we expect both together to give a five pound increase? Not necessarily; it may be that the first fertilizer supplies all the nutrients that the plant needs, in which case the addition of the second may have little or no effect. As another example, calculus instruction by television may reduce the students' test scores if it is used with a conventional textbook, but perhaps be beneficial if used with a novel textbook written with television in mind. Medical practice often calls for the joint use of two *synergistic* drugs, which together have several times the effect of either used separately. In all of these examples the effects do not simply add; rather, the presence of one treatment will influence the action of the other. In such cases the treatments are said to *interact*.

When the treatments A and B interact, the effect of either of them is not a constant, but depends on whether or not the other is present. As before, let α denote the effect of A when B is absent, and denote by α' the effect of A when B is present. Then $\overline{Y}_A - \overline{Y}_C$ is an unbiased estimate for α, while $\overline{Y}_{AB} - \overline{Y}_B$ is an unbiased estimate for α'. The estimate (5) therefore has the expected value $\frac{1}{2}(\alpha + \alpha')$, which is known as the *main effect* of A, and of which we may think as the average effect of A. Similarly, β and β' are the effects of B in the absence and presence of A, and (6) estimates the main effect of B, which is $\frac{1}{2}(\beta + \beta')$.

Since α' is the effect of A in the presence of B and α the effect of A in the absence of B, the quantity $\alpha' - \alpha$ indicates how much the presence of B improves the effect of A, and hence measures the strength of the *interaction* of A and B. (In advanced texts, interaction is defined for technical reasons as $\frac{1}{2}(\alpha' - \alpha)$ by some authors and as $\frac{1}{4}(\alpha' - \alpha)$ by others.) Alternatively, the interaction of A and B could be defined in terms of $\beta' - \beta$, which measures how much the presence of A improves the effect of B. It turns out (Problem 9) that these two measures of interaction coincide, and that they are zero if and only if the additivity assumption (4) holds. Combining the unbiased estimates of α and α' given above, it is seen that the quantity $\alpha' - \alpha = \beta' - \beta$ is estimated by

$$(7) \qquad \overline{Y}_{AB} - \overline{Y}_B - \overline{Y}_A + \overline{Y}_C.$$

It is clear that design 4 is better than any of the others if there is no interaction, as it provides equally good estimates at lower cost. But if interaction is present, the situation is not clearcut. In some cases one is equally interested in all four effects α, α', β, and β', wanting to know the effect of each treatment both with and without the other. In such cases, the factorial design is called for, as the others do not provide estimates for all four parameters. In other cases one is mainly interested in knowing the average effects, or in estimating the magnitude of interaction, and again the factorial design is indicated. But in some problems, where one is considering the use of one treatment or the other but not both, design 3 will be better than 4, as it provides estimates of α and β of given precision at lower cost. To provide estimates of α and β with variance $.04\ \tau^2$, when there is interaction, design 4 will require 50 subjects in each group or a total of 200, 54 more than required by design 3. In this case, as in all statistical problems, the correct answer depends very heavily on the circumstances.

PROBLEMS

1. If $\sigma'^2 = 9\sigma^2$, determine the best division of (i) 40, (ii) 60, (iii) 100 observations in Example 1.

2. Show that in Example 1 with $\sigma' = 3\sigma$ and equal allocation, one requires 50 observations to achieve variance $\frac{2}{5}\sigma^2$.

3. Prove directly that in Example 1 with $\sigma' = 3\sigma$ the best allocation of $n + k$ observations is $n = 10$, $k = 30$, as follows. Since with this allocation the variance of the estimate is $\frac{2}{5}\sigma^2$ while with allocation n and $k = 40 - n$ it is $\sigma^2\left[\dfrac{1}{n} + \dfrac{9}{40 - n}\right]$, it is only necessary to prove that $\dfrac{1}{n} + \dfrac{9}{40 - n} \geqq \dfrac{2}{5}$. Multiplying through by $5n(40 - n)$, this is seen to be equivalent (for $0 < n < 40$) to the true inequality $n^2 - 20n + 100 = (n - 10)^2 \geqq 0$.

4. The observations $X_1, X_2, \ldots$ cost \$1 each, while the observations $X_1', X_2', \ldots$ cost \$2 each. A budget of \$20 is available for observation. What allocation of this budget to the two kinds of measurement will give the smallest variance to $\bar{X}' - \bar{X}$ if the two kinds of measurement are equally precise ($\sigma'^2 = \sigma^2$)?

5. Suppose that the observations $X_1, X_2, \ldots$ cost c dollars each and have variance σ^2, while the observations $X_1', X_2', \ldots$ cost c' dollars each and have variance σ'^2. Use result (2) to show that with a fixed total budget the best allocation is such that

$$\frac{k}{n} = \frac{\sigma'\sqrt{c}}{\sigma\sqrt{c'}}.$$

[Hint: We must minimize $\mathrm{Var}(\bar{X}' - \bar{X}) = \dfrac{\sigma'^2}{k} + \dfrac{\sigma^2}{n} = \dfrac{(\sigma\sqrt{c})^2}{cn} + \dfrac{(\sigma'\sqrt{c'})^2}{c'k}$, with $cn + c'k$ being fixed.]

6. Twenty experimental subjects are to be used to estimate the effects of three treatments. Suppose that n subjects are to be allocated to each of the three treatments and that k are to serve as controls so that $3n + k = 20$ and that the division into four groups is at random. Assuming a constant effect model so that all observations may be taken to have the same variance and using (9.4.6), find (by trial and error) the value of k that minimizes the variances of $\bar{Y}_1 - \bar{Y}_c$, $\bar{Y}_2 - \bar{Y}_c$ and $\bar{Y}_3 - \bar{Y}_c$ where $\bar{Y}_c$ is the average response of the k controls, and $\bar{Y}_1, \bar{Y}_2, \bar{Y}_3$ are the average responses of the three treatment samples.

7. Suppose in Example 1 that the total number $n + k$ of observations is fixed at m. Show that the optimum allocation (2) implies

$$n = m\sigma/(\sigma + \sigma'), \quad k = m\sigma'/(\sigma + \sigma').$$

8. Use the result of the preceding problem to show that when the total number $n + k$ of observations is fixed at m, the minimum value of $\mathrm{Var}(\bar{X}' - \bar{X})$ is $(\sigma + \sigma')^2/m$.

9. Show that the quantities α, β, α', β' in design 4 of Example 3 are related by the equation $\alpha + \beta' = \alpha' + \beta$. [Hint: α' is the expected difference in response when A and B are applied and when B is applied alone, and the other quantities are defined analogously.]

10. In design 4 of Example 3, if v is the untreated response of a subject, show that the response is

$v + \alpha$ when A alone is applied,
$v + \beta$ when B alone is applied,
$v + \alpha' + \beta = v + \beta' + \alpha$ when both A and B are applied.

11. Use Problem 10 to show that the interaction of A and B is zero if and only if the additivity assumption (4) holds.

12. Let A, B, C, D be four objects whose weights α, β, γ, δ we wish to determine. In any one weighing, we can weigh any sum or difference of the objects by placing them either on the same or on opposite sides of a chemical balance. Suppose that the variance of any individual weighing is σ^2 and that a measurement model is appropriate. Consider the following three designs:

Design 1 (16 weighings). Object A is weighed four times, yielding readings X_1, X_2, X_3, X_4 with

$$E(X_1) = E(X_2) = E(X_3) = E(X_4) = \alpha.$$

Similarly each of the objects B, C, D is weighed four times.

Design 2 (8 weighings). Two weighings Y_1, Y_2 are obtained of $\alpha + \beta$ and two weighings Y_3, Y_4 of $\alpha - \beta$ so that

$$E(Y_1) = E(Y_2) = \alpha + \beta; \quad E(Y_3) = E(Y_4) = \alpha - \beta.$$

Similarly two weighings each are obtained of $\gamma + \delta$ and of $\gamma - \delta$.

Design 3 (4 weighings). Four weighings Z_1, Z_2, Z_3, Z_4 are obtained such that

$$E(Z_1) = \alpha + \beta + \gamma + \delta, \quad E(Z_2) = \alpha - \beta + \gamma - \delta$$
$$E(Z_3) = \alpha + \beta - \gamma - \delta, \quad E(Z_4) = \alpha - \beta - \gamma + \delta.$$

For all three designs, find unbiased estimates of α, β, γ, δ and compare their variances.

13. Suppose that in design 4, $4s$ subjects are randomly divided into four treatment groups of s each. Using the fact that each term in the numerator of (5) has variance $(N - s)\tau^2/(N - 1)s$, and Problems 9.4.6(i) and 9.4.3(i) show that the variance of (5) is $N\tau^2/(N - 1)s$.

10.3 STRATIFIED SAMPLING

In Section 9.1 we considered a method of obtaining an estimate for the average value $\bar{v}$ of some quantity v in a population, by drawing a simple random sample of s items from the population, observing their v values $Y_1, \ldots, Y_s$, and then computing the average $\bar{Y}$. As a further illustration of the concept of experimental design, we shall now discuss an alternative design that sometimes produces an equally good estimate for $\bar{v}$ with far fewer observations.

It is often possible to divide the population into a number of subpopulations, called *strata*, within each of which the value of v is more nearly constant than it is in the population as a whole. For example, a city may be divided into districts within each of which the expenditure on housing is nearly the same for all households; or a state may be divided into counties or blocks of counties within each of which the farms have a nearly constant percentage of their area planted to wheat. When such a stratification is possible, marked savings in the cost of a sample survey can usually be achieved by drawing unrelated small samples from each stratum separately instead of taking a single large sample from the entire population. This procedure is known as *stratified sampling*. (Examples 3.1.1, 4.2.3, 5.5.3.)

The possibility of saving through stratification can be seen intuitively by considering the extreme case that the quantity v is exactly constant within each stratum. Then a single item drawn from each stratum will enable us to determine the population average value $\bar{v}$ exactly, while a much larger simple random sample from the entire population would not provide a perfect estimate. This suggests that even in less extreme cases a stratified sample might be advantageous.

Let us suppose that there are L strata numbered from 1 to L, containing $N_1, \ldots, N_L$ items respectively so that

$$N_1 + \ldots + N_L = N.$$

We shall assume that we know $N_1, \ldots, N_L$; in practice, this places a severe limitation on the kinds of stratification that can be used. Let the average value of the quantity v with which we are concerned be $\bar{v}_1$ in the first stratum, $\bar{v}_2$ in the second, ... Then $\bar{v} = (N_1\bar{v}_1 + \ldots + N_L\bar{v}_L)/N$, or

$$(1) \qquad \bar{v} = \frac{N_1}{N}\bar{v}_1 + \ldots + \frac{N_L}{N}\bar{v}_L.$$

Suppose that from the first stratum a random sample of s_1 items is drawn, from the second stratum a random of s_2 items, . . . , and denote by $\overline{Y}_1$, $\overline{Y}_2, \ldots$ the average v values in these samples. Then by (9.1.6)

$$(2) \qquad E(\overline{Y}_1) = \bar{v}_1, \quad \ldots, \quad E(\overline{Y}_L) = \bar{v}_L$$

and by (9.1.8)

$$(3) \qquad \mathrm{Var}(\overline{Y}_1) = \frac{N_1 - s_1}{N_1 - 1} \cdot \frac{\tau_1{}^2}{s_1}, \quad \ldots, \quad \mathrm{Var}(\overline{Y}_L) = \frac{N_L - s_L}{N_L - 1} \cdot \frac{\tau_L{}^2}{s_L}$$

where $\tau_1{}^2$ is the variance of a single observation from the first stratum, . . . , $\tau_L{}^2$ the variance of a single observation from the Lth stratum. If our attempt was successful to define strata within which v is nearly constant, the strata variances $\tau_1{}^2, \ldots, \tau_L{}^2$ will be small compared with the variance τ^2 of v within the entire population.

The separate estimates $\overline{Y}_1$ for $\overline{v}_1, \ldots, \overline{Y}_L$ for $\overline{v}_L$ may now be combined to produce an estimate Y^* for $\overline{v}$ given by

(4) $$Y^* = \frac{N_1}{N} \cdot \overline{Y}_1 + \ldots + \frac{N_L}{N} \cdot \overline{Y}_L.$$

This is a weighted average of the means of the samples from the various strata, the weights being proportional to the stratum sizes. Unless the stratum sizes $N_1, \ldots, N_L$ were known, the estimate Y^* could not in practice be computed.

By taking expectations on both sides of (4), the estimate Y^* is seen to be unbiased. Furthermore, if $\overline{v}_1, \ldots, \overline{v}_L$ are completely unknown, as we assume them to be, Y^* is the only weighted average of $\overline{Y}_1, \ldots, \overline{Y}_L$ that is unbiased for estimating $\overline{v}$ (Problem 3). If the drawings from the different strata are unrelated, a product model with L factors is appropriate for the experiment as a whole, and $\overline{Y}_1, \ldots, \overline{Y}_L$ are defined on different factors of this product. Hence the variance of Y^* is

$$\mathrm{Var}(Y^*) = \frac{N_1{}^2}{N^2} \mathrm{Var}(\overline{Y}_1) + \ldots + \frac{N_L{}^2}{N^2} \mathrm{Var}(\overline{Y}_L).$$

To simplify the calculations, we shall now suppose that the finite correction factors can be neglected in the variances of $\overline{Y}_1, \ldots, \overline{Y}_L$—which is reasonable if the sampling fractions $s_1/N_1, \ldots, s_L/N_L$ are small—so that

$$\mathrm{Var}(\overline{Y}_1) = \tau_1{}^2/s_1, \quad \ldots, \quad \mathrm{Var}(\overline{Y}_L) = \tau_L{}^2/s_L.$$

Then we have

(5) $$\mathrm{Var}(Y^*) = [N_1{}^2\tau_1{}^2/s_1 + \ldots + N_L{}^2\tau_L{}^2/s_L]/N^2.$$

(For the exact formula see Problem 4.)

This formula shows that the accuracy of Y^* will depend on the numbers $s_1, \ldots, s_L$. If the total sample size

$$s = s_1 + \ldots + s_L$$

is fixed, for example by the available funds, the problem arises of deciding how to allocate the sample among the strata. Perhaps the most natural procedure is to divide the sample among the strata in proportion to the sizes of the strata. This implies (Problem 5) that

(6) $$s_1 = s \cdot \frac{N_1}{N}, \quad \ldots, \quad s_L = s \cdot \frac{N_L}{N} \qquad \text{(proportional allocation)}.$$

(The number of observations from each stratum must of course be an integer; in addition, there should be at least one observation from each stratum so that the quantities $\overline{Y}_1, \ldots, \overline{Y}_L$ can be computed. Therefore in practice one cannot use the exact values for $s_1, s_2, \ldots$ given by (6); instead one might take positive integral values, adding up to s, and as close

to (6) as possible.   To simply further calculations, however, we shall ignore
these small rounding corrections.)

When the values (6) are substituted in (5), we obtain the variance for
proportional allocation

(7) $$\mathrm{Var}(Y^*) = \left[\frac{N_1}{N} \tau_1{}^2 + \ldots + \frac{N_L}{N} \tau_L{}^2 \right] \Big/ s.$$

We shall show at the end of the section that this is always less than or equal
to the variance of the estimate $\overline{Y}$ available from a single random sample
of size s.   Since by (9.1.7) the variance of the average $\overline{Y}$ of the values in a
simple random sample of size s is approximately given by

(8) $$\mathrm{Var}(\overline{Y}) = \tau^2/s,$$

we see that the estimate Y^* based on stratified sampling with proportional
allocation and the same total sample size s will be substantially better
than $\overline{Y}$ provided

$$\frac{N_1}{N} \tau_1{}^2 + \ldots + \frac{N_L}{N} \tau_L{}^2$$

is substantially smaller than

$$\tau^2 = \frac{N_1}{N} \tau^2 + \ldots + \frac{N_L}{N} \tau^2;$$

i.e., when we have succeeded in finding strata that are considerably less
variable than the population as a whole.   We shall however show at the
end of the section that (7) is in any case never larger than (8).

EXAMPLE 1.   *Smoking.*   The college physician in a men's college wishes
to estimate the average daily consumption $\bar{v}$ of cigarettes by undergraduates
of his college.   He can obtain such an estimate by drawing a random
sample of, say, $s = 100$ from the registrar's list of undergraduates, and
then interview each of the selected students to determine the number v of
cigarettes he smokes per day.   If then the standard deviation of v in the
population is for example $\tau = 6$, the estimate for $\bar{v}$ obtained by averaging
the 100 consumption figures would have variance $6^2/100 = .36$.

Let us suppose that cigarette smoking tends to increase with age of
student, so that the freshmen smoke relatively little compared with the
seniors.   This means that the students within any given class tend to
resemble each other in this regard, and hence that the variance within any
one class would be smaller than the variance within the entire under-
graduate population.   To get an idea of the numerical effect of stratified
sampling with proportional allocation, suppose that the actual standard
deviations within the four classes are as shown in column 3 of the following
table.

Class	Number of students	Standard deviation
Freshman	3100	2
Sophomore	2700	3
Junior	2300	4
Senior	1900	5
Total	10,000	6

The variance of the estimate Y^* based on proportional allocation given by (7) can now be computed, using the class sizes given in the table:
$$\text{Var}(Y^*) = [(.31)4 + (.27)9 + (.23)16 + (.19)25]/s = 12.1/s.$$
We can therefore attain the variance .36, provided by a sample of 100 drawn at random from the whole school, by using a stratified sample whose size is given by the equation $12.1/s = .36$, that is, $s = 33.61$. The allocation of a sample of 34 as nearly as possible in proportion to class size is as follows:

Freshman	Sophomore	Junior	Senior
10	9	8	7

The variance of Y^* with this allocation, as computed from (7), is in fact .346, even slightly better than .36. Through stratification, we have thus obtained a slightly better estimate with only about one third as many observations!

In some cases, even if a simple random sample has been drawn, it will be possible to compute an estimate nearly as good as the estimate Y^* based on stratified sampling with proportional allocation. After a simple random sample is taken from the whole population, the items in the sample may be classified according to the strata from which they happen to have come, and one may compute the average values, say $\tilde{Y}_1, \ldots, \tilde{Y}_L$, of the observations from the various strata. These may be combined to produce the estimate

$$\tilde{Y} = \frac{N_1}{N} \tilde{Y}_1 + \ldots + \frac{N_L}{N} \tilde{Y}_L$$

for $\bar{v}$. This procedure is of course possible only if there happened to be at least one observation from each stratum, which is very likely to happen if the sample size s is large. It can be shown that, with large s, the estimate $\tilde{Y}$ is nearly as accurate as the estimate Y^* based on proportional allocation of the same total sample size. However, Y^* is always at least somewhat better, and it also exists (and is unbiased) in all cases since one would always allocate at least one observation to each stratum.

So far, we have considered only a stratified sample with proportional allocation, but perhaps further improvements could be obtained by a different scheme of allocation. Intuitively, it would seem reasonable to take into account not only the size of the strata but also their variability. We shall show in the next section that the variance (5) is minimized, for a given total sample s, by making $s_1, \ldots, s_L$

proportional to $N_1\tau_1, \ldots, N_{LTL}$, so that

(9) $\qquad s_1 = s \dfrac{N_1\tau_1}{N_1\tau_1 + \ldots + N_{LTL}}, \quad \ldots, \quad s_L = s \dfrac{N_{LTL}}{N_1\tau_1 + \ldots + N_{LTL}}.$

With this optimum or Tschuprow-Neyman allocation, the variance (5) reduces to

(10) $\qquad \mathrm{Var}(Y^*) = \dfrac{(N_1\tau_1 + \ldots + N_{LTL})^2}{s N^2} = \dfrac{1}{s}\left(\dfrac{N_1\tau_1 + \ldots + N_{LTL}}{N}\right)^2.$

The table below shows for Example 1 the division of a total sample of 34 according to optimal allocation, and for comparison that according to proportional allocation considered earlier.

	Freshman	Sophomore	Junior	Senior
Optimum allocation	6	8	10	10
Proportional allocation	10	9	8	7

In this example, optimum allocation gives an estimate with variance .321, somewhat better than the .346 of proportional allocation.

The use of formula (9) for optimum allocation requires knowledge of the stratum standard deviations $\tau_1, \ldots, \tau_L$. If these are radically different, optimum allocation may give a substantial improvement over proportional allocation. However, as the numerical illustration suggests, the differences between $\tau_1, \ldots, \tau_L$ must be extreme before the gain is appreciable. Since this seldom happens, and since knowledge of $\tau_1, \ldots, \tau_L$ is often uncertain, proportional allocation is frequently used in practice.

The idea of stratification is exceedingly useful also in connection with experiments for estimating the average effect $\bar{\Delta}$ of a treatment in a population of subjects. Recall design B of Section 9.4, where $\bar{Y}$ is the mean response of a sample of s untreated subjects, $\bar{Z}$ is the mean response of a distinct sample of t treated subjects, and $\bar{Z} - \bar{Y}$ is an unbiased estimate of $\bar{\Delta}$ with (approximately)

$$\mathrm{Var}(\bar{Z} - \bar{Y}) = \frac{\tau^2}{s} + \frac{\omega^2}{t}.$$

In many cases, the responses of the subjects in the population differ considerably among each other, so that τ^2 and ω^2 are large, and hence large sample sizes s and t are required to produce an accurate estimate with this design.

However, the experimenter is frequently able to divide the subjects into subpopulations that are relatively homogeneous in responsiveness. For example, the plots to be used in an agricultural experiment for comparing two varieties of wheat may be grouped into those of high fertility, medium fertility, and low fertility. Again, in the clinical trial of a new drug, the patients may be divided according to age, sex, and grade of disease into subgroups that are similar with regard to their time required for recovery.

These groups or subpopulations play the role of the strata. As before, let L denote the number of groups and suppose that they contain

$N_1, \ldots, N_L$ subjects. Let $\tau_1^2, \ldots, \tau_L^2$ denote the variances of the v values within the L groups, while $\omega_1^2, \ldots, \omega_L^2$ denote the corresponding variances of the w values. If the grouping has been successful, these values will be much smaller than τ^2 and ω^2. As a modification of design B, suppose that distinct samples of size s_1 and t_1 are drawn from the first group providing an estimate $\overline{Z}_1 - \overline{Y}_1$ for the average effect $\overline{\Delta}_1$ of the treatment in the first group, and so forth. These estimates may, in analogy with (4), be combined to produce the estimate

$$(11) \qquad \frac{N_1}{N}(\overline{Z}_1 - \overline{Y}_1) + \ldots + \frac{N_L}{N}(\overline{Z}_L - \overline{Y}_L)$$

for $\overline{\Delta}$. This estimate is unbiased (Problem 7), and its variance may easily be computed (Problem 8). If the grouping has produced homogeneity, the estimate (11) will be much more precise than the estimate $\overline{Z} - \overline{Y}$ of design B applied to the whole population with the same total number of subjects.

To conclude this section, let us prove that the variance (8) of the estimate $\overline{Y}$ based on a single sample of size s is always greater than or equal to the variance (7) of the estimate Y^* based on proportional allocation. This is in fact an immediate consequence of the identity (see Problem 9)

$$(12) \qquad \mathrm{Var}(\overline{Y}) = \mathrm{Var}(Y^*) + \frac{1}{s}\left[\frac{N_1}{N}(\overline{v}_1 - \overline{v})^2 + \ldots + \frac{N_L}{N}(\overline{v}_L - \overline{v})^2\right]$$

where $\overline{v}_1, \ldots, \overline{v}_L$ are the strata means and where $\overline{v}$ is given by (1), since the second term on the right side of (12) cannot be negative.

Equation (12) throws some light on the effect of different stratifications. Since the left side of (12) is fixed, that is, does not depend on how the population is stratified, any decrease in $\mathrm{Var}(Y^*)$ results in a corresponding increase of the quantity

$$(13) \qquad \frac{N_1}{N}(\overline{v}_1 - \overline{v})^2 + \ldots + \frac{N_L}{N}(\overline{v}_L - \overline{v})^2$$

and vice versa. Now it is clear from (7) that $\mathrm{Var}(Y^*)$ will be small (and hence stratification successful) if the strata variances $\tau_1^2, \ldots, \tau_L^2$ are small, that is, if the strata are homogeneous. On the other hand, the quantity (13) will be large if the strata means $\overline{v}_1, \ldots, \overline{v}_L$ differ widely from the over-all mean $\overline{v}$, and hence if they differ widely among each other. The expression (13) therefore shows that to obtain homogeneity within strata one must have wide differences between the strata.

PROBLEMS

1. What stratification would you suggest for estimating the average weight of children in a public grammar school?

2. A population of ten subjects is divided into two strata of five subjects each, having these v values:

Stratum 1: 16, 13, 24, 17, 15
Stratum 2: 34, 22, 30, 28, 31

(i) Find the variance of the estimate (4) of $\bar{v}$ if $s_1 = s_2 = 1$.

(ii) How large a sample would be required to produce as small a variance if simple random sampling from the whole population were used?

3. Let $E(\bar{Y}_1) = \bar{v}_1, \ldots, E(\bar{Y}_L) = \bar{v}_L$ where $\bar{v}_1, \ldots, \bar{v}_L$ are completely unknown. Show that (4) is the only weighted average which provides an unbiased estimate of (1). [Hint: Write down the expected value of an arbitrary weighted average $a_1\bar{Y}_1 + \ldots + a_L\bar{Y}_L$ and equate this to (1). Then suppose for example that $\bar{v}_1 = 1$, $\bar{v}_2 = \ldots = \bar{v}_L = 0$.]

4. Show that the variance of the estimate (4) is

$$\frac{N_1^2}{N^2} \frac{N_1 - s_1}{N_1 - 1} \frac{\tau_1^2}{s_1} + \ldots + \frac{N_L^2}{N^2} \frac{N_L - s_L}{N_L - 1} \frac{\tau_L^2}{s_L}.$$

5. Show that the numbers $s_1, \ldots, s_L$ with $s_1 + \ldots + s_L = s$ that are proportional to the numbers $N_1, \ldots, N_L$ with $N_1 + \ldots + N_L = N$ are given by (6). [Hint: Proportionality means the existence of a number k such that $s_1 = kN_1, \ldots, s_L = kN_L$.]

6. Show that the allocations given in Example 1 as "proportional" and "optimal" are as close to these properties as possible.

7. Prove that (11) is an unbiased estimate of $\bar{\Delta}$.

8. Show that if $s_1/N_1, \ldots, s_L/N_L$ and $t_1/N_1, \ldots, t_L/N_L$ are small, then the variance of (11) is approximately

$$\frac{N_1^2}{N^2}\left(\frac{\tau_1^2}{s_1} + \frac{\omega_1^2}{t_1}\right) + \ldots + \frac{N_L^2}{N^2}\left(\frac{\tau_L^2}{s_L} + \frac{\omega_L^2}{t_L}\right).$$

9. Prove equation (12). [Hint: Denote the v values in the first stratum by $v_1, \ldots, v_{N_1}$. The contribution of the first stratum to τ^2 is then

$$\frac{1}{N}\left[(v_1 - \bar{v})^2 + \ldots + (v_{N_1} - \bar{v})^2\right] = \frac{N_1}{N}\tau_1^2 + \frac{N_1}{N}(\bar{v}_1 - \bar{v})^2$$

where the equality follows from (5.7.18). Adding this contribution to the corresponding ones from the other strata gives (12).]

10. Compare proportional and optimum allocation in the special case when all stratum variances are equal, $\tau_1^2 = \tau_2^2 = \ldots = \tau_L^2$.

10.4 AN INEQUALITY AND ITS APPLICATIONS

In the present section we shall prove three results stated earlier: (a) that the best weighted average of several independent unbiased estimates uses weights proportional to the reciprocals of the variances (1.6); (b) that the optimum allocation for estimating a difference is given by (2.2); (c) that the best allocation in stratified sampling is the Tschuprow-Neyman allocation (3.9). The reader willing to accept these theorems without proof may omit this section. All three proofs rest on the following inequality: for any positive numbers $a, b, c, \ldots$ and $A, B, C, \ldots$

(1)
$$\frac{(a + b + c + \ldots)^2}{A + B + C + \ldots} \leqq \frac{a^2}{A} + \frac{b^2}{B} + \frac{c^2}{C} + \ldots.$$

To prove (1) we recall from (5.6.2) that the variance of any random variable Z is given by $E(Z^2) - [E(Z)]^2$. Since a variance is never negative, it follows that for any random variable Z

(2)
$$[E(Z)]^2 \leqq E(Z^2).$$

Let us now consider a random variable Z with the following probability distribution. To simplify the notation, we write S for $A + B + C + \ldots$.

Value taken by Z	a/A b/B ...
Probability of this value	A/S B/S ...

Then
$$E(Z) = \frac{a}{A} \cdot \frac{A}{S} + \frac{b}{B} \cdot \frac{B}{S} + \ldots = \frac{a + b + c + \ldots}{S}.$$

Similarly,
$$E(Z^2) = \frac{a^2}{A^2} \frac{A}{S} + \frac{b^2}{B^2} \frac{B}{S} + \ldots = \left(\frac{a^2}{A} + \frac{b^2}{B} + \ldots \right) \frac{1}{S}.$$

Substituting these expressions in (2) and cancelling a common factor $1/(A + B + C + \ldots)$ from both sides gives exactly (1).

As the first application of (1), recall from Example 1.1 the problem of finding those weights $a, b, \ldots$ with

(3)
$$a + b + \ldots = 1$$

which minimize the variance of $aX + bY + \ldots$, where $X, Y, \ldots$ represent unrelated measurements of μ with variances $\text{Var}(X) = \sigma^2$, $\text{Var}(Y) = \tau^2, \ldots$. If we define

$$A = 1/\sigma^2, \quad B = 1/\tau^2, \ldots$$

then by (9.2.4)

(4)
$$\text{Var}(aX + bY + \ldots) = a^2/A + b^2/B + \ldots.$$

We wish to show that this variance is smallest for the values $a, b, \ldots$ given by (1.6), i.e., for $a = A/S$, $b = B/S$, ... With these weights, (4) becomes

(5)
$$\frac{A}{(A + B + \ldots)^2} + \frac{B}{(A + B + \ldots)^2} + \ldots = \frac{1}{A + B + \ldots}.$$

Using (3), the fact that (4) is always at least equal to (5) follows from (1).

Let us consider next the problem of optimum allocation discussed in Example 2.1. The variance of the unbiased estimate $\bar{X}' - \bar{X}$ of $\mu' - \mu$ is

$$(6) \qquad \mathrm{Var}(\overline{X}' - \overline{X}) = \frac{\sigma'^2}{k} + \frac{\sigma^2}{n}$$

If the total sample size $k + n$ is fixed, equal to m say, the problem is to find those values of n and k with $n + k = m$, for which (6) is as small as possible. We shall show that the optimum allocation consists in devoting to each of the two quantities a number of measurements proportional to the standard deviation of those measurements, that is, by putting

$$(7) \qquad n = \frac{\sigma}{\sigma + \sigma'} m \quad \text{and} \quad k = \frac{\sigma'}{\sigma + \sigma'} m.$$

When these values are used for n and k, the variance of the estimate reduces to $(\sigma + \sigma')^2/m$. To prove that (7) is the optimum allocation we must show that any other allocation will make the variance larger. That is, we must show

$$\frac{(\sigma + \sigma')^2}{n + k} \leqq \frac{\sigma^2}{n} + \frac{\sigma'^2}{k}$$

whatever be the positive numbers n and k. This follows from inequality (1) with σ, σ', n, k in place of a, b, A, B.

We conclude by proving that the best allocation in stratified sampling is provided by the Tschuprow-Neyman allocation, defined by (3.9). To this end we must show that the variance (3.9) never exceeds the variance (3.5) of an arbitrary allocation of the same total sample size, i.e., that

$$\frac{1}{s} \left(\frac{N_1\tau_1 + N_2\tau_2 + \dots}{N} \right)^2 \leqq \left(\frac{N_1^2\tau_1^2}{s_1} + \frac{N_2^2\tau_2^2}{s_2} + \dots \right) \frac{1}{N^2}.$$

If in (1) we set $a = N_1\tau_1$, $b = N_2\tau_2$, $\dots$ and $A = s_1$, $B = s_2$, $\dots$ and then divide by N^2, the result follows.

From the fact (Problem 5.7.16) that $\mathrm{Var}\, Z > 0$ unless Z is constant, it follows that the left side of (1) is strictly smaller than the right side unless $a/A = b/B = \dots$. From this it can be shown that in the three applications of (1) considered above, the optimum solution is always unique.

CHAPTER 11

TESTS OF SIGNIFICANCE

11.1 FIRST CONCEPTS OF TESTING

By a statistical hypothesis we mean a statement about the way a random variable is distributed. In this section we shall explain how the observed value of the random variable can be used to test a statistical hypothesis.

The use of a random experiment to test a statistical hypothesis is an extension of the method of testing scientific hypotheses by means of non-random experiments. The following experiment, apocryphally attributed to Galileo, provides a simple illustration of this method. In order to test the hypothesis that two objects of different weights will fall at the same speed, Galileo is reported to have dropped from the leaning tower of Pisa one large and one small cannon ball. His hypothesis was supported by the observation that the cannon balls did indeed strike at nearly the same instant.

To see the general structure of this method, suppose that an observable quantity x according to the hypothesis should have the value x_0, while alternative theories predict values different from x_0. (In the example, x would be the difference in the times of impact of the two cannon balls, and x_0 would be zero.) If on performing the experiment we find that x is different from x_0, we must reject the hypothesis since its predictions are not fulfilled. The observation that $x = x_0$, on the other hand, serves to support the hypothesis, though it cannot be regarded as proving it since there may be other theories that would also predict $x = x_0$.

When the hypothesis is statistical, the situation is not quite so clearcut. The observable quantity is then a random variable, say X, and the hypothesis states that X is distributed in a certain way. Usually, the hypothesis will assign positive probability to all possible values of X; no matter which value is observed it could therefore have occurred under the hypothesis, so that no observation can rule out the hypothesis with certainty. However, certain values of X may cast such serious doubt on the hypothesis as to persuade us to abandon it. This will be illustrated by the following examples.

EXAMPLE 1. *Changing the sex ratio.* In the dairy industry, a male calf is much less valuable than a female one. A biochemist claims to have found a treatment that increases the chance of females above its usual value of $\frac{1}{2}$. The association of dairy farmers is skeptical regarding this claim, and proposes an experiment to test their suspicion that the treatment is worthless. The treatment is applied in 20 cases, and the number F of females among the 20 calves is observed. According to the association's hypothesis, F has the binomial distribution ($n = 20$, $p = .5$). Since each of the possible values of F, 0, 1, . . . , 20, has positive probability under this distribution, no observation can definitely disprove the hypothesis. However, an observation of $F = 20$ might well persuade the association to "reject its hypothesis," that is, to abandon its skepticism and grant the biochemist's claim, since such an observation would be extremely unlikely if the hypothesis were true but is the type of result expected if the new treatment is highly effective.

The example suggests a general approach to the problem of testing a statistical hypothesis H. Suppose that an experiment is performed resulting in the observation of a random variable T and that certain values, say large values, of T are very unlikely under H but quite likely under some alternative theory. Then if T is observed to have a sufficiently large value, we would be inclined to reject H in favor of such an alternative theory. The random variable T is called a *test statistic;* the hypothesis H is called the *hypothesis tested,* or the *null hypothesis* since it is often an assertion that there is nothing to certain claims.

EXAMPLE 2. *Extrasensory perception.* A spiritualist claims to have clairvoyant powers that permit her to perceive the arrangement of a deck of cards in the next room. If she were to claim complete accuracy, the problem would be nonstatistical and there would be little difficulty in checking her claim. However, she pleads that the identification of each card requires great concentration and admits that she is only partly successful. A psychologist who doubts that even this partial claim has any validity wishes to test the null hypothesis that she has no such powers at all. He numbers eight cards from 1 to 8, shuffles them thoroughly, and invites the lady to arrange a similar deck in what she believes to be the same order. He will then count the number M of cards that occupy the same position in the two decks. Under his hypothesis, matchings will occur only by chance and M will have a distribution of the type illustrated in Example 5.5.4. It is intuitively clear that large values of M will be more likely if the lady's claim is justified, than if the matching is at random. Therefore M would be a reasonable test statistic, with large values of M throwing doubt on the null hypothesis.

So far, we have considered qualitatively how observations may influence our attitude toward a hypothesis being tested, by raising or quieting doubts concerning its validity. However, frequently something more specific is required: a decision between two possible recommendations or actions, one of them appropriate if the null hypothesis is true, the other if it is false. Thus in Example 1, the association may have to decide whether or not to recommend the new treatment. In Example 2, the psychologist may be faced with the choice of either dismissing the clairvoyant or retaining her for further study. The decision to act as if the null hypothesis H were true will be called *acceptance* of H; that appropriate if H is false will be called *rejection* of H. In the two examples, rejection would stand respectively for recommending the new treatment and retaining the clairvoyant.

If a statistical hypothesis is to be rejected when a test statistic T is sufficiently extreme, it is necessary to decide just where to draw the line between acceptance and rejection. (To simplify the discussion, let us suppose that large values of T are the ones that are "significant," i.e. that cause us to doubt H.) If it is agreed to reject the null hypothesis when $T \geq c$ and to accept it for $T < c$, the boundary value c is called the *critical value*. It is the smallest value of the test statistic that leads to rejection. For example, if the dairy association decided to reject the null hypothesis (and hence to recommend the treatment) if $F \geq 18$, and to accept it (i.e. not recommend the treatment) if $F < 18$, it would be using 18 as the critical value of F.

There are several issues involved in the choice of a critical value, which will be discussed in Chapter 13. In the present section we shall consider only one issue: the possibility of choosing the critical value to control the risk of *false rejection* of the null hypothesis, that is, of rejecting the null hypothesis when it is actually true. This risk is measured by the probability of false rejection $P_H(T \geq c)$, where P_H denotes a probability computed under the assumption that the null hypothesis H is true. It is customary to denote the probability of false rejection by α, and

(1) $$\alpha = P_H(T \geq c)$$

is also known as the *level of significance* of the test. When a null hypothesis is tested at significance level α and is rejected, the outcome of the experiment is said to be "significant at level α." Instead of selecting directly a critical value c, it is customary to select a value of α—the probability that we wish to control—and then determine c through equation (1). Before showing how this is done, we discuss briefly the choice of α.

In deciding how to choose α, it is important to be clear about the meaning of this probability. To be specific, suppose that $\alpha = .01$. What implications does this have in each of the examples? In Example 1, the probability is then $\frac{1}{100}$ of recommending the treatment when it is actually completely ineffectual in raising the proportion of females. Perhaps it is one of the functions of the dairy association to test various new discoveries, treatments, etc., that might be of benefit to its members. Adoption in each such test of the level $\alpha = .01$ means that the association will mistakenly recommend to its members about 1% of the useless "discoveries" submitted to its scrutiny, successfully screening out the remaining 99%. Similarly, in Example 2, adoption of the level $\alpha = .01$ means that if the clairvoyant has no special powers of extrasensory perception, there is only a chance of $\frac{1}{100}$ that she will uselessly be retained for further study.

It is seen from (1) that α decreases as the critical value c increases. We shall now illustrate with the two examples how to determine c so as to reduce α to a satisfactorily low value.

EXAMPLE 1. *Changing the sex ratio (continued)*. According to the null hypothesis, F has the binomial distribution ($n = 20$, $p = .5$). Using Table C we can compute $P_H(F = c)$ for $c = 20, 19, \ldots$, and then add these terms to find successively $P_H(F \geq c)$ for $c = 20, 19, \ldots$, as shown below.

c	20	19	18	17	16	15	14
$P_H(F = c)$	.0000	.0000	.0002	.0011	.0046	.0148	.0370
$P_H(F \geq c)$	.0000	.0000	.0002	.0013	.0059	.0207	.0577

This table shows that α can have only certain values. There is no critical value c for which α is exactly .01. If we wanted one chance in a hundred of false rejection, we would have to settle for $c = 15$ and get $\alpha = .0207$, or use $c = 16$ and get $\alpha = .0059$. Similarly, if we wanted five chances in a hundred of false rejection, the choice of critical value lies between 14 and 15 and that for α correspondingly between .0577 and .0207

In this example we formulated the null hypothesis as $p = \frac{1}{2}$ under the assumption that this is the probability of a female calf when the treatment has no effect. Actually, for most animals as for humans the probability of a female birth is not exactly $\frac{1}{2}$. Suppose that for the kind of cattle under consideration this probability (under the assumption of no treatment effect) is .48. Then the rejection probability $P_H(F \geq c)$ should be computed for the binomial distribution corresponding to $p = .48$ rather than that corresponding to $p = .5$, giving rather different values than those listed in the table.

Actually, it is better in such circumstances not to speculate about a value of p which is either not exactly known or which, even if it is known for cattle in general, may not apply to the particular herd which is used for the experiment. Instead, there should be included in the experiment control cases to which the treatment being tested is not applied. (Controls are needed here for essentially the same reasons as when estimating a treatment effect; see Section 9.4.) In this way, it is possible to study the effect of the treatment directly by a comparison of treated and untreated cases. The problem of testing the effect of the treatment for this type of experiment will be studied in detail in Chapter 12.

EXAMPLE 2. *Extrasensory perception (continued).* For a second numerical illustration of the relation between α and c, we use Table 6.8.1, which gives the distribution of the number M of matchings when N items are arranged in random order. (It is not necessary to read Section 6.8 to follow this illustration.) In that table, $(N)_N$ is the number of equally likely orderings of N items, and $C(m, N)$ is the number of these which give exactly $M = m$ matchings. For example, for $N = 8$ the table shows that there are $(8)_8 = 40{,}320$ possible orderings, of which $C(4, 8) = 630$ have exactly $M = 4$ matchings. Thus, if matching is done at random, $P_{\mathrm{H}}(M = 4) = 630/40{,}320 = .0156$. Proceeding similarly, the following table can be obtained.

c	8	7	6	5	4
$P_{\mathrm{H}}(M = c)$	.0000	.0000	.0007	.0028	.0156
$P_{\mathrm{H}}(M \geqq c)$	.0000	.0000	.0007	.0035	.0191

Supposing the psychologist decides to reject the null hypothesis (and therefore retain the clairvoyant) provided she correctly places 5 or more cards, his risk of retaining her is .0035 if she has no powers of extrasensory perception. (In this case, $c = 5$ and $\alpha = .0035$.)

The distinction between statistical and nonstatistical hypotheses made at the beginning of the section is not as sharp as it may seem, since all observations involve at least some degree of uncertainty. Thus it would, for example, not be possible to observe exactly the difference in time of impact of Galileo's two cannon balls. A more careful model for this experiment would treat this time difference as a random variable, to which Galileo's hypothesis assigns the expected value zero. However, in this case the random element is so minor that it can be neglected, and the observed quantity treated as nonrandom.

PROBLEMS

1. An instructor asks each member of his class of ten students to write down "at random" one of the digits $0, 1, \ldots, 9$. Since he believes that people incline to select the digits 3 and 7, he counts the number S of students selecting 3 or 7.
   (i) What is the distribution of S if the students do select at random?
   (ii) How large must S be before the hypothesis of randomness would be rejected at significance level $\alpha = .1$?

2. Solve the preceding problem if the instructor believes that people incline to select an odd digit in preference to an even one and if S denotes the number of students selecting an odd digit.

3. Solve the preceding problem if the number of students is 15 and $\alpha = .05$.

4. A surgeon believes he has found an improved technique for performing a certain heart operation for which the mortality has been 30%. He performs the operation on 10 patients all but one of whom survive. At the 3% significance level, is he justified in claiming to have proved the success of his innovation?

5. In the preceding problem, suppose the mortality had been 40%. If all but one of the ten patients survive, is the surgeon justified in his claim at the 1% level?

6. A manufacturer of a certain food product believes that he can use less expensive ingredients without noticeably lowering the quality of his product. To test the hypothesis that the cheapening cannot be detected, he presents 25 customers with samples of both products and asks them to state which they prefer. If 17 prefer the old and only 8 the new product, is this result significant at the 1% level, at the 5% level, at the 10% level? (Assume that a customer who cannot distinguish the products chooses at random.)

7. In the preceding problem, suppose the number of customers is 20, of which 14 prefer the old and 6 the new product. Is this result significant at the 1% level, at the 5% level, at the 10% level?

8. A delegation of three is selected from the council of Example 2.1.1 consisting of four Conservatives and three Liberals. All three members chosen for the delegation (supposedly at random) turn out to be Conservatives. At significance level 5%, would the Liberals reject the hypothesis that the delegation was selected at random?

9. In the preceding problem, suppose that the council consists of nine members, five Conservatives and four Liberals, and that a delegation of five is to be selected. Let D denote the number of Conservatives included in the delegation.
   (i) What is the distribution of D under the hypothesis that the delegation has been selected at random?
   (ii) For what values of D should the Liberals reject the hypothesis of random selection at approximate significance level 5%?

10. In Example 2, suppose that the spiritualist is shown eight cards (face down and in random order) and is told that four are black and four red. She is asked to identify the four red cards. Let D be the number of red cards among the four cards she selects.

(i) What is the distribution of D under the hypothesis that she has no extrasensory powers at all?

(ii) For what values of D would you reject the hypothesis of no extrasensory powers—in favor of recognizing the lady's claim at approximate significance level .02?

11. What significance levels between .005 and .1 are available if, under the hypothesis, T has the binomial distribution with n trials and success probability p, and if the hypothesis is rejected for large values of T, in the following cases:

(i) $n = 7, p = .2,$ (ii) $n = 9, p = .1,$ (iii) $n = 14, p = .5.$

12. Under the assumptions of the preceding problem, find the significance level closest to .05 when

(i) $n = 6, p = .3,$ (ii) $n = 10, p = .4,$ (iii) $n = 25, p = .5.$

13. What significance levels between .005 and .1 are available if, under the hypothesis, T has the hypergeometric distribution (1) of the random variable D in Section 6.2 and if the hypothesis is rejected for large values of T, when

(i) $N = 15, r = 10, s = 6;$ (ii) $N = 20, r = 10, s = 8?$

14. Let X denote the number of heads obtained when a fair penny is tossed n times.

(i) Suppose you observe X and believe that n was equal to 18 but are not sure that you did not lose count and that actually n was greater than 18. Would you reject H: $n = 18$ against the alternatives $n > 18$ for small values of X or for large values of X and why?

(ii) At significance level $\alpha = .1$, would you reject it if $X = 15$?

15. Solve the preceding problem if the hypothesis H: $n = 15$ is to be tested against the alternatives $n > 15$, if $X = 12$ and $\alpha = .05$.

16. Let B denote the number of successes in n binomial trials with success probability p. For testing $p = .3$ against the alternatives that $p < .3$, would you reject H for small values of B or for large values of B?

17. You wish to test the hypothesis H that in a certain species of animals the probability p of an offspring being male is $\frac{1}{2}$. Since you believe it possible that p may be either $> \frac{1}{2}$ or $< \frac{1}{2}$, you wish to test H against the alternatives $p \neq \frac{1}{2}$. If M denotes the number of male offspring in $n = 16$ births, for what values of M would you reject H at level $\alpha = .02, .05, .1$?

18. Solve the preceding problem if $n = 14$.

19. Suppose H is rejected when T is sufficiently large, and that in particular when $T = 6$, H is rejected at level $\alpha = .05$. Determine for each of the following levels α' whether for $T = 6$, H should definitely be rejected, definitely accepted, or whether not enough information is given to tell:

(i) $\alpha' = .01,$ (ii) $\alpha' = .02,$ (iii) $\alpha' = .1.$

[Hint: Suppose H is rejected when $T \geq c$. What is the relation between c and 6, and how is $P_H(T \geq c)$ related to α?]

20. In Problem 1, suppose that S is observed to be equal to 5. What is the smallest significance level α for which the hypothesis of random selection would be rejected?

21. In Example 2, suppose the psychologist uses a deck of N cards, and rejects the null hypothesis (of no clairvoyant powers) only if the spiritualist arranges all N in the correct order ($M = N$). How large must N be for α to be less than .01? [Hint: Use Table 6.8.1.]

22. If one wishes to test $p = .4$ against $p < .4$ with a value of n in the range $12 \leqq n \leqq 15$, and if α is to be as near .05 as possible, what values of n and c should be used?

11.2 METHODS FOR COMPUTING SIGNIFICANCE

In the preceding section we illustrated the exact computation of the relation between the critical value c of a test statistic T (large values of which are significant) and the corresponding significance level $\alpha = P_H(T \geqq c)$. We shall now develop a short-cut method and an approximation for this computation.

To emphasize the fact that the probability of the event $T \geqq t$ is computed under the assumption that the null hypothesis H is true, we have written $P_H(T \geqq t)$. However, to simplify the notation, we shall omit the subscript H in the remainder of this chapter and in Chapter 12. The reader should keep in mind that in this part of the book all probabilities, expectations, and variances are computed under H. In Chapter 13, where probabilities are computed under various hypotheses, a subscript notation will have to be reintroduced.

EXAMPLE 1. *Matching.* On return from a sabbatical year in France, a professor astonishes his friends by claiming to have become a connoisseur of French clarets. To test his claim, they get clarets from nine famous chateaux and ask him to match the wines with the list of chateaux in a blindfold test. He correctly matches seven out of the nine. It is of course possible that he knows nothing about wine and is just guessing at random (null hypothesis). Is his achievement significant evidence against the null hypothesis at the 1% level?

If M denotes the number of correct matchings, large values of M are significant. The distribution of M when N items are matched at random was discussed in Example 1.2. However, Table 6.8.1 unfortunately extends only up to $N = 8$. Is it necessary to extend that table to $N = 9$ in order to answer the question of the significance of the professor's performance? Fortunately not, since we know (Problem 6.8.12) that $c(7, 9) = 36$, $c(8, 9) = 0$ and $c(9, 9) = 1$, and hence that the number among the $(9)_9$ equally likely arrangements leading to seven or more correct matchings is 37. Since $(9)_9 = 362,880$ by Table 2.3.1, it follows that under the null hypothesis, $P(M \geq 7) = \frac{37}{362,880} = .000102$.

If the entire distribution of M were available, we would know exactly what significance level α corresponds to each possible critical value c. The computation we have just made shows that level $\alpha = .000102$ corresponds to $c = 7$. Clearly, a larger level would correspond to any c smaller than 7, and a smaller level to any c larger than 7. Thus, we know that the observation $M = 7$ is significant at all possible levels greater than or equal to $.000102$ and hence in particular at the one percent level, without the labor of computing the values of α corresponding to $c = 6, 5, \ldots$.

The short-cut device just illustrated is often helpful. Let T be any test statistic, large values of which are significant, and suppose that in the experiment we observe $T = t$. We compute the probability $P(T \geq t)$ of observing a value of T as significant as, or even more significant than, the value t actually observed. This probability is known as the *significance probability* of the observed value t of T. By the argument given above, the significance probability tells us just at which levels α the hypotheses would be rejected, without the necessity of computing the entire distribution of T. The observation $T = t$ will be significant at each level greater than or equal to $P(T \geq t)$, and not significant at any level less than $P(T \geq t)$.

When reporting the result of an experiment designed to test a null hypothesis, it is better practice to give the significance probability than merely to assert whether or not the experiment is significant at a prechosen level. This has the advantage that a reader of the report can decide for himself what significance level he wishes to use. For example, suppose $P(T \geq t) = .04$, and the report merely says "the experiment was significant at the 5% level." A user of the report, who would prefer to use the 2% level, does not know whether to reject the null hypothesis at that level; but publication of the value $.04$ of $P(T \geq t)$ tells him to accept the hypothesis at level $.02$.

The significance probability may be thought of as giving, in a single convenient number, a measure of the degree of surprise which the experiment should cause a believer of the null hypothesis. Behind this interpretation lies the assumption that large values of the test statistic are just what would be expected under the alternative hypothesis. Thus, if the professor really has acquired a fine taste in French clarets, it would be expected that he can correctly identify the chateaux in most cases. For this reason, the very small probability $P(M \geq 7) = .000102$ casts very strong doubt on the null hypothesis that he is just guessing.

In many cases, the significance level corresponding to a given critical value, or the critical value appropriate for a desired significance level, can be approximated to a satisfactory degree of accuracy by use of the normal approximation. This requires only knowledge of the expectation and variance of the test statistic under the null hypothesis.

EXAMPLE 2. *Triangular taste tests.* A manufacturer of powdered coffee is considering a change in the production process which would reduce his costs, but he does not wish to make the change if it would result in an alteration in flavor that could be detected by his customers. He arranges to conduct taste trials in a supermarket. Each of 500 customers will be offered in random order three cups of coffee, one of which is made by the new process and the other two by the old process. He is then asked to identify which cup is different from the other two. If the customers are in fact unable to distinguish the change in flavor, each will have one chance in three of correctly guessing which cup is different. Let B denote the number of customers who correctly identify the cup made by the new process. Under the null hypothesis of no change in flavor, B has the binomial distribution with $n = 500$, $p = \frac{1}{3}$. If the flavor has been altered, B would tend to have large values. What is the critical value for B corresponding to the 5% level?

We seek that value c for which $P(B \geqq c) = .05$. Formulas (6.1.6) and (6.1.7) give $E(B) = \frac{500}{3} = 166.67$ and $\mathrm{Var}(B) = \frac{1000}{9} = 111.11$ so that $\mathrm{SD}(B) = 10.54$, where of course the expectation and variance are computed assuming H. The normal approximation gives

$$.95 = P(B < c) = \Phi\left(\frac{c - .5 - 166.67}{10.54}\right).$$

Comparing this with the equation

$$.95 = \Phi(1.645),$$

we see that

$$1.645 = \frac{c - .5 - 166.67}{10.54} \quad \text{or} \quad c = 184.5.$$

Now c must be an integer. For $c = 184$, the normal approximation gives $\alpha = P(B \geqq 184) = .0552$, while for $c = 185$, $\alpha = P(B \geqq 185) = .0453$. (The correct values, obtained from a large table of the binomial distribution, are in fact .0560 and .0462.) Thus, if 185 or more customers correctly identify the cup made by the new process, the experiment is significant at the five percent level.

PROBLEMS

1. Find the significance probability in the following cases: (i) in Problem 1.1 if $S = 4$; (ii) in Problem 1.4; (iii) in Problem 1.6; (iv) in Problem 1.9 if $D = 4$;

(v) in Problem 1.10 if $D = 3$.

2.   In Problem 1.1 use the normal approximation to find the critical value of S if the number n of students and the significance level are (i) $n = 64$, $\alpha = .1$; (ii) $n = 64$, $\alpha = .05$; (iii) $n = 100$, $\alpha = .1$; (iv) $n = 100$, $\alpha = .05$; (v) $n = 100$, $\alpha = .01$.

3.   In Problem 1.4 suppose that the operation is performed on n patients, S of whom survive. At significance level $\alpha = .01$, use the normal approximation to find the critical value of S if (i) $n = 80$, (ii) $n = 120$, (iii) $n = 160$.

4.   In the preceding problem, use the normal approximation to find the significance probability if (i) $n = 90$, $S = 65$; (ii) $n = 90$, $S = 70$; (iii) $n = 900$, $S = 650$; (iv) $n = 900$, $S = 700$.

5.   In Problem 1.6 suppose that n customers are asked to state their preference and that S of the n prefer the old product. Use the normal approximation to find the critical value of S if $n = 100$ at significance levels

(i) $\alpha = .01$ (iii) $\alpha = .05$
(ii) $\alpha = .02$ (iv) $\alpha = .10$.

6.   To test the hypothesis H that a coin is fair, suppose it is tossed n times, the number S of heads is observed, and H is accepted if $-c \leqq S - \frac{1}{2}n \leqq c$. Use the normal approximation to find the critical value c corresponding to significance level $\alpha = .05$ if (i) $n = 50$, (ii) $n = 100$.

7.   In Example 2, suppose that the taste test is given to 200 customers. Use the normal approximation to find the significance probability if B takes on the value (i) 75, (ii) 90.

8.   A teacher of a class of 100 children is instructed to select a treatment group of 50 at random to receive a daily supplement of orange juice in their school lunches, the other 50 to serve as controls. When the results of the experiment are analyzed, it is found that 30 of the 100 children came from an orphanage and that 23 of these 30 were included in the treatment group. At significance level .05, would you reject the hypothesis that the teacher followed instructions and instead entertain the suspicion that he tended to favor the orphans with the juice? [Hint: If D is the number of orphans included in the treatment group, what is the distribution of D when the hypothesis of random assignment is true?]

9.   In the preceding problem, find the significance probability if the class consists of 150 children, of which 75 receive the juice and 75 serve as controls, and if 40 come from the orphanage of which 25 are included in the treatment group.

10.   Find the significance probability in the following cases: (i) Problem 1.14; (ii) Problem 1.15; (iii) Problem 1.16 when $B = 5$.

11.   In Problem 1.17, find the significance probability when $M = 5$. [Hint: Put $T = |M - 8|$ and note that large values of T are significant.]

12.   In Problem 1.18, find the significance probability when $M = 3$.

13. In Problem 6, find the significance probability when (i) $n = 100$, $S = 40$; (ii) $n = 200$, $S = 75$.

14. A new drug cures nine of 200 patients suffering from a type of cancer for which the historical cure rate is two percent. How significant is this result? [Hint: Use the Poisson approximation developed in Section 6.7.]

11.3 THE CHI-SQUARE TEST FOR GOODNESS OF FIT

The basic concept underlying this book is that of a probability model representing a random experiment, as described in Chapter 1. Before applying any such model one should ask: is the model realistic? Deciding whether a proposed model is realistic may be viewed as a problem of testing a statistical hypothesis, and we shall now present a test appropriate for this purpose. The theoretical basis for this test is the multinomial distribution discussed in Section 7.3.

To be specific, suppose a model is being considered with k simple events, to which are assigned probabilities $p_1, p_2, \ldots, p_k$. This model is supposed to represent a random experiment with k simple results. A test of realism requires the collection of suitable data. Suppose to this end we perform the experiment n times and observe the numbers of times, say $B_1, B_2, \ldots, B_k$, that the various simple results occur. Intuitively, one would judge the model to be realistic if the observed frequencies $B_1/n, B_2/n, \ldots, B_k/n$ are close enough to the probabilities $p_1, p_2, \ldots, p_k$ that are supposed to represent them.

EXAMPLE 1. *Random digit generator.* The idea of a random digit generator was introduced in Example 1.2.2, and was discussed further in Section 3.4. The probability model for the experiment of operating the generator to produce a random digit has ten simple events ($k = 10$) corresponding to the ten digits, and assigns equal probability to each ($p_1 = p_2 = \ldots = p_{10} = \frac{1}{10}$). The results of $n = 250$ trials of the experiment are given in the first five rows of Table 3.4.1. Inspection of this table shows that the digit zero occurred 32 times, or with frequency $\frac{32}{250} = .128$. The other digits may be counted similarly, giving the following results:

(1)

Digit	0	1	2	3	4	5	6	7	8	9
Number of occurrences	32	22	23	31	21	23	28	25	18	27
Frequency	.128	.088	.092	.124	.084	.092	.112	.100	.072	.108

"In the long run," each of the ten frequencies is supposed to be .1. While the frequencies are not expected to be exactly equal to this value in the short run, are the observed discrepancies perhaps too large to be attributed to chance fluctuations, and hence cast doubt on the operation of the generator? To help answer this question, we need a reasonable test statistic for which we can compute the significance probability.

Many different reasonable measures could be suggested for the discrepancy between the frequencies B_1/n, B_2/n, . . . and the corresponding probabilities p_1, p_2, . . . ; and each such measure could be used to test the proposed model. For example, one might consider the maximum of the absolute differences $|(B_1/n) - p_1|$, $|(B_2/n) - p_2|$, ... Another possibility is the sum of the squared differences $[(B_1/n) - p_1]^2 + [(B_2/n) - p_2]^2 + ...$ The measure most commonly used in practice is

$$(2) \qquad Q = \frac{(B_1 - np_1)^2}{np_1} + \frac{(B_2 - np_2)^2}{np_2} + \ldots + \frac{(B_k - np_k)^2}{np_k}.$$

Large values of Q correspond to large discrepancies, so that large values of Q are significant. This test was proposed by Karl Pearson in 1900, and is known as the *chi-square test of goodness of fit* (of the observed frequencies to the probabilities of the model).

In order to carry out the test it is necessary to know the distribution of Q under the null hypothesis, i.e., according to the model being tested. If the model is correct, and if the n trials of the experiment are unrelated so that a product model may be used, the random variables $(B_1, B_2, \ldots, B_k)$ have a k-nomial distribution as discussed in Section 7.3. From this result, the null distribution of Q can be obtained by listing all possible sets of values of $(B_1, B_2, \ldots, B_k)$, and for each set computing both its null probability according to the method of Section 7.3, and also the associated value of Q from (2).

EXAMPLE 2. *Computation of null distribution of Q.* To illustrate the method, let us take $k = 3$, $p_1 = .5$, $p_2 = .3$, $p_3 = .2$ and $n = 4$. The fifteen possible sets (b_1, b_2, b_3) are shown in Table 1 arranged in order of decreasing value of Q. Consider for example the row $(b_1 = 1, b_2 = 3, b_3 = 0)$. For this case, the value of Q is by (2)

$$Q = \frac{(1 - 2.0)^2}{2.0} + \frac{(3 - 1.2)^2}{1.2} + \frac{(0 - 0.8)^2}{0.8} = 4$$

and the probability that $(B_1 = 1, B_2 = 3, B_3 = 0)$ is by (7.3.4)

$$\binom{4}{1,\,3} (.5)^1 (.3)^3 (.2)^0 = .0540.$$

The column $P(Q \geq q)$ of Table 1 gives the significance probabilities, obtained by adding the probabilities of the sets with Q as large or larger than the observed value q. Suppose for example that one observed $B_1 = 0$, $B_2 = 1$, $B_3 = 3$. These values are represented in the third row of the table, and give $Q = 8.08$. The significance probability is $P(Q \geq 8.08) = .0193$.

The procedure illustrated in Example 2 is clearly not practical unless n is quite small, and we shall therefore consider an approximation. It is a remarkable fact, discovered by Pearson, that when n is large the null distribution of Q is very nearly independent of the values of $p_1, p_2, \ldots, p_k$, depending essentially only on the value of k. Since most applications do involve a large value of n, this means that in practice it is usually sufficient to employ Pearson's approximation, values of which are shown in Table G.

It can be shown (Problem 2) that $E(Q) = k - 1$, and we shall for brevity denote this value by $k - 1 = \nu$, where ν is often called the "degrees of freedom." Large values of Q are significant, and it is values of Q larger than the expected value that are of primary interest. Accordingly, Table G gives Pearson's approximation to $P(Q \geq q)$ only for values of q that exceed ν. In fact, the columns of the table are headed with values of $q - \nu$, the amount by which the observed value q exceeds the expected value ν of Q. The use of Table G is illustrated by the following examples.

TABLE 1. NULL DISTRIBUTION OF Q

b_1	b_2	b_3	q	$P(Q = q)$	$P(Q \geq q)$	Approx. to $P(Q \geq q)$
0	0	4	16.00	.0016	.0016	.000
0	4	0	9.33	.0081	.0097	.010
0	1	3	8.08	.0096	.0193	.017
1	0	3	7.75	.0160	.0353	.021
0	3	1	4.75	.0216	.0569	.094
0	2	2	4.33	.0216	.0785	.115
4	0	0	4.00	.0625		
1	3	0	4.00	.0540	.1950	.135
2	0	2	3.00	.0600	.2550	.224
1	1	2	2.33	.0720	.3270	.313
3	0	1	1.75	.1000	.4270	
3	1	0	1.33	·.1500		
2	2	0	1.33	.1350	.7120	
1	2	1	1.08	.1080	.8200	
2	1	1	.08	.1800	1.0000	

EXAMPLE 2. *Computation of null distribution of Q (continued).* Let us return to Example 2 and consider the problem of finding the approximate significance probability of the observation $(B_1 = 1, B_2 = 0, B_3 = 3)$. For this observation, Q has the value 7.75, so that the desired probability is $P(Q \geq 7.75)$. This probability was computed exactly and found to be .0353; let us see what value is given by Pearson's approximation. Since $k = 3$, and hence $\nu = k - 1 = 2$, the appropriate row of Table G is the first. The columns of Table G are headed with values of $q - \nu$, which in

this case means $q - 2$. Since we are concerned with the value $q = 7.75$, we wish to enter the table at $7.75 - 2 = 5.75$. There is no column headed 5.75, but we can interpolate between the values $q - \nu = 5.5$ and $q - \nu = 6.0$. The entry in the 5.5 column is .024, while the entry in the 6.0 column is .018. Interpolation gives $P(Q \geqq q) = .021$, which may be compared with the correct value .0353.

This method was used to compute the final column of Table 1, which should be compared with the corresponding correct entries given in the penultimate column. The reader will see that the approximation is far from perfect, as is only to be expected since in this case $n = 4$ while the approximation is based on the assumption that n is large. Nevertheless, even with this extremely small n, the approximation gives a fair idea of the magnitude of the significance probabilities. Pearson's approximation improves as the numbers $np_1, \ldots, np_k$ increase, and it is a common rule-of-thumb that the approximation is reliable when all of these numbers are 5 or more. Our example, in which these numbers are $np_1 = 2.0$, $np_2 = 1.2$ and $np_3 = 0.8$, suggests that the approximation may be useful even for values of $np_1, \ldots, np_k$ less than 5.

EXAMPLE 1. *Random digit generator (continued).* Let us now apply the chi-square test to answer the question raised in Example 1: are the discrepancies between the observed frequencies of the ten digits and their probabilities too great to be explained by chance? Applying formula (2) to the data given in (1), we have $k = 10$, $n = 250$, $p_1 = \ldots = p_{10} = .1$, and

$$Q = \frac{(32 - 25)^2}{25} + \frac{(22 - 25)^2}{25} + \ldots + \frac{(27 - 25)^2}{25} = 7.2.$$

In this case $\nu = 10 - 1 = 9$, so that $E(Q) = 9$. The observed value is actually smaller than the expected value, so that the discrepancies are certainly not significant. Table G gives $P(Q \geqq \nu)$ to be .433 for $\nu = 8$ and .440 for $\nu = 10$; thus by interpolation it is .436 for $\nu = 9$. Since our observed q is less than ν, $P(Q \geqq 7.2)$ exceeds .436. In fact, from a more complete table of Pearson's approximation* it appears that the value is .616.

The chi-square test can also be used for a more general problem than that discussed here, namely for testing the goodness of fit of a model whose probabilities are not completely specified. The application of the test in such cases is explained in more advanced books.

---

* *Biometrika Tables for Statisticians, Vol. 1* (edited by E. S. Pearson and H. O. Hartley), Cambridge University Press (1954). Table 7.

PROBLEMS

1. Check the values of q and $P(Q = q)$ for the following entries in Table 1:
  (i) $b_1 = 0$, $b_2 = 1$, $b_3 = 3$;   (ii) $b_1 = 1$, $b_2 = 1$, $b_2 = 2$.

2. Show that $E(Q) = k - 1$.
[Hint: $E(B_1 - np_1)^2$ is just the variance of the binomial random variable B_1 and similarly for $B_2, \ldots, B_k$.]

3. Use Table G to check the entries in the last column of Table 1 for the cases
  (i) $b_1 = 0$, $b_2 = 4$, $b_3 = 0$;   (ii) $b_1 = 2$, $b_2 = 0$, $b_3 = 2$.

4. Use Table G to find an approximate value for $P(Q \geq q)$ when $k = 10$ and $q = 12.2$. [Hint: This value requires interpolation in both directions, which should be done as follows: since $q = 12.2$ and $\nu = 9$, $q - \nu = 3.2$ so we must interpolate between the columns headed 3.0 and 3.5. This can be done both for rows $\nu = 8$ and $\nu = 10$, and the average of these values used for the desired $\nu = 9$.]

5. Use Table G and the method of the preceding problem to find an approximate value for $P(Q \geq q)$ when $k = 34$ and $q = 33.1$.

6. Use the last ten rows of Table 3.4.1 to test the fit of the model discussed in Example 1.

7. Carry out the experiment of throwing a die 60 times, and record the frequencies of the results "one point," $\ldots$, "six points." For testing the hypothesis that the die is fair, compute Q and use Table G to determine whether Q is significant at the 10% significance level.

8. Divide the 200 tosses with a coin performed as Problem 1.2.1 into 100 pairs of two consecutive tosses and count the numbers of cases with two heads, one head and one tail, and two tails. For testing the hypothesis that the coin is fair and the tosses are unrelated, so that the probabilities of the three simple events are, respectively, $\frac{1}{4}$, $\frac{1}{2}$ and $\frac{1}{4}$, compute Q and use Table G to determine whether Q is significant at the 8% significance level.

9. (i) For the case $k = 2$ (and hence $\nu = 1$) which is not covered by Table G, show that the quantity Q defined by (2) reduces to $Q = (B_1 - np_1)^2/np_1(1 - p_1)$.
  (ii) Use the normal approximation to the binomial distribution for computing approximately the probability $Q \geq q$ when $k = 2$, $n = 100$, $p_1 = \frac{1}{4}$ and $q = 4$.
[Hint: (ii) B_1 has a binomial distribution, and Q in this case is the square of the standardized binomial variable $B_1{}^*$.]

10. Obtain the exact distribution of Q and sketch its histogram for $k = 3$, $p_1 = p_2 = p_3 = \frac{1}{3}$ and
  (i) $n = 3$,   (ii) $n = 6$,   (iii) $n = 9$.

[Hint: Since $p_1 = p_2 = p_3$, both the value of Q and its probability do not depend on the order of the values b_1, b_2 and b_3, which greatly reduces the number of cases.]

11. Use Table 1 to find the significance probability of the observation $B_1 = 1$, $B_2 = 3$, $B_3 = 0$ if instead of Q the following measures of discrepancy are used as test statistics:

(i) the maximum of the absolute differences

$$|(B_1/n) - p_1)|, \quad |(B_2/n) - p_2|, \quad |(B_3/n) - p_3|;$$

(ii) the sum of the squared differences

$$[(B_1/n) - p_1]^2 + [(B_2/n) - p_2]^2 + [(B_3/n) - p_3]^2;$$

(iii) the sum of the absolute differences

$$|(B_1/n) - p_1| + |(B_2/n) - p_2| + |(B_3/n) - p_3|.$$

12. Suppose that in Example 2 the observed values are $B_1 = 1$, $B_2 = 3$, $B_3 = 0$. Then the largest of the absolute differences $|B_1 - np_1|$, $|B_2 - np_2|$, $|B_3 - np_3|$ is $|B_2 - np_2| = |3 - 1.2| = 1.8$. Since B_2 has the binomial distribution with $n = 4$, $p_2 = .3$, it is tempting to compute the significance probability of the value 1.8 from this distribution, using Table B.

(i) Compute the probability in this way.

(ii) Explain why this does *not* give the correct significance probability of the observed result.

(iii) Determine without computation whether the correct value is larger or smaller than that computed in (i).

(iv) Find the correct significance probability.

[Hint: (iv) See Problem 11(i).]

CHAPTER 12

TESTS FOR COMPARATIVE EXPERIMENTS

12.1 COMPARATIVE EXPERIMENTS

In nearly all areas of human endeavor, efforts are constantly being made to find better methods of doing things. When a new method is proposed it becomes necessary to decide whether it really is superior to the method currently used, and if so by how much. The problem of estimating the magnitude of the improvement provided by a new method was discussed in Section 9.4 as the problem of "estimating the effect of a treatment." In the present chapter, we shall develop methods for testing whether there is an improvement at all. In spite of the fact that essentially the same experimental designs are used to provide answers for both questions, the present development will be independent of the earlier one, at the price of some repetition.

EXAMPLE 1. *Salk vaccine.* To find out whether the Salk vaccine would give protection against polio, a large field trial was conducted in the United States in 1954. In this trial, 200,745 children were given the vaccine, while 201,229 other children received no vaccine. The numbers of cases of polio in the two groups were then compared (see Example 2.2).

EXAMPLE 2. *Rainmaking.* Experiments have been carried out in various parts of the world to test the effectiveness of techniques for artificial rainmaking by cloud seeding. In a typical experiment, some storms were seeded and the others left alone, and the frequency and amount of rainfall were compared for the two groups of storms.

EXAMPLE 3. *Fertilizer.* To find out whether a new fertilizer gives better results than the type currently in use with cotton crops, a farmer applies each type to a number of plots of land, and weighs the lint cotton produced from each plot to obtain a comparison.

EXAMPLE 4. *Physics by TV.* A research foundation sponsors an experiment to determine whether physics can be taught as well by television as in the classroom. A number of students are instructed by television while others are taught in the ordinary way. At the end of the course, all students take the same final examination and their scores are compared.

Despite the variety of applications, all of these examples have the same basic structure. In each case, two ways of doing something, to which we shall refer as two *treatments*, are being compared. One of these is the standard treatment, the one now in use (which may as in Examples 1 and 2 consist in doing nothing), the other is the new treatment that has been proposed as a possible improvement. The experiment requires the use of experimental *subjects* (children, storms, plots, students), each of which receives one treatment or the other. The experiment consists in observing the *response* of each subject (polio or no polio, amount of rainfall, pounds of cotton, test score), and then comparing the responses of the subjects receiving the new treatment with those receiving the standard treatment.

We wish to emphasize the importance in such experiments of having a group of subjects who receive the standard treatment. These subjects are called the *controls* and are used as a background against which to compare the results of the experimental subjects. Frequently experimenters will omit the controls, feeling that their inclusion would be a waste of experimental facilities. They then compare the responses of the experimental subjects with past experience with the standard treatment, or with the response they would have expected from theoretical considerations if the standard treatment had been used. This is in most cases a very hazardous practice, as a little consideration of the examples will show. In the rainmaking experiment, for instance, we might seed all the storms this year, and compare the rainfall with that obtained in previous years. However, some years are much wetter than others, and if this should happen to be a wet year we would conclude that cloud seeding was effective even if it were not. Similarly, the incidence of polio and the yield of crops varies from one year to another. Another instance of the same difficulty occurs in the evaluation of a new experimental treatment for mental patients. Here the inclusion of a patient in the experiment and the resulting additional attention he receives may produce an improvement which will be attributed to the new treatment, unless a control group of patients is also included in the experiment. In almost all cases of this kind, it is better practice to build a control group into the experiment.

Even worse than the omission of controls is the use of a control group that is not comparable to the experimental group in its responsiveness to treatment, since this lends to the experiment an entirely misleading aura of objectivity. If the control group consists of subjects who are on the whole less able to respond, then the new treatment may look good even if in fact it is worthless. On the other hand, the new treatment would not have a fair chance if it is applied to subjects less responsive than those who serve as controls. A doctor once "demonstrated" the ability of a vaccine to prevent tuberculosis by giving it to his private patients living in a suburb, and using as controls the patients at a clinic in a slum area.

It can even happen that biases are introduced as a consequence of a well-intentioned effort to assure comparability of the control and experimental groups. In a famous experiment on the nutritive value of milk for school children, carried out in 67 Scottish schools in 1930, half of the children in each school were "randomly" selected to be given milk for a four-month period beginning in February, with the other half serving as controls. It was found that the weight gain of the treated children was dramatically greater (about 40%) than that of the controls. The validity of this plausible finding was however later challenged. The control children were definitely taller and heavier at the beginning of the experiment than the treated children, so that bias appears to have entered into the selection. The original report stated: "In any particular school where there was any group to which these methods [random selection] had given an undue proportion of well-fed or ill-nourished children, others were substituted in order to obtain a more level selection." It may be presumed that the teachers, acting from the most humane motives, would be more inclined to make such a substitution when it resulted in giving the milk to children who obviously needed it; thus, needy children would be over-represented in the treatment group. Since the children were weighed in their clothes, and since the reduction in clothing weight from February to June was presumably greater for wealthy children than for needy ones, the overrepresentation of the latter group would tend to exaggerate the apparent effect of the milk.

In order to insure comparability of the experimental and control groups, one may attempt to use experimental subjects who are alike; then we could be sure that the observed differences in response are attributable to the differences in treatment rather than to differences in the subjects. Although this ideal is unattainable, experimenters should try to obtain subjects as homogeneous as possible. Thus, the children being compared in a vaccine trial should perhaps be of the same age, live in the same part of town, and attend the same school; the plots of land used in a fertilizer test should be contiguous and should all have been treated alike in recent years; the students in an experiment to compare teaching methods should have similar aptitude scores. In order to produce homogeneous material for animal experiments, pure strains of mice and other laboratory animals have been bred which have very similar genetic composition.

In general, in spite of the best efforts to the contrary, there will of course remain differences between the experimental subjects. This opens the door to the introduction of bias for purely accidental reasons, as in the milk experiment mentioned above. Also, frequently an experimenter has a strong desire for the new treatment to show up well, and there is the possibility that this will cause him unconsciously to assign to the control group the subjects less likely to give a good response. Finally, even if such a bias does not exist, critical observers may believe that it does. There is one very simple way for an experimenter to remove both the possibility and the suspicion of any such assignment bias: to assign the subjects to the treatment groups at random. One can rely on the impartiality of a set of random number tables!

In addition to providing a safeguard against bias or suspicion of bias, randomization has another important function: it furnishes the basis for the probabilistic evaluation of the results. The situation is quite analogous to that in sampling where the random character of the sample drawn gave insurance against a biased selection of the sample and at the same time assigned to the sample a probability distribution which provided a basis for estimation. This analogy between randomization and sampling is not accidental; as we shall see below, randomization is actually carried out by means of sampling, with different sampling designs leading to different methods of randomization. We shall now briefly outline three such methods.

(i) *Complete randomization.* Suppose that N subjects are available for the experiment. Of these, t are to receive the treatment, with the remaining $s = N - t$ serving as controls. The simplest randomized method of dividing the N subjects into s controls and t treatment subjects, consists in selecting the t treatment subjects completely at random; that is, in such a manner that all $\binom{N}{t}$ possible choices of the t subjects have the same probability, which is therefore

$$(1) \qquad\qquad 1 \Big/ \binom{N}{t}.$$

With this procedure, the $s = N - t$ remaining subjects who will serve as controls also constitute a random sample (of size s) from the given set of N subjects. (See Section 2.3.)

(ii) *Randomization within blocks.* As remarked above, it is desirable that the subjects used in a comparative experiment be as homogeneous as possible so that any observed difference in response to the two treatments may be attributed to difference in treatment, rather than to difference in the subjects. Homogeneity may often be attained by imposing conditions on the experimental subjects, to make them more nearly alike. For example, in the clinical trial of a new drug, one may require that all patients be of the same sex, of nearly the same age, and be afflicted by the disease to a similar degree. Unfortunately, such conditions will often reduce the number of available subjects so drastically that not enough will remain for the experiment.

An important technique devised to get around this difficulty is *blocking*. The available subjects are divided into homogeneous blocks, those within each block being like each other, while subjects in different blocks may be quite dissimilar. Within each block, a certain number of subjects will then be treated, with the rest serving as controls. The subjects in each block are randomly divided into treatment and control groups. As the random divisions of the various blocks are performed separately, we may regard them as unrelated parts of the whole experiment, and use a product model each factor of which has structure (1).

To be specific, suppose there are n blocks. The first, consisting of N_1 subjects, is to be divided into s_1 controls and $t_1 = N_1 - s_1$ treatment subjects; the second, consisting of N_2 subjects, is to be divided into s_2 and $t_2 = N_2 - s_2$; and so on. The probability that the $t_1 + t_2 + \ldots + t_n$ treatment subjects will consist of a particular set of t_1 from the first block, a particular set of t_2 from the second block, etc. is just the product of the corresponding probabilities (1), i.e.

$$(2) \qquad 1 \bigg/ \binom{N_1}{t_1}\binom{N_2}{t_2} \cdots \binom{N_n}{t_n}$$

The idea of blocking is completely parallel to that of stratification discussed in Section 10.3, with the blocks playing the role of the strata. The $t_1 + t_2 + \ldots + t_n$ treatment subjects constitute in fact a stratified sample.

(iii) *Paired comparisons.* An extreme case of blocking occurs when pairs of subjects form natural blocks, as is the case for example when identical twins are available for the experiment or when the "subjects" are human hands. Then in each block of two, one subject receives the treatment and the other serves as control, the assignment being made at random. In this case $N_1 = N_2 = \ldots = N_n = 2$ and $t_1 = t_2 = \ldots = t_n = 1$. Therefore $\binom{N_1}{t_1} = \binom{N_2}{t_2} = \ldots = \binom{N_n}{t_n} = 2$ and the probability of any particular set of n subjects receiving the treatment is seen from (2) to be

$$(3) \qquad (\tfrac{1}{2})^n.$$

Although methods exist for testing the effectiveness of a treatment in the general case of blocking, based on the equal probabilities (2), we shall in this book for simplicity restrict ourselves to the two extreme cases of designs (i) and (iii). Tests for design (i) are discussed in Sections 2–4, while Sections 5 and 6 present tests for design (iii).

Strictly speaking, conclusions reached from comparative experiments of types (i)–(iii) are valid only for the particular subjects that were used, and extrapolation of the conclusions to other subjects must always be somewhat questionable. It is not unreasonable, for example, to assume that a chemical that appears worthless when tried on one set of cancer patients will also prove worthless for another. On the other hand, such an assumption may be unjustified, and a valuable discovery may thus be missed. However, scientific work is possible only if one proceeds on the basis of such assumptions.

From the point of view of extrapolating a finding made on a few experimental subjects to a large population of subjects, it is obviously desirable to use for the experiment subjects which are as representative as possible, and correspondingly dangerous to use subjects selected for their uniformity. On the other hand, as discussed above, uniformity is desirable in order to avoid masking the possible effect of the treatment by the variability of the subjects. A happy compromise is provided by the blocking design (ii).

The tests presented in Sections 3 and 6 of this chapter (and also a test to be given in Section 6 of the next chapter) are called "rank tests" because they employ the ranks of the observations with respect to each other, rather than using the observations directly. Rank tests were first introduced for their ease of use, as short-cut methods. It then came to be realized that they have another advantage: their validity does not require the specific assumptions needed by the tests that had been used previously. These often unreliable assumptions are expressed in terms of probability models that are mathematical functions precisely specified except for the values of some parameters. (In contrast to the older tests based on such parametric formulation, the rank tests are called "nonparametric.") In recent years it has been learned that rank tests, in addition to the advantages of simplicity and freedom from parametric assumption, are often more sensitive than the parametric tests for detecting treatment effects of the kind that arise in practical work.

PROBLEMS

1. In a primitive country in which on the average it rains on only 3% of the summer days, it was observed that it rained on 27% of the days following a special rain dance. Examine the experiment for possible bias in the assignment of treatment (dance or no dance) to subjects (summer days).

2. In an agricultural field trial, ten plots are available. of which five are selected at random to receive a new variety of wheat. Find the probability that
   (i) the five most fertile plots are selected;
   (ii) at least four of the five most fertile plots are selected.

3. In the preceding problem suppose that the ten plots are divided into five pairs, one of each pair being randomly chosen for the new variety. Find the probability that the new variety will be assigned to the more fertile plot

(i) in each pair;

(ii) in at least four of the five pairs.

4. Among the designs (i)–(iii), the number of possible choices of the subjects is largest for (i) and smallest for (iii). To illustrate this, find the number of possible choices of t subjects from a population of $2t$ both when the t subjects are chosen at random and when the population is divided into t pairs and one subject is chosen at random from each pair, in the cases (i) $t = 4$, (ii) $t = 6$, (iii) $t = 8$. [Hint: Use Table A.]

5. In an experiment to evaluate instruction by television, the lectures are given in one classroom and are carried by television into another room. Students are permitted to choose at the beginning of the term which room they will attend. At the end of the term, all take the same examination, and the marks of the two rooms are compared.

(i) Discuss the possibility of the conclusions being invalidated by bias;

(ii) Suggest a bias-free design which does not require the assignment of unwilling students to one room or the other.

6. In Problem 5, suggest reasonable ways of forming blocks for the use of design (ii).

12.2 THE FISHER-IRWIN TEST FOR TWO-BY-TWO TABLES

Experiments for the comparison of two treatments may be classified not only according to the design used but also according to the nature of the observed response. In an important class of problems, one is primarily interested in which of two possible responses occurs: one observes for example whether a child does or does not contract polio, whether or not it rains, or whether or not a surgical patient survives an operation. In these experiments we say the response is all-or-none or *quantal*. Alternatively, one may observe a *graded* response, which can take on many different numerical values, for example, the degree of paralysis, amount of rainfall, or length of life following an operation. Any graded response can be converted into a quantal one by recording only whether or not it exceeds a specified value. Thus, studies of cancer therapy often report merely how many patients survive five years, rather than giving the exact time of survival of the patients. We shall discuss now the analysis of comparative experiments with quantal response, taking up the consideration of graded response in the next two sections.

The data from a comparative experiment with complete randomization and quantal response can be summarized by giving the numbers of subjects with each of the two possible responses among those treated and among the controls. These can be arranged in a *two-by-two* table, as is shown in the following example.

EXAMPLE 1. *A heart treatment.* A physician believes that a certain treatment may prolong the life of persons who have suffered a coronary attack. To test his belief, an experiment is carried out in which five patients receive the treatment and five do not. The ten patients are carefully selected to be similar with regard to age, severity of attack, and general health, and then five are selected at random from the ten to receive the treatment, the remaining five serving as controls. Five years later, it is found that of the patients who received the treatment four are still alive, while of the control patients only two have survived. The data can be exhibited in the following two-by-two table.

	Alive	Dead	Total
Treated	4	1	5
Control	2	3	5
Total	6	4	10

The table suggests that the treatment helps, but the numbers are very small and perhaps it is only a result of chance that more of the treated than of the untreated patients lived for five years. In order to see whether this is a real possibility, we may test the null hypothesis that the treatment is completely without effect. Under this hypothesis the fate of each patient will be the same whether or not he is treated. We may think of his fate as determined even before it was decided (by lot) which patients would receive the treatment and which would serve as controls. Regardless of the outcome of this lottery, the six patients who did survive were "saved" (that is, were destined to survive the next five years) and the four who died were "doomed" to die before the termination of this period.

We may thus think of four of the patients as labeled "doomed," the remaining six carrying the label "saved." From these ten patients, five are chosen at random to constitute the treatment group. This means that all of the $\binom{10}{5} = 252$ possible choices of a treatment group are equally likely: a random sample of five was drawn from a lot of size ten of whom four are doomed and six saved. The number D of doomed patients found in the sample, that is, in the treatment group, is a random variable capable of taking on the values 0, 1, 2, 3, 4. Under the null hypothesis, D has the hypergeometric distribution ($N = 10$, $r = 4$, $s = 5$).

The variable D is a reasonable test statistic with small values of D furnishing evidence against the null hypothesis of no treatment effect, since D will tend to be smaller when the treatment is effective than when it is without value. The significance probability of the experiment is therefore by (6.2.1)

$$P(D \leqq 1) = P(D = 1) + P(D = 0)$$

$$= \frac{\binom{4}{1}\binom{6}{4}}{\binom{10}{5}} + \frac{\binom{4}{0}\binom{6}{5}}{\binom{10}{5}} = \frac{4 \cdot 15 + 1 \cdot 6}{252} = \frac{66}{252} = 0.26.$$

The evidence offered by the experiment against the null hypothesis is not very strong, since even if the treatment is without effect there is a chance of more than $\frac{1}{4}$ that $D \leqq 1$. Actually, the experiment is too small to provide a satisfactory test: even the most extreme value $D = 0$ has a probability of .024 under the hypothesis, so that at significance level $\alpha = .01$ we would accept the hypothesis regardless of how the observations come out.

The test we have just illustrated is called the Fisher-Irwin test since it was proposed independently by R. A. Fisher and J. O. Irwin. We shall now give another illustration of the test where however the numbers are so large that we shall use the normal approximation to the hypergeometric distribution instead of calculating the significance probability exactly.

EXAMPLE 2. *Salk vaccine.* The data from part of the 1954 trial of the Salk polio vaccine (see Example 1.1) may be summarized in the following two-by-two table.

	Polio	No Polio	Total
Vaccinated	33	200,712	200,745
Control	115	201,114	201,229
Total	148	401,826	401,974

The situation is quite analogous to that discussed in the preceding example. The null hypothesis of no treatment effect implies that the vaccine gave no protection, and hence that regardless of which children were vaccinated, the same 148 children would have contracted polio, the other 401,826 being spared. We may thus, under the null hypothesis, think of the children used in the experiment (the experimental subjects), as a "lot" of size $N = 401,974$ of which $r = 148$ are marked "polio," the remaining 401,826 being labeled "no polio." From this lot,* a sample of size $s = 200,745$ is selected at random to constitute the treatment group (that is, be vaccinated). It then follows, under the null hypothesis, that the number D of polio cases in the vaccinated group has the hypergeometric

---

* The method of randomization actually used was more complicated, but to simplify the discussion we shall analyze the experiment as if complete randomization had been employed. An analysis based on the actual randomization would yield essentially the same results.

distribution ($N = 401{,}974$, $r = 148$, $s = 200{,}745$). Since small values of D are to be expected if the vaccine is effective, we are inclined to reject the null hypothesis when D is small. The exact calculation of the significance probability $P(D \leqq 33)$ of the experiment is laborious, but we may use the normal approximation. Since by (6.2.3) and (6.2.5)

$$E(D) = \frac{sr}{N} = 73.9 \quad \text{and} \quad \text{Var}(D) = \frac{N-s}{N-1} s \frac{r(N-r)}{N^2} = 36.99$$

we find that approximately

$$P(D \leqq 33) = \Phi\left(\frac{33 + .5 - 73.9}{6.08}\right) = \Phi(-6.64).$$

This value is extremely small. From the auxiliary entries of Table E, and the symmetry of $\Phi(z)$, we see that it is less than one in ten million. One need not trust the normal approximation to seven decimal places to be quite confident that the apparent reduction of polio in the vaccinated group is not due to chance. In particular, we would reject the null hypothesis for example at level $\alpha = .0001$.

Frequently the experimental subjects enter the experiment, and must be dealt with, one at a time. For example, we cannot collect 20 rainstorms, choose 10 at random, and seed them. Again, in Example 1, a heart patient must be treated as soon as his disease has been diagnosed. How, in such circumstances, can we choose the treated group at random? This is in fact not difficult. Suppose in a rainmaking experiment it has been decided to observe twelve storms of which six are to be seeded. From the first twelve integers, we choose six at random, and seed the storms occurring at the corresponding six positions in time. For example, if our sample consists of the integers 1, 3, 4, 9, 10, and 12, we will seed the first storm that comes along, use the second as a control, seed the third and fourth, use the fifth as a control, and so on.

It is often necessary for the experimenter to exercise judgement in deciding whether a subject qualifies for inclusion in the experiment, and whenever there is an exercise of judgement there is also a possibility of bias. Suppose for example that the cloud seeding experiment is to be conducted by a professional rainmaker who has a heavy financial stake in proving that his service is effective. He insists that he select the storms that are to be included in the study, since not all storms are suitable for treatment by this method. Thus, when a storm approaches the test area, the rainmaker studies the weather map and announces whether this storm is suitable for seeding.

Now suppose he knows that of the storms selected by him as suitable, those numbered 1, 3, 4, 9, 10, and 12 are to be seeded, with storms 2, 5, 6, 7, 8, and 11 serving as controls. Since he then knows that the first storm he selects will be seeded, he is under great temptation to wait for a heavy

storm to serve as number 1; since the second is to be a control, he may
thereafter wait for a rather small storm to select as number 2; and so forth.
As a result, the treated storms would give heavier rain than the controls
even if the treatment is completely without effect: the experiment is biased
in favor of the treatment.   We may refer to bias that enters through the
selection of subjects as *selection bias*.

A simple device will greatly reduce the risk of selection bias: we may
keep secret from the experimenter the serial numbers of the subjects to be
treated.   This can be achieved by having one person (for example, a
statistician) do the random sampling, and another (the rainmaker) select
the storms.   When the rainmaker has decided that a storm is suitable, he
informs the statistician, who only then reveals whether the storm is to be
treated or used as a control.

Even this device will not completely eliminate the bias if the rainmaker knows
that, in all, 6 storms are to go into each group.   After storm 4 in the example
above, he knows that of the remaining 8, only 3 are to be treated.   It is therefore
more likely (probability = $\frac{5}{8}$) that the next storm will be a control, and by picking
a small one to serve as number 5 the rainmaker can again bias the experiment in his
favor, although to a much lesser degree than before.   Similarly, after storm 8 the
odds favor the treatment of the next storm, so that the rainmaker will be tempted
to pick a juicy one.   If this possibility were used deliberately and to full advantage,
the bias would be large enough completely to destroy the validity of the experiment.

The possibility of selection bias can be eliminated completely by changing the
design of the experiment.   Instead of deciding in advance to treat 6 storms and
have 6 controls, the decision may be made independently each time on the toss of
a penny.   The storm will be treated or used as control as the penny falls heads or
tails.   In this case the probability is always $\frac{1}{2}$ that the next selected storm will be
treated, and there is no way for selection bias to enter.

This design involves the new difficulty that the numbers of treated and control
storms now are random variables instead of being predetermined constants.   This
requires a different analysis, which we shall not discuss here.

As we remarked earlier in the section, the random assignment of subjects
to the two groups gives protection from biased assignment, but it should
not be thought that no further concern about bias is necessary.   Bias is
also possible in the conduct and interpretation of the experiment, and may
enter in an almost unbelievable variety of ways.   Consider, for example,
an experiment to compare two teaching methods.   If the person grading
the final examination papers has an interest in the outcome of the experiment, this may unconsciously influence his grading.   In a psychiatric experiment involving two different methods of treatment, the assessment of
the improvement of the patients is clearly highly subjective, and may quite
easily be influenced, at least to some degree, if the psychiatrist strongly believes in one of the treatments.   Such *assessment bias* may enter even if the

results of the experiment are measurements of length, weight readings, or counts of particles or stars on a photographic plate. Even with great care, it easily happens when counting a large number of dots that some are omitted or some counted twice. This phenomenon of assessment bias has often been demonstrated by presenting to an observer two groups of subjects in fact all treated alike, but telling the observer that only one group has received the treatment. It then frequently turned out that the observers reported higher responses in the group they believed to have been treated. To prevent such bias it is desirable that whenever possible the experimenter himself be kept in ignorance of the identity of the treated subjects until after he has recorded his assessment of the responses, or that this assessment be carried out by someone else who is not aware of the identity of the treated subjects.

A similar bias arises when the subjects themselves are people, who may believe in the treatment. Physicians are well acquainted with the fact that patients feel better merely because they are receiving attention, and frequently some innocuous prescription (called a placebo) is made when patients demand medication that the doctor considers unnecessary. The existence of a *placebo effect* has frequently been shown by giving sugar pills to an experimental group and nothing to a control group: usually, the patients in the experimental group respond significantly better. It is clear that the placebo effect may introduce a powerful bias. If the experimental group receives an active drug and the control group gets nothing, how can we be sure that the better response of the treated patients is due to the action of the drug and not merely to the fact that the patients are receiving attention? A simple way to avoid this bias is to give the control patients a placebo indistinguishable from the drug (or injection) given to the treated subjects, and to keep all patients in ignorance as to which ones are in which group. A clinical trial that is conducted in this manner is called *blind*. If in addition, in order to avoid assessment bias, the doctor evaluating the success of the treatment in each case is also ignorant as to which patients are the controls, the experiment is said to be *double blind*. For example, in the 1954 trials of the Salk vaccine, half the vials contained a placebo instead of the vaccine, and the vials were labeled with code numbers whose meaning was not revealed, even to the physicians conducting the experiment, until after their diagnoses of polio were on file.

PROBLEMS

1. In Example 1, what would be the available significance levels less than $\frac{1}{4}$ if
   (i) 5 of the 10 patients had died within 5 years;
   (ii) 3 of the 10 patients had died within 5 years?

2. In Example 1, suppose the actual survival times, in years, were

Treated	6.5,	4.2,	17.8,	7.9,	13.2
Control	6.7,	.4,	2.9,	1.2,	5.6

Find the significance probability if the only available information for each patient is whether or not he is alive after seven years.

3. In Example 1, find the significance probability if there are twelve patients and if the data are given by the following table.

(i)	Alive	Dead		(ii)	Alive	Dead
Treated	5	2		Treated	5	2
Control	2	3		Control	3	2

4. To determine whether an engine oil additive decreases the chances of needing an engine overhaul within two years, a truck company with 25 new trucks selects ten of these at random to be given the additive. The results of the experiment are given in the following table.

	No overhaul	Overhaul	Total
Additive	5	5	10
No additive	3	12	15
Total	8	17	25

Find the significance probability, and determine whether the results are significant at the 5% level.

5. A group of 200 students is used to test whether a certain law course improves the chances of passing the bar examination. One hundred of the students are selected at random to attend the course, the other hundred serve as controls, with the following results.

	Pass	Fail	Total
Attend course	75	25	100
Do not attend	57	43	100
Total	132	68	200

Use the normal approximation to find the significance probability, and determine whether the results are significant at the 2% level.

6. In the preceding problem find the significance probability if among those attending the course 70 passed and 30 failed.

7. The results of a comparative experiment are given by the following table.

	Success	Failure	Total
Treated	73	727	800
Control	72	928	1000
Total	145	1655	1800

Is the number of successes in the treated group significantly high at the 5% level? (Use the normal approximation.)

8. In order to determine whether a more expensive method of storing would, as claimed, give better protection against spoilage, 1000 items are assigned at random, 500 to each of the two methods. If the number of spoiled items turns out to be 30 among those stored in the standard way and 20 among those stored in the more expensive way, is this result significant at the 3% level against the null hypothesis that there is no difference between the two methods?

9. In the preceding problem, let there be $1000k$ items of which half are assigned to each of the two methods, and suppose that the number of spoiled items under the standard and more expensive way are $30k$ and $20k$ respectively. Find the significance probability if (i) $k = .5$; (ii) $k = 1$; (iii) $k = 1.5$; (iv) $k = 2$.

10. Suppose that in a comparative experiment with 100 treatment subjects and 100 controls, 50 of the controls are successes and 50 failures. How many successes must there be among the treated subjects for the experiment to be significant at the 1% level? [Hint: Denote the number of successes among the treated subjects by D and use the normal approximation.]

12.3 THE WILCOXON TWO-SAMPLE TEST

In the preceding section we saw how to test the hypothesis of equality of two treatments for the case of quantal response; we shall now consider the same problem when the response is graded (that is, able to take on many different numerical values). As before, we assume a completely randomized design.

EXAMPLE 1. *A heart treatment.* Let us suppose that in Example 2.1 the response observed is the length of life of each patient instead of merely whether the patient survives five years. Let the actual survival times, in years, be as follows:

Treated	4.2,	6.5,	7.9,	13.2,	17.8
Control	.4,	1.2,	2.9,	5.6,	6.7

The two-by-two table of Example 2.1 is obtainable from these data by ignoring the actual values and recording only that four of the treated and two of the control patients survived five years. We may refer to this process as "quantizing" the data.

It is intuitively clear that quantizing will in general involve a loss of information, since it suppresses the relationship among the observed values that are grouped together. Thus, in the example, not only did four treated patients survive five years as compared with but two control patients, but the four tended to live longer than the two. Similarly, among those that did not survive five years, the one treated patient lived longer (4.2 years) than any of the three control patients. These interrelationships can be made clear visually by plotting the two sets of data on a common time scale (Fig. 1).

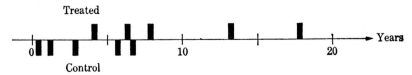

FIGURE 1.

It is seen that the survival times of the treated patients tend to lie to the right of those of the control patients, and it is clear that this fact tends to undermine the null hypothesis of no treatment effect. We shall now present a test that pays attention to all of the order relationships between the two groups of survival times.

Let us arrange the ten survival times in order and number them from the smallest to the largest:

.4	1.2	2.9	4.2	5.6	6.5	6.7	7.9	13.2	17.8
1	2	3	4	5	6	7	8	9	10

The number that each observation receives in this way is called the *rank* of the observation. Thus we say that the smallest observation .4 has rank 1, the largest 17.8 has rank 10, and so forth. (Note that all ten survival times are different, which permits them to be ranked unambiguously. A method for dealing with "tied ranks" will be discussed in Section 12.7.) We have underscored the ranks of the treated survival times. The fact that the treated patients, as a group, lived longer is reflected by their having, in general, higher ranks (4, 6, 8, 9, 10) than do the control patients (1, 2, 3, 5, 7).

A simple but highly effective test statistic is the sum W_T of the ranks of the treated subjects, which in our example would be $W_T = 4 + 6 + 8 + 9 + 10 = 37$. If the treatment has the desired effect of increasing a subject's response, the response of the treated subjects will tend to have large values and hence large ranks. The sum of the ranks of the treated subjects will then tend to be larger than if the treatment had no effect, so that one would tend to reject the null hypothesis when W_T is large.

We could equally well have used as test statistic the sum W_C of the ranks of the control subjects, rejecting the null hypothesis when W_C is small. Since $W_C + W_T$ is the sum of all 10 ranks, it must be true that

$$W_C + W_T = 1 + 2 + \ldots + 10 = 55$$

however the experiment turned out; thus, each of the two rank sums determines the other. With large values of W_T and small values of W_C considered significant, the two statistics determine in fact the same test, which was proposed by Frank Wilcoxon in 1945 and which is known as the *Wilcoxon two-sample test.*

To determine the relationship between the critical value of W_C and the level of significance, we must find the distribution of W_C under the null hypothesis. The basic argument here is very similar to that used for the Fisher-Irwin test. If the null hypothesis is correct, the treatment has no effect and the five treated patients would have lived just as long without treatment. The ten observed survival times may be regarded as ten fixed

numbers, which are determined even before it is decided which patients
are to be treated and which to serve as controls. At the time this
assignment is made, each patient may thus be considered as being labeled with his survival time and with the rank of this time (one patient
for example with the time .4 and rank 1, another patient with the time
1.2 and rank 2, etc.). The selection of five patients (and hence of five
ranks between 1 and 10) to constitute the control group is then made at
random, so that all $\binom{10}{5} = 252$ possible choices of the control ranks are
equally likely. Each such choice gives to W_C a value equal to the sum of
the five chosen ranks. The situation may be represented schematically
by a lottery in which ten tickets are available paying prizes of $1, 2, \ldots, 10$.
We select five of the tickets at random and W_C denotes the total prize won
by these five tickets. (This problem was considered in Example 5.3.5.)
The entire distribution of W_C can be found by listing the 252 sets of five
of the first ten positive integers, forming the sum of each set, and counting
the number of times each of the possible values of W_C occurs in this process.

Since this is rather laborious, it is fortunate that we do not need the
entire distribution to determine the significance probability of the result
$W_T = 37$, or equivalently $W_C = 18$. To find the probability that
$W_C \leq 18$, it is only necessary to determine in how many ways five distinct
positive integers between 1 and 10 may be chosen to produce a sum less
than or equal to 18. It is easy to list all such choices by inspection, as
follows:

$$1 + 2 + 3 + 4 + 5 = 15$$
$$1 + 2 + 3 + 4 + 6 = 16$$
$$1 + 2 + 3 + 4 + 7 = 17$$

$$1 + 2 + 3 + 5 + 6 = 17$$
$$1 + 2 + 3 + 4 + 8 = 18$$
$$1 + 2 + 3 + 5 + 7 = 18$$
$$1 + 2 + 4 + 5 + 6 = 18$$

Out of the 252 equally likely cases, seven are favorable to the event
$W_C \leq 18$, so that $P(W_C \leq 18) = 7/252 = .026$. This is considerably
smaller than the value .26 obtained by applying the Fisher-Irwin test to
the same data. The Fisher-Irwin test does not always give a larger significance probability, but when both tests are applicable, the Wilcoxon
test may in general be expected to be the more sensitive. The basic
reason for this is that quantization of the data throws away useful information.

In the above example, the numbers of subjects in the treated and control
groups were equal. The Wilcoxon test is however not restricted to this

case. If there were, for example, five controls and eight treated subjects, the ranks of the five controls would be randomly selected from the positive integers 1, 2, ..., 13 and the significance probability of the sum of these ranks could be computed as before. In the next section we shall discuss a table which gives the null distribution of W_C (and of W_T) for small numbers of treated and control subjects.

When the group sizes are too large to be covered by the table, it has been found that a normal approximation gives good results. For this approximation we need the expectation and variance of W_C and W_T, which at the end of the section will be shown to be given by

$$(1) \qquad E(W_C) = \frac{s(N+1)}{2}, \quad E(W_T) = \frac{t(N+1)}{2}$$

and

$$(2) \qquad \mathrm{Var}(W_C) = \mathrm{Var}(W_T) = \frac{st(N+1)}{12}$$

where, as in Section 1, t and s denote the numbers of treated and control subjects, and where $N = s + t$.

EXAMPLE 2. *Rainmaking.* In a rainmaking experiment (Example 1.2), there were 13 storms of which eight were selected at random and treated by seeding. The average amounts of rainfall in a system of rain gauges were as follows:

> Treated storms: .06, .13, .15, .28, .41, .62, .83, 1.26
> Control storms: .02, .09, .21, .29, 1.09

Here the control storms have ranks 1, 3, 6, 8, 12 while the treated storms have ranks 2, 4, 5, 7, 9, 10, 11, 13 so that $W_C = 30$ and $W_T = 61$, where of course $W_C + W_T = 1 + 2 + \ldots + 13 = 91$. If the treatment were effective, the treated ranks and hence W_T would tend to be large, while the control ranks and hence W_C would tend to be small. Thus one could either use large values of W_T or small values of W_C to indicate significance. Choosing for example the former, we may ask: is the result $W_T = 61$ significant at the 5% level? We find $E(W_T) = (8 \cdot 14)/2 = 56$, $\mathrm{Var}(W_T) = (5 \cdot 8 \cdot 14)/12 = \frac{140}{3}$, and $\mathrm{SD}(W_T) = \sqrt{\frac{140}{3}} = 6.83$. Since the upper 5% point on the normal curve is at 1.645, the corresponding critical value c for W_T would be approximately given by

$$\frac{c - .5 - 56}{6.83} = 1.645 \quad \text{or} \quad c = 67.7.$$

Thus, the observed value $W_T = 61$ is far from being significant at the 5% level. In fact, $P(W_T \geq 61)$ is approximately

$$1 - \Phi\left(\frac{61 - .5 - 56}{6.83}\right) = 1 - \Phi(.659) = .255.$$

(In Problem 4.1(v) the exact value will be found to be $P(W_T \geq 61) = P(W_C \leq 30) = \frac{337}{1287} = .262$, so that the normal approximation is reasonably close.)

In both of our examples, the treatment, if effective, would tend to increase the response (lifetime, amount of rain) and hence to increase the ranks of the treated subjects. Accordingly, it is large values of W_T (and small values of W_C) which are significant. However, in other experiments, the treatment effect being tested may tend to decrease response. In those cases, it will of course be *small* values of W_T and *large* values of W_C that are significant.

Suppose for example that the response is the number of points obtained on the final examination in Example 1.4. Here we wish to know whether the treatment (TV) has the effect of lowering the response. Thus low values of W_T and high values of W_C are significant. On the other hand, if the final is a multiple-choice examination, and the observed response is the number of wrong answers, the situation is reversed. In this case, high values of W_T and low values of W_C are significant.

We shall now derive formulas (1) and (2). Since W_C is the sum of s integers chosen at random from the integers $1, 2, \ldots, N = s + t$, we can apply the results obtained for the sampling model in Section 9.1. In the notation of that section, $v_1, \ldots, v_N$ denote the integers from 1 to N, and $Y_1, \ldots, Y_s$ the s chosen integers, so that

$$W_C = Y_1 + \ldots + Y_s \quad \text{and} \quad \bar{v} = (1 + 2 + \ldots + N)/N = \tfrac{1}{2}(N + 1).$$

Hence from formula (9.1.6)

$$E(W_C) = s \cdot \bar{v} = \tfrac{1}{2}s(N + 1)$$

as was to be proved.

From formula (9.1.8) it follows that

$$\text{Var}(W_C) = \text{Var}(Y_1 + \ldots + Y_s) = \frac{N - s}{N - 1} s\tau^2,$$

where by (5.6.5)

$$\tau^2 = \frac{1^2 + 2^2 + \ldots + N^2}{N} - \bar{v}^2$$

Since $\bar{v} = \tfrac{1}{2}(N + 1)$ and the sum of the squares of the first N integers is $\tfrac{1}{6}N(N + 1)(2N + 1)$ (see (6.8.4)), τ^2 simplifies to

$$\tau^2 = \frac{(N+1)(2N+1)}{6} - \left(\frac{N+1}{2}\right)^2 = \frac{N^2-1}{12}.$$

Substitution of this in the formula for $\mathrm{Var}(W_C)$ completes the proof of (3.2).

Since W_T is the sum of a random sample of t of the integers, instead of s, the formulas apply to W_T with s and t interchanged.

PROBLEMS

1. Let N subjects be divided at random into t treatment subjects and $s = N - t$ controls, and let a graded response be observed for each subject.

   (i) What is the smallest possible value of W_C?

   (ii) Assuming the null hypothesis of no treatment effect, find the probability that W_C takes on this smallest possible value.

[Hint: See (6.8.2).]

2. Under the assumptions of Problem 1, find the largest possible value of W_C.

3. Under the assumptions of Problem 1 and assuming the hypothesis of no treatment effect, determine the distribution of W_C when $s = 1$.

4. Use the distribution of W_C determined in the preceding problem to find (i) $E_H(W_C)$; (ii) $\mathrm{Var}_H(W_C)$ and check your results with formulas (1) and (2).

5. Under the assumptions of Problem 1, find a formula for $W_C + W_T$ and check it against the value $W_C + W_T = 55$ found in Example 1 for $s = t = 5$.

6. Under the assumptions of Problem 1, find

   (i) the second smallest value of W_C;

   (ii) the probability of this value, assuming the hypothesis of no treatment effect.

7. Suppose that in Example 1, we have $s = 4$, $t = 5$ and the following nine survival times:

$$\begin{array}{llllll}
\text{Treated} & 1.8, & 2.7, & 6.3, & 8.1, & 8.8 \\
\text{Control} & .6, & 1.1, & 4.5, & 5.1
\end{array}$$

Use the method of Example 1 to find the significance probability of W_C.

8. To find out whether preparation can raise one's IQ as measured by an intelligence test, five of eleven available subjects are selected at random, and permitted to see the questions and answers on a similar test. The other six serve as controls and receive no preparation. All eleven are then given the test with the following results:

$$\begin{array}{lllllll}
\text{Preparation} & 108, & 112, & 114, & 120, & 126 \\
\text{No preparation} & 98, & 101, & 102, & 105, & 110, & 111
\end{array}$$

(i) Would high or low values of W_T be significant?

(ii) Find the significance probability of W_T.

9. In the preceding problem, suppose that $N = 29$ students are involved in the study and that the results are as follows:

Preparation

97, 108, 111, 112, 114, 118, 120, 121, 123, 125, 126, 128, 131, 139

No Preparation

94, 95, 98, 100, 101, 102, 105, 107, 108, 109, 113, 117, 119, 122, 127

Compute the significance probability using the normal approximation.

10. In Example 2, suppose eight storms were treated and six served as controls with the following results:

Treated   .07,   .26,   .29,   .68,   .73,   .89,   1.02,   1.14
Control   .03,   .05,   .11,   .23,   .67,   .81

Use the normal approximation to find the significance probability of W_C.

11. In the preceding problem, find the significance probability of the data under the Fisher-Irwin test if it is only known for each storm whether the amount of rainfall was above or below (i) .1, (ii) .2, (iii) .3.

12. In Problem 9, use the normal approximation to find the significance probability of the data under the Fisher-Irwin test if it is only known for each student whether the result of the test was above or below (i) 103; (ii) 110; (iii) 116; (iv) 124.

13. Suppose that of 20 students volunteering for the television experiment of Example 1.4, ten are selected at random to receive instruction by TV, the other ten receiving their instruction directly and thereby serving as controls. Low values of W_T being significant, use the normal approximation to find the significance probability of W_T if the number of points received by the students on the final examination are as follows:

TV instruction   3,   6,   18,   25,   37,   48,   49,   53,   81,   89
Direct instruction 9,   34,   57,   61,   64,   75,   91,   93,   98,   100

14. Solve the preceding problem if $N = 24$ and the results were as follows:

TV instruction   3,   5,   6,   13,   19,   23,   25,   38,   44,   49,   74,   98
Direct instruction 9,   10,   34,   40,   57,   62,   65,   72,   88,   95,   96,   100

15. Construct other survival times for Example 1 satisfying the restrictions of Example 2.1 and such that the Wilcoxon test has

(i) the maximum significance probability possible under these restrictions;

(ii) the minimum significance probability possible under these restrictions.

16. Solve the two parts of the preceding problem for the data of Problem 2.3(i) and Problem 2.3(ii).

12.4 THE WILCOXON DISTRIBUTION

As is the case with the normal approximation to other distributions such as the binomial and hypergeometric, the normal approximation to the null distribution of the Wilcoxon rank-sum test statistic illustrated in the preceding section tends to be accurate only if the sample sizes are not too small. In this section we shall discuss a table, given as Table H at the end of the book, which gives the exact null probabilities when neither s nor t exceeds eight; this table, together with the normal approximation, covers most practical uses of the test.

We can best present Table H by asking the reader first to work through a simple illustration. Suppose there are $t = 2$ treated subjects and $s = 3$ control subjects. Then the two treated subjects can be chosen from the total number $N = 5$ of subjects in $\binom{5}{2} = 10$ ways. If the five responses are ranked, say from smallest to largest, the sum W_T of the ranks of the treated subjects can have any value from $1 + 2 = 3$ to $4 + 5 = 9$. Let $\#(W_T = w)$ be the number of choices of two of the five ranks for which W_T has the value w. Since under the null hypothesis the 10 choices are equally likely, the probability that $W_T = w$ is then

$$P(W_T = w) = \#(W_T = w)/10.$$

Tableau (1) shows $\#(W_T = w)$ for $w = 3, 4, \ldots, 9$.

(1)

w	3	4	5	6	7	8	9	10	11	12
$\#(W_T = w)$	1	1	2	2	2	1	1			
$\#(W_C = w)$				1	1	2	2	2	1	1

The reader should check these entries by actually listing the ten possible choices of two out of the five ranks together with the values of W_T resulting from each choice.

Also shown in tableau (1) is the distribution of the sum W_C of the three control ranks. Each choice of two subjects for treatment leaves three as controls, so there are again $10 = \binom{5}{3}$ possible cases, and

$$P(W_C = w) = \#(W_C = w)/10.$$

The reader will notice that the distributions of W_C and W_T are the same, except that W_C runs from 6 ($= 1 + 2 + 3$) to 12 ($= 3 + 4 + 5$) rather than from 3 to 9. The two distributions can be made to coincide, thereby permitting the table to be cut in half, by subtracting from each statistic its minimum value, so that after the subtraction each begins at zero. Let the resulting statistics be $U_T = W_T - 3$ and $U_C = W_C - 6$. Then U_T

and U_C have the same distribution, shown in tableau (2), so that the subscript may be omitted. We also give in (2) the number of cases for which $U \leq u$, which is more convenient for obtaining significance probabilities.

(2)

u	0	1	2	3	4	5	6
$\#(U = u)$	1	1	2	2	2	1	1
$\#(U \leq u)$	1	2	4	6	8	9	10

Another saving is possible in the size of the table. Suppose there had been $t = 3$ treated and $s = 2$ control subjects, instead of the other way around. Then W_T would have been the sum of three of the five ranks, and would have therefore had the distribution of W_C shown in (1). Similarly, W_C would have had the distribution of W_T. It follows that the distribution (2) serves not only for either statistic, but also for group sizes 2 and 3 regardless of which is treated and which is control. The simple distribution (2) thus covers four different cases. This welcome consolidation is not an accident of the particular group sizes 2 and 3, but holds for any pair of group sizes, as we shall show later in the section.

We are now ready to present Table H. Each row of the table corresponds to a pair of group sizes—one treated, the other control, it does not matter which is which. The table can be used for either W_T or W_C, but from the chosen statistic one must first subtract its minimum possible value. By (6.8.2) this is

$$1 + 2 + \cdots + t = \tfrac{1}{2}t(t + 1) \qquad \text{for } W_T$$

and

$$1 + 2 + \cdots + s = \tfrac{1}{2}s(s + 1) \qquad \text{for } W_C,$$

so that the resulting statistics are

(3) $U_T = W_T - \tfrac{1}{2}t(t + 1)$ and $U_C = W_C - \tfrac{1}{2}s(s + 1)$.

Since U_T and U_C have the same distribution, as we shall show below, we can omit the subscript. The entries in the table give, for each value of u, the number of equally likely cases $\#(U \leq u)$ that have that value of u or a smaller value. Division of this number by the total number $\binom{N}{s} = \binom{N}{t}$ of possible cases gives the probability $P(U \leq u)$. These binomial coefficients are of course obtainable from Table A, but for convenience they are also given as the second column of Table H. (When expressed in terms of the statistic U_T or U_C, the test of the preceding section is known as the *Mann-Whitney* test.)

It is not necessary in Table H to provide also the probabilities $P(U \geq u)$; for by proper choice of the statistic W_T or W_C, one can always arrange it so that small values of the chosen statistic will be significant. (Alternatively, but less conveniently, the probability $P(U \geqq u)$ can be obtained from those given in the table by using the fact that the distribution of U is symmetric about the point $\frac{1}{2}st$, as will be shown below.)

To illustrate the use of the table, consider once more Example 3.1. Here $s = t = 5$ and $W_C = 18$, so that by (3), $U_C = 18 - 15 = 3$. Since in this example small values of W_C are significant, the significance probability is $P(W_C \leqq 18) = P(U_C \leqq 3)$. Entering Table H in the row corresponding to the group sizes (5, 5) and the column corresponding to the value $u = 3$, we find that $\#(U \leqq u) = 7$. The denominator $\binom{N}{s} = \binom{10}{5} = 252$ is given in the second column and the desired significance probability is therefore $P(U \leqq 3) = 7/252 = .026$, in agreement with the result found in Example 3.1 by counting cases.

Let us now show that U_T and U_C always have the same distribution. To this end, we shall show first that the distribution of W_T is symmetric about the point $\frac{1}{2}t(t + 1) + \frac{1}{2}st = \frac{1}{2}t(N + 1)$. (Thus, when $t = 2$, $s = 3$, any value of w above $6 = \frac{1}{2} \cdot 2 \cdot (5 + 1)$ is as likely as the value equally far below 6.) The responses of the subjects are usually ranked from smallest to largest, but suppose that they are instead assigned the *inverse ranks* from largest to smallest. Thus the subject with rank 1 has inverse rank N, the subject with rank 2 has inverse rank $N - 1$, etc. For each subject, the sum of his rank and his inverse rank is always $N + 1$. If W_T' is the sum of the inverse ranks of the t treatment subjects, then the total of the ranks and inverse ranks of these subjects is $t(N + 1)$, so that $W_T + W_T' = t(N + 1)$, or

(4) $$W_T - \tfrac{1}{2}t(N + 1) = -[W_T' - \tfrac{1}{2}t(N + 1)].$$

However W_T', like W_T, is just the sum of t integers chosen at random from $1, 2, \ldots, N$, so that it has the same distribution as W_T. Therefore the right-hand side of (4) has the same distribution as $-[W_T - \frac{1}{2}t(N + 1)]$. It follows that the left-hand side has the same distribution as its negative and hence is symmetrically distributed about zero. (See Problem 5.4.16.)

We shall now use this result to prove that U_T and U_C have the same distribution. Note first that $W_C + W_T$ is the sum of all the integers from 1 to N and hence

$$W_C + W_T = \tfrac{1}{2}N(N + 1) = \tfrac{1}{2}s(N + 1) + \tfrac{1}{2}t(N + 1),$$

or

(5) $$W_C - \tfrac{1}{2}s(N + 1) = -[W_T - \tfrac{1}{2}t(N + 1)].$$

By the symmetry just established, the right-hand side of (5) has the same distribution as its negative. It follows that

$$W_T - \tfrac{1}{2}t(N + 1) = U_T + \tfrac{1}{2}st$$

has the same distribution as

$$W_C - \tfrac{1}{2}s(N + 1) = U_C + \tfrac{1}{2}st.$$

Hence U_T and U_C have the same distribution, as was to be shown. The common distribution runs from 0 to st, and is symmetric about $\tfrac{1}{2}st$ (Problem 11).

We now have to explain the last row of Table H.   An inspection of the columns of Table H shows that as the group sizes are increased, the entries in a given column at first also increase but eventually achieve a stable maximum.   In fact, this maximum is attained as soon as both group sizes reach the value of u that heads the column.   For example, the entries in the column $u = 5$ are equal to 19 for all pairs of group sizes both of which are at least 5.   This phenomenon, the reason for which is given below, makes it possible to present in the last row of Table H values of $\#(U \leq u)$ which hold for any pair of group sizes both of which are at least equal to u.

To illustrate this last row, let us show how to find the probability $P(W_C \leq 43)$ when $s = 8$, $t = 9$, in spite of the fact that the pair (8,9) is outside of the range of the table.   Since $\tfrac{1}{2}s(s + 1) = 36$, the desired probability is equal to $P(U_C \leq 7)$.   Neither group size is less than 7, and we may therefore use the last row of Table H to find $\#(U_C \leq 7) = 45$. It now follows from Table A that the desired probability is $P(U_C \leq u) = 45/24310 = .00185$.

To see why the entries in the u column of Table H are the same for all rows for which both $s \geq u$ and $t \geq u$, consider W_T for a fixed number t of treated subjects and a fixed value of u, but with varying s.   Since the minimum value of W_T is $1 + 2 + \cdots + t$, the value of W_T is then fixed at $w = [1 + \cdots + (t - 1) + t] + u$. The desired entry is the number of ways of finding t distinct positive integers with sum w, subject to the restriction that none of the summands may exceed $N = t + s$. Without this restriction, what is the largest integer that could enter into such a sum?   It is obtained from the arrangement

$$w = [1 + 2 + \cdots + (t - 1)] + (t + u),$$

in which the first $t - 1$ summands are as small as possible and hence the last one as large as possible.   Without the restriction, the largest usable integer is therefore $t + u$.   It follows that the restriction, that no summand may exceed $t + s$, does not cut down the number of possibilities as soon as $s \geq u$.   The analogous consideration of W_C shows that the restriction has no effect on the distribution of U_C as soon as $t \geq u$, and this completes the proof.

PROBLEMS

1. Use Table H to find the following probabilities, and check each against its normal approximation.

(i) $P(W_C \leqq 27)$ when $s = 6$, $t = 8$;
(ii) $P(W_C \leqq 55)$ when $s = t = 8$;
(iii) $P(W_T \leqq 36)$ when $s = 8$, $t = 7$;
(iv) $P(W_T \leqq 34)$ when $s = t = 6$;
(v) $P(W_C \leqq 30)$ when $s = 5$, $t = 8$.

2. Use Table H to find the following probabilities.
   (i) $P(W_C = 31)$ when $s = 5$, $t = 8$;
   (ii) $P(W_T = 38)$ when $s = 4$, $t = 7$.

3. (i) Compute by enumeration the entries required to extend Table H to the cases $s = 2$, $t = 2$; $s = 2$, $t = 4$; $s = 2$, $t = 6$.
   (ii) Conjecture a general formula for $s = 2$ and t even.
   (iii) Make a similar analysis for $s = 2$ and t odd.

4. Find the probability $P(W_C = 11)$ when $s = 3$ and $t = 5$ by listing cases.

5. Use Tables A and H to find the following probabilities.
   (i) $P(W_T \leqq 9)$ when $s = 9$, $t = 3$;
   (ii) $P(W_T \leqq 9)$ when $s = 12$, $t = 3$;
   (iii) $P(W_C \leqq 26)$ when $s = 6$, $t = 10$;
   (iv) $P(W_C \leqq 35)$ when $s = 7$, $t = 10$.

6. From a class of 15 students, eight are selected at random to receive a daily vitamin pill, the other seven serving as controls. At the 5% level, can the vitamins be said to have reduced significantly the number of school days missed on account of illness, if the results are as follows:

	Number of Days Missed							
Treated	0,	2,	3,	7,	8,	10,	13,	18
Control	4,	11,	12,	15,	20,	21,	27	

Use Table H to find the significance probability of W_C and compare this probability with its normal approximation.

7. Solve the preceding problem if the data are as follows:

	Number of Days Missed							
Treated	0,	1,	4,	7,	8,	10,	16,	18
Control	3,	11,	14,	17,	20,	21,	47	

8. For $s = 4$, $t = 7$ find (i) $P(W_C = 35)$, (ii) $P(W_C \geq 35)$, using Table H and the fact that the distribution of W_C is symmetric about $\frac{1}{2}s(N + 1) = 24$.

9. For $s = 8$, $t = 5$ find (i) $P(W_T = 48)$; (ii) $P(W_T \geq 48)$, using Table H and the fact that the distribution of W_T is symmetric about $\frac{1}{2}t(N + 1)$.

10. Find (i) the expectation and (ii) the variance of the random variable U (i.e., U_C or U_T) given by (3).

11. (i) Show that the value set of U is $0, 1, \ldots, st$.
    (ii) Show that the distribution of U is symmetric about $\frac{1}{2}st$.

12. Suppose that $s = t$ and that the ranks of the control responses are $1, 3, \ldots, 2s - 1$ and those of the treatment responses $2, 4, \ldots, 2s$. If the hy-

pothesis is rejected for small values of W_c,

   (i) find the significance probability for $s = 2, 4, 6$;

   (ii) use the normal approximation to find the (approximate) significance probability for $s = 10, 20, 30$;

   (iii) determine from the normal approximation what happens as s tends to infinity.

13. Suppose the ordered responses are assigned the scores $2, 4, 6, \ldots$ instead of the ranks $1, 2, 3, \ldots$. How would the critical value of the Wilcoxon test be altered by the use of these "even ranks"? What if the odd ranks $1, 3, 5, \ldots$ were used instead?

14. Is it possible for the Wilcoxon test to show that the treatment of Example 2.1 significantly prolongs life, even though on the average the control patients lived longer?

15. Suppose the ranks in Example 1 had been assigned inversely, giving time 17.8 years the rank 1, time 13.2 years the rank 2, etc. How would this have changed the significance probability of the data?

16. Suppose the ordered subject responses are numbered from both ends toward the middle in this way:

$$1 \quad 4 \quad 5 \quad 8 \quad 9 \quad \ldots \qquad \ldots \quad 7 \quad 6 \quad 3 \quad 2.$$

The sum S for the treated subjects of these scores is the *Siegel-Tukey* test statistic. What is the null distribution of S? Qualitatively, for what sort of treatment effect would S tend to have a small value? A large value?

17. Let $\#(w; s, t)$ be the number of possible choices of s integers from the integers $1, \ldots, s + t$ such that the sum óf the s chosen integers is equal to w. Prove the recursion formula

$$(2) \qquad \#(w;\ s, t) = \#(w - N;\ s - 1, t) + \#(w;\ s, t - 1).$$

[Hint: How many choices are there if the s integers are to be selected from $1, \ldots, s + t - 1$ (i.e., do not include the integer $s + t$)? How many choices are there if one of the s integers to be selected is $s + t$?]

18. Use (2), (which was used to compute Table H), and Problem 3.3 to add to Table H the rows $(2, 9)$, $(3, 9)$, and $(4, 9)$.

19. Let the s control responses be denoted by $X_1, \ldots, X_s$ and the t treatment responses by $Y_1, \ldots, Y_t$ and let U denote the number among all possible pairs (X, Y):

$$(X_1, Y_1), (X_1, Y_2), \ldots, (X_s, Y_t)$$

for which X is less than Y. Prove (3), by noting that the number of X's less than the smallest Y (which has rank s_1) is $s_1 - 1$; the number of X's less than the second smallest Y is $s_2 - 2$; etc.

12.5 THE SIGN TEST FOR PAIRED COMPARISONS

In Section 1 we discussed three designs for comparative experiments:

(i) the completely randomized design, (ii) randomization within blocks, and (iii) the design for paired comparisons, also called the "matched pairs" design, to which (ii) specializes when the blocks are of size two. We shall now discuss tests that may be used with design (iii). As distinguished from (i), the assignment of subjects to treatment or control is not completely at random in design (iii) but rather the randomization is restricted to occur separately within each pair. This design is particularly appropriate when the subjects are naturally paired as in Examples 1 and 4 below. It may however also be used in other cases by grouping the subjects into artificial pairs, as illustrated in Examples 2 and 3.

EXAMPLE 1. *Twins.* In an experiment to compare the synthetic (phonetic) and analytic ("look-say") methods of teaching children to read, the comparison is complicated by the fact that different children learn at such different rates, even when taught the same way. Identical twins form an ideal pair for this and many other experiments. They have the same genetic composition, are the same age, and usually have very similar environments. The reading methods may be compared by using a number of such pairs of twins, one twin being instructed by each method.

EXAMPLE 2. *Agricultural field trials.* To compare two methods of irrigating cotton, a field is divided into 40 parallel strips of equal area. The strip is the subject, and the response is the yield of cotton from the strip. The yield will be influenced not only by the irrigation method, but by many other factors, such as soil fertility and insect damage. Two strips that are close together are likely to be similar with respect to these other factors. Therefore it seems reasonable to match two adjacent strips as a pair, considering the field as divided into 20 blocks, each block consisting of two adjacent strips.

EXAMPLE 3. *Clinical trials.* When two drugs or methods of treating a disease are to be compared, the clinician must take into account the great variability of patients in their rates of recovery. The disturbing effect of this variability may be reduced by arranging the patients who are to be the experimental subjects into pairs of patients with similar prognosis. For example, if age is known to influence the speed of recovery, it will be desirable to put into the same pair two patients of about the same age. Sex, severity of the disease, and medical history are other factors that may be considered when forming the pairs. The design is successful to the degree that the two patients within each pair will behave alike in their response to the same therapy, so that differences in response are attributable to treatment differences.

The aim of matching is to pair together subjects as alike as possible, but of course no two subjects are identical and even if they were treated alike they would not give identical responses. As was pointed out in Section 1, the fairness of the comparison can be insured by randomization. In the case of matched pairs, this can be done simply by tossing a coin for each pair to decide which subject of the pair gets which treatment. When the assignment of treatments to subjects within each pair is random, there is no possibility of the experimenter favoring one treatment over the other.

In addition to preventing assignment bias the act of random assignment, as in the case of complete randomization, has another important property. As we shall see below, it provides the probabilistic basis for a test of the null hypothesis of no treatment effect. Let us illustrate the procedure by an example.

EXAMPLE 4. *Sun-tan lotions.* The manufacturer of a sun-tan lotion wishes to know whether a new ingredient increases the protection his lotion gives against sunburn. Seven volunteers have their backs exposed to the sun, with the old lotion on one side and the new lotion on the other side of the spine. Here the block is the volunteer, and the two matched "subjects" are the two sides of the volunteer's back.

Suppose that the experimental data on degree of sunburn (as measured on some scale) for the seven volunteers was as follows:

Volunteer	1	2	3	4	5	6	7
Burn on side with old lotion	42	51	31	61	44	55	48
Burn on side with new lotion	38	53	36	52	33	49	36
Difference	4	−2	−5	9	11	6	12

Consider in particular Volunteer 1, and suppose that the fall of the coin happened to assign the new lotion to his right side. As it turned out, his right side was burnt less (a reading of 38) than his left side (a reading of 42). Perhaps this is attributable to greater protection given by the new lotion. However, it is also possible that the lotions are really equally good, and that the difference is attributable to other factors: perhaps his right side is a little less sensitive, or happened to get a little less exposure. The null hypothesis, i.e., that the lotions are equivalent, asserts that the readings 38 and 42 are not influenced by the way the lotions happened to be assigned. Under the null hypothesis, if the coin had chanced to fall the other way so that the new lotion went on the left, still the left would have given the reading 42 and the right would have given 38. Then the numbers 42 and 38 in our table would have been reversed, and the difference would have been −4 instead of 4. The sign of the difference, plus or minus, is determined by whether the coin falls heads or tails, if the null hypothesis is correct. A similar analysis holds for the other six volunteers.

(It might of course have happened that a difference is 0. How to handle such cases will be taken up in Section 12.7.)

As the data are given, there are five positive differences and two negative differences. If the null hypothesis is correct, each difference had equal chances of being positive or negative, and the signs of the differences are unrelated, since the coin is assumed to be fair and the tosses unrelated. The number, say B, of positive signs is accordingly binomially distributed with $n = 7$ and $p = \frac{1}{2}$.

Now let us consider how B would tend to behave if the new lotion is really superior to the old. Imagine first an extreme possibility. If its superiority were great enough, so great as to override the other factors such as differences in sensitivity or in exposure, then the side protected by the new lotion would be less burned than the other side in every case. That is, every sign would be positive and B would equal 7. In a less extreme case, in which the new lotion is superior but not overwhelmingly so, there would still be a tendency for the side with the new lotion to be the less burned, and accordingly a tendency for the signs to be positive and for B to be large.

This analysis suggests that B can serve as a statistic for testing the hypothesis that the treatments are equally good, with large values of B significant evidence against the null hypothesis, in favor of the alternative that the new treatment is superior to the old. This test is known as the *sign test*, because B is the number of positive signs. The significance probability $P(B \geqq 5)$ can be obtained from Table C by the method discussed in Section 6.1. In fact, $P(B \geqq 5) = P(B \leqq 2) = .2266$. (An alternative argument consists in observing that the number B' of negative signs is just $B' = 7 - B$, so that the test based on rejecting H for large B is equivalent to the test based on rejecting H for small B'. Since B' also has the binomial distribution ($n = 7$, $p = .5$) under H, we may obtain $P(B' \leqq 2) = .2266$ directly from the table.) The data given above is therefore not significant at the twenty percent level, when analyzed with the sign test.

In general, the sign test can be used whenever an experiment is conducted to compare a treatment with a control on a number, say n, of matched pairs; provided the two treatments are assigned to the members of each pair at random (for example by tossing a fair coin). To test the null hypothesis that the new treatment is indistinguishable from the control, against the alternative that the new treatment is better, we simply observe the number B of pairs for which the new treatment gives better results. Large values of B are significant, and the null distribution of B is binomial $(n, p = \frac{1}{2})$. Fortunately, the normal approximation to the binomial distribution is especially accurate when $p = \frac{1}{2}$.

The sign test is not the only reasonable test for the matched pairs design, and we shall in the next section suggest an alternative test that has certain advantages. However, the sign test is often used, not only because it is so very simple and easy to apply, but also because it is applicable when the comparisons are qualitative rather than quantitative. For instance, suppose that in Example 4 we did not have actual readings of degree of burn, but that a judgement was made for each volunteer as to which side of his back was less severely burned. If B is the number of volunteers having less burn on the side with the new lotion, then B again has (under the null hypothesis) the binomial distribution ($n = 7$, $p = \frac{1}{2}$).

The dangers of bias in interpretation, discussed in Section 1, are of course also present with the matched pairs design, and are particularly serious when the observations are subjective. If the person who judges which side is less burned knows how the lotions were assigned, it will be difficult for him to avoid having his judgement influenced by this knowledge. It is therefore very desirable to keep the judge in ignorance of the assignment. Similarly, in a clinical trial, the physician who decides which of two patients is making the better recovery should if possible not know which therapy was used on each.

The model of matched pairs is often used in situations where there is not a random assignment of treatments, and where in consequence the conclusions are questionable. For example, a study of the relation of cigarette smoking to lung cancer could be made as follows. A number of heavy cigarette smokers are located. To each of them there is matched another person who is a nonsmoker but who is similar with regard to various factors, such as age, sex, occupation, and area of residence. Then the two matched persons are treated as a pair, and the responses (i.e., development or nondevelopment of lung cancer) are compared. The trouble is that there is no random assignment. The subjects decided themselves whether or not to become smokers. Perhaps there is a tendency for persons whose lung tissue is sensitive to cancer-causing agents in the atmosphere, to find inhaling cigarette smoke pleasurable. This would be an example of *self-selection* bias.

Another difficulty that arises in practice is the necessity of insuring the unrelatedness of the subjects. Suppose a military installation is to be used to test the influence of room temperature on susceptibility to the common cold. Fifty soldiers are matched into 25 pairs, according to tendency to colds and general health, and one of each pair is randomly assigned to a barracks that will be kept at very low temperature. If it turns out that colds are very frequent in this barracks, we are not justified in using the sign test, because the 25 soldiers there are not unrelated with regard to colds, which they may catch from each other.

PROBLEMS

1. In Example 4, suppose that the number of subjects is $n = 18$ of which 13 show less burn on the side receiving the new lotion. Find the significance probability of this result.

2. To see whether a new material for contact lenses reduces irritation as claimed, 23 volunteers are given a standard lens for one of the eyes and one made of the new material for the other; which eye receives the standard is determined by tossing a coin for each volunteer.

   (i) Let B denote the number among the volunteers for whom the eye receiving the standard lens shows more irritation. For what values of B will the hypothesis H of no difference between the lenses be rejected at (approximate) significance level .05?

   (ii) Compare the exact significance level of part (i) with the approximate level obtained by using the normal approximation.

3. Use the normal approximation to give an approximate solution to the preceding problem if, instead of 23, the number of volunteers is (i) 100, (ii) 200, (iii) 300.

4. In a paired comparisons experiment involving n pairs of subjects, what significance levels between .05 and .1 are available if
$$\text{(i) } n = 10, \quad \text{(ii) } n = 20, \quad \text{(iii) } n = 30?$$

5. In the preceding problem find the significance level closest to .05 when
$$\text{(i) } n = 5, \quad \text{(ii) } n = 15, \quad \text{(iii) } n = 25.$$

6. In Problem 2 suppose that B is observed to be equal to 17. What is the smallest significance level α for which H would be rejected?

7. Suppose that in a paired comparison experiment involving n pairs of subjects, the treated subject comes out ahead in 60% of the pairs. Find the significance probability of this result when
$$\text{(i) } n = 5, \quad \text{(ii) } n = 10, \quad \text{(iii) } n = 20.$$

8. In the preceding problem, use the normal approximation to find the approximate significance probability when
$$\text{(i) } n = 50, \quad \text{(ii) } n = 100, \quad \text{(iii) } n = 150.$$

9. In the preceding problem, make a conjecture as to the behavior of the significance probability as n tends to infinity and prove your result.

10. Suppose that in a paired comparison experiment involving n pairs of subjects, the treated subject comes out ahead in (i) 55%, (ii) 51% of the pairs. Find the significance probability when $n = 100$, 200 and 300 and determine what happens as n tends to infinity.

11. Suggest three experiments in which there are natural pairs, other than those already considered.

12. A paired comparisons experiment is performed on 28 pairs of subjects to see whether there is any difference in the effectiveness between two new treatments A and A'. Let B be the number of pairs for which the effect observed on A' exceeds that observed on A. Suppose the hypothesis that the treatments do not differ is rejected both when $B \leq k$ and when $B \geq 29 - k$. Find the value k for which the resulting test has (approximately) significance level .05.

Why is it reasonable in such a case to reject both when B is too large and too small?

12.6   WILCOXON'S TEST FOR PAIRED COMPARISONS

In the preceding section we discussed the sign test for testing the hypothesis H of no treatment effect with paired comparisons. This test, just as the Fisher-Irwin test, involves quantization and neglects an important aspect of the data. Even if the values of the differences for all the pairs are available, the sign test takes notice only of whether they are greater or less than zero.

EXAMPLE 1.   *Sun-tan lotions.*   Consider once more the differences between the treatment and control observations in Example 5.4. Of these differences, two are negative $(-2, -5)$ and five positive $(4, 6, 9, 11, 12)$. However, not only are most of the differences positive, but the positive differences are on the whole of larger absolute size than the negative ones. Intuitively, this seems to provide further evidence that the new treatment is superior to the old. It seems plausible that a test which also pays attention to the magnitudes of the absolute differences would be more sensitive in detecting a treatment effect than a test which considers only the signs. We shall now present such a test.

Let us arrange the seven differences in order of size, without regard to their signs, and assign to each its rank:

Differences	-2	4	-5	6	9	11	12
Ranks		1	2	3	4	5	6	7

Let V_- denote the sum of the ranks of the negative differences, so that in the example $V_- = 1 + 3 = 4$. This small value of V_- reflects both that only few of the differences are negative, and also that these negative differences tend to be smaller in absolute size than the positive ones. Thus, small values of V_- are evidence against the null hypothesis of no treatment effect, in favor of the alternative that the new lotion tends to give greater protection than the old. We could equally well have considered the sum V_+ of the ranks of the positive differences. In fact, $V_- + V_+$ is the sum of all the ranks, i.e., the fixed number $1 + 2 + \ldots + 7 = 28$. Therefore the test based on rejecting the null hypothesis when V_- is small (large) is equivalent to rejecting when V_+ is large (small).

In order to determine the significance probability of the observed value of V_-, we require its distribution under H. When the hypothesis is true, we have seen in the preceding section that the assignment of plus and minus

signs to the seven differences constitute seven binomial trials with probability $p = \frac{1}{2}$. By (3.3.1), all $2^7 = 128$ possible assignments of signs therefore have the same probability $(\frac{1}{2})^7$. To find $P(V_- \leq 4)$ we need only
count how many of the 128 equally likely cases give a value of V_- not
greater than 4. The statistic V_- takes on its smallest value 0 if all differences are positive; its next smallest value 1 if only the smallest difference
(i.e., the difference with rank 1) receives a negative sign; etc. We list
below, for each value of V_- up to 4, all the sets of ranks whose sum gives
that value of V_-.

Value of V_-	Sets of ranks totaling V_-
0	Empty set
1	$\{1\}$
2	$\{2\}$
3	$\{3\}$, $\{1, 2\}$
4	$\{4\}$, $\{1, 3\}$

There are seven "favorable" cases out of 128, so that the significance
probability of the observed value of V_- is $P(V_- \leq 4) = \frac{7}{128} = .0547$.

The test based on the rank sum V_- or V_+, as the corresponding test for
complete randomization, was proposed by Wilcoxon in 1945; it is known
as the Wilcoxon paired-comparison test.

We give in Tables I, J, and K values from which the significance probabilities can be obtained provided the number n of matched pairs is not too
large. Since V_- and V_+ have the same null distribution (Problem 16), it
is not necessary to distinguish between them in the tables, where V stands
for either V_- or V_+. With n matched pairs, there are 2^n equally likely
choices of signs for the n ranks; the values of 2^n, for $1 \leq n \leq 20$, are shown
in Table I. The null distribution of V is given by

$$P(V \leq v) = \#(V \leq v)/2^n$$

where $\#(V \leq v)$ is the number of the 2^n equally likely choices of signs
giving a value of V not exceeding v.

A table of $\#(V \leq v)$ splits naturally into two parts, according as $v \leq n$
or $v > n$. The reason for this is that, when $v \leq n$, the number $\#(V \leq v)$
does not depend on n, while when $v > n$, $\#(V \leq v)$ does depend on n
(Problem 13). Table J covers the former case for all values of v up to 20,
while Table K covers the latter case for $v - n \leq 30$ and $n \leq 20$.

To illustrate the use of Table J, suppose there are $n = 12$ matched pairs,
and the observed value of V is $v = 10$. Since $10 < 12$ we use Table J, to
find $\#(V \leq 10) = 43$. From Table I we read $2^{12} = 4096$. Therefore,
under the null hypothesis $P(V \leq 10) = \frac{43}{4096} = .0105$.

To illustrate the use of Table K, suppose again that $n = 12$, but that now the observed value is $v = 16$, so that $v > n$ and $v - n = 4$. From the column headed $n = 12$ and the row headed $v - n = 4$, we read $\#(V \leq 16)$ $= 158$, so that $P(V \leq 16) = \frac{158}{4096} = .0386$. (For larger values of n and v than those covered in Table K, a normal approximation may be used as discussed below.)

Since the tables cover only the lower tail of the distribution, it is convenient to make the choice between V_- and V_+ in such a way that small values of the chosen statistic are significant. This choice is illustrated in the next example.

EXAMPLE 2. *College entrance tests.* To find out whether hypnosis can improve performance on a standard college entrance examination, 12 pairs of twins are used. One twin of each pair, chosen at random, is hypnotized and told that he will do well; the other twin serves as control. The differences between the scores of hypnotized and nonhypnotized twins are

$$-10 \quad 12 \quad -8 \quad -23 \quad 7 \quad -5 \quad 2 \quad 14 \quad 19 \quad 6 \quad 27 \quad 16.$$

Large values of V_+ indicate success for the treatment, and in the present case $V_+ = 1 + 3 + 4 + 7 + 8 + 9 + 10 + 12 = 54$. The significance probability is $P(V_+ \geq 54)$. However, since Tables J and K give only the lower tail of the distribution, it is more convenient to use the equivalent test based on V_-, small values of which are significant. Here $V_- = 2 + 5 + 6 + 11 = 24$, and we need $P(V_- \leq 24)$. Entering Table K at $n = 12$, $v = 24$, we find $P(V_- \leq 24) = \frac{545}{4096} = .133$. An improvement by hypnosis would therefore not be clearly established by this data.

For values of n and v not covered by the table, it is again found that a normal approximation gives good results. The expectation and standard deviation of V under the null hypothesis, which are needed for this approximation, are (see Problems 19 and 20)

(1) $E(V) = n(n + 1)/4$

and

(2) $\text{SD}(V) = \sqrt{n(n + 1)(2n + 1)/24}.$

Table I gives the values of $E(V)$ and $\text{SD}(V)$ for $11 \leq n \leq 30$.

EXAMPLE 2. *College entrance tests (continued).* In Example 2, $n = 12$ so that from Table I, $E(V) = 39.0$ and $\text{SD}(V) = \sqrt{162.5} = 12.75$. Hence $P(V_- \leq 24)$ is approximately

$$\Phi\left(\frac{24.5 - 39}{12.75}\right) = \Phi(-1.137) = .128,$$

which may be compared with the correct value .133.

The null distribution arising here also provides the basis for a rank test in a different testing problem (the "one-sample" problem) to be taken up in Section 13.6.

PROBLEMS

1. If n pairs of subjects are used, find
   (i) the largest possible value of V_-,
   (ii) the probability that V_- takes on this largest value under the null hypothesis of no treatment effect.

2. If n pairs of subjects are used, find $P(V_- = 1)$.

3. If n pairs of subjects are used, find a formula for $V_- + V_+$ and check it against the value $V_- + V_+ = 28$ found in Example 1 for $n = 7$.

4. Suppose that in Example 1, we have $n = 8$ and the following degrees of burn

Old lotion	37	34	50	53	41	61	47	59
New lotion	41	29	47	45	47	51	35	52

   (i) Use the method of Example 1 to find the significance probability of V_-.
   (ii) What significance probability does the sign test attach to the data?

5. If in three out of four pairs of subjects the treated subject gives higher (improved) response than the control, find
   (i) the largest possible value of V_-,
   (ii) the significance probability attached to the data by the sign test,
   (iii) the significance probability of the value of V_- found in part (i), where the differences are obtained by subtracting the control responses from the treated responses.

6. Construct other values for the measurements in Example 5.4 such that the new lotion comes out ahead in 5 of the 7 cases and the Wilcoxon test has
   (i) the maximum significance probability possible under these restrictions;
   (ii) the minimum significance probability possible under these restrictions.

7. Solve the preceding problem when the new lotion comes out ahead in 4 of the 7 cases.

8. Use Table I, J and K to find the following four probabilities and check each against its normal approximation.
   (i) $P(V \leq 19)$ when $n = 20$,   (iii) $P(V \leq 45)$ when $n = 15$,
   (ii) $P(V \leq 19)$ when $n = 7$,   (iv) $P(V \leq 35)$ when $n = 12$.

9. For $n = 4$ find the distribution of V by enumeration, and graph its histogram. Use your results to check the entries in column $n = 4$ of Table K.

10. Use the distribution of V determined in the preceding problem to find (i) $E(V)$, (ii) $\text{Var}(V)$ and check your results against formulas (1) and (2).

11. Find the distribution of V in the case $n = 10$ from Tables J and K and the fact that the distribution is symmetric about $\frac{1}{4}n(n + 1)$.

12. Suppose the 20 students of Problem 3.14 are not divided at random into two

groups of 10 each but instead are divided into 10 matched pairs on the basis of their performance in a previous course. Of the two members of each pair, one is assigned at random to treatment and the other to control. Suppose the numbers of points received by the 10 pairs of students on the final examination are as follows:

| TV instruction | 35 | 6 | 18 | 25 | 37 | 48 | 49 | 53 | 81 | 89 |
| Direct instruction | 34 | 57 | 9 | 64 | 61 | 75 | 91 | 100 | 98 | 93 |

Find the significance probability attached to the data (i) by the sign test; (ii) by the Wilcoxon paired-comparison test.

13. (i) Show that for $v \leq n$, $\#(V \leq v)$ is the number of ways in which v can be represented as the sum of distinct integers chosen from $1, 2, \ldots, v$.
   (ii) Use (i) to show that for $\leq n$, $\#(V \leq v)$ is independent of n.
[Hint: (i) Note that if any of the ranks $v + 1, \ldots, n$ is included in V, then $V > v$.]

14. For $n = 8$, find $P(V \geq 24)$ using Table K and the fact that the distribution of V is symmetric about $\frac{1}{4}n(n + 1) = 18$.

15. For $n = 13$, find $P(V \geq 151)$ using Table K and the fact that the distribution of V is symmetric about $\frac{1}{2}n(n + 1)$.

16. (i) Explain why V_- and V_+ have the same null distribution.
   (ii) Use (i) and Problem 3 to show that the distribution of V_- is symmetric about $\frac{1}{4}n(n + 1)$.

17. Suppose that $n = 2k$ and that the ranks of the absolute values of the negative differences are $1, 3, \ldots, 2k - 1$ and those of the positive differences $2, 4, \ldots, 2k$. If small values of V_- are significant, find the significance probability of the results for $n = 2, 4, 6, 8$
   (i) for the sign test,
   (ii) for the Wilcoxon paired-comparison test.

18. In the preceding problem, use the normal approximation to
   (i) find the (approximate) significance probability for $n = 60, 80, 100$;
   (ii) determine what happens as n tends to infinity.

19. For a paired comparisons experiment involving n pairs of subjects, let I_1 indicate the event that a positive sign is attached to the difference with rank 1, let I_2 indicate that a positive sign is attached to rank 2, etc. Show that

$$(3) \qquad V_+ = I_1 + 2I_2 + \ldots + nI_n.$$

20. Use (3) to prove (1) and (2). [Hint: Under the hypothesis of no treatment effect, $I_1, \ldots, I_n$ indicate n binomial trials with success probability $p = \frac{1}{2}$. Use formulas (6.8.2) and (6.8.4).]

21. Let $\#(v; n)$ be the number of choices of n signs giving to V_+ the value v. Prove the recursion formula

$$(4) \qquad \#(v; n) = \#(v; n - 1) + \#(v - n; n - 1).$$

[Hint: How many choices are there such that the difference with rank n receives a minus sign? How many such that it receives a plus sign?]

22. Work Problem 12.4.13 for the Wilcoxon paired-comparison test.

12.7 THE PROBLEM OF TIES

Throughout this chapter we have tacitly assumed that it will be clear from the data which of two responses (or two response differences) is the larger. Unfortunately, in practice, ambiguities may arise. Suppose for instance that the longest surviving control patient in Example 3.1 had lived 7.9 instead of 6.7 years. He would then be *tied* with the treated patient who survived 7.9 years. The two patients would share the ranks 7 and 8, but we would not know which had rank 7 and which rank 8 and would thus be unable to compute the Wilcoxon two-sample statistic. As another illustration, consider the data of Example 5.4. If volunteer 4 had received score 63 instead of 61 on the side with the old lotion, his difference would be 11 instead of 9, and in the ranking of absolute differences he would be tied with volunteer 5 for ranks 5 and 6. A somewhat similar difficulty may arise in the sign test (if we do not know whether a difference is positive or negative) and in the Fisher-Irwin test (if the response category of some subjects may be unclear).

The present section is concerned with methods of dealing with such ties, but these should be prefaced with the advice: avoid ties if you can! Prospects for doing so vary with the origin of the tie, and one may distinguish between four main types.

(i) *Imprecise record or observation.* The most common source of ties is the failure to record observations with sufficient precision. Many experiments involve a continuously variable response, such as time, height, weight, breaking strength or temperature. If the measuring instrument is read with sufficient precision, it is unlikely that two subjects will have exactly the same response. (It is of course possible that the instrument is so crude that sufficiently precise values cannot be obtained; in this case, the situation is essentially that described in (ii).)

(ii) *Discrete response.* In some experiments the response is an integer, such as the number of eggs laid by a hen, or the number of wrong answers on a true-false examination. Here it is often easy to break ties in a reasonable way by considering an auxiliary variable. For example, if two hens have laid the same number of eggs, they may be ranked according to the average weight of the eggs. Similarly, two students tied on the final examination may be ranked by the total score on their homework.

(iii) *Judged response.* It is a great virtue of the tests discussed in this chapter that they can be used in cases where the responses are not measured but the subjects are ranked by an exercise of subjective judgement. (These judgements should of course be "blind" in the sense of Section 2.) If a judge finds it difficult to distinguish between two or more subjects, it may be possible to break the resulting tie by bringing in a second judge.

(iv) *Truncated observations.* Quite often all the ranks will be distinct except for those at one end or the other. For example, the observed survival times of surgical patients will be distinct except for those who die during the operation, or who are still alive when the study is being published. Similarly, the breaking strengths of steel rods are distinct except for those which withstand the maximum stress of which the testing machine is capable. Fortunately, when all the ties occur at one end, Table H can still be used to compute the significance probability of the Wilcoxon two-sample test, by means of a device explained at the end of the section.

One method of breaking ties is always available: by means of a random decision. If two subjects are tied for ranks 6 and 7, a penny can be tossed to decide which subject gets which rank. While we are ardent advocates of randomization *before* the experiment (this being necessary to protect against bias and to provide the probabilistic basis of the analysis), we do not recommend randomization *after* the experiment as an aid to analyzing the data. It is in a sense the purpose of the analysis to extract the true situation from data that reflect random disturbances, and it then seems illogical to inject into the data an extraneous element of randomness. In addition, it is troublesome that two workers, applying the same test to the same data, could reach opposite conclusions.

While ties can in many cases be avoided by some foresight, data will often reach us with ties which we are unable to resolve. How to handle the ties will then depend on the problem and on the test to be used, and we begin by considering the Wilcoxon two-sample test. Suppose the data are those of Example 3.1, except that treated and control subject are tied for ranks 7 and 8. It is then natural to assign to both of these subjects the *mean rank* 7.5 (and, in general, to each of a number of tied subjects the average of the ranks for which they are tied). The resulting ranks are then

	Ranks					Sum
Treated	4,	6,	7.5,	9,	10	36.5
Control	1,	2,	3,	5,	7.5	18.5
						55

Let us continue to denote the sum of control and treated (mean) ranks by W_C and W_T. Small values of W_C will still be significant, and the significance probability is then $P(W_C \leq 18.5)$.

This probability must be calculated under the null hypothesis, according to which the 10 survival times are fixed numbers, unaffected by the random assignment of 5 patients to the treatment. Since the survival times are fixed, so are their ranks or mean ranks. The statistic W_C is therefore the sum of 5 numbers chosen at random from the set

$$1, \quad 2, \quad 3, \quad 4, \quad 5, \quad 6, \quad 7.5, \quad 7.5, \quad 9, \quad 10.$$

(This random choice is another example of the lottery model.) To obtain the desired probability, we need only count the choices yielding a sum of 18.5 or less, namely

$$
\begin{aligned}
1 + 2 + 3 + 4 + 5 \; &= 15 & 1 + 2 + 3 + 5 + 6 \; &= 17 \\
1 + 2 + 3 + 4 + 6 \; &= 16 & 1 + 2 + 3 + 5 + 7.5 &= 18.5 \\
1 + 2 + 3 + 4 + 7.5 &= 17.5 & 1 + 2 + 3 + 5 + 7.5 &= 18.5 \\
1 + 2 + 3 + 4 + 7.5 &= 17.5 & 1 + 2 + 4 + 5 + 6 \; &= 18.
\end{aligned}
$$

There are 8 "favorable" cases among $\binom{10}{5} = 252$ equally likely cases, so that

$$P(W_C \leq 18.5) = 8/252 = .0317.$$

Calculation of the significance probability by enumeration is in principle always possible, but it is feasible only if there are not too many favorable cases. Unfortunately, there is no hope of publishing a table like Table H for the mean ranks, since each of the many possible patterns of ties would require a separate table. However, if the group sizes are sufficiently large, the normal approximation may again be used. Replacing the ranks by mean ranks does not change the expectation of W_C or W_T (Problem 7(i)), but for each set of d tied ranks it reduces the variance by

$$(1) \qquad\qquad st\, d(d^2 - 1)/12N(N - 1)$$

below the value given by (3.2). (See Problem 7(ii).)

As an illustration, let us compute the normal approximation to the probability $P(W_C \leq 18.5)$ computed above. The expectation of W_C is 27.5 and application of the correction (1) to the variance formula (3.2) shows the variance to be 22.778 and hence the standard deviation 4.7726 (Problem 8). Due to the presence of the mean rank 7.5, W_C can take on not only integral values but also half-integer values. As a result (Problem 6.5.19), the normal approximation to $P(W_C \leq 18.5)$ is

$$\Phi\left[\frac{18.75 - 27.5}{4.7726}\right] = \Phi(-1.833) = .0334,$$

while the correct value is .0317. The normal approximation is thus reasonably satisfactory. In general, the presence of ties tends to reduce its accuracy since the distribution of a rank sum becomes less smooth and regular when the ranks are not equally spaced.

Let us next consider the problem of ties in the sign test and the Wilcoxon

paired-comparison test.   Here two kinds of ties may arise: (a) ties among
the absolute differences being ranked; and (b) ties among the two responses within a pair, which result in zero differences and leave us in doubt
even as to which of the two subjects in the pair came out ahead.

Suppose that there are eight pairs of subjects and that within each pair
one is chosen at random, with the following observations:

Pair	1	2	3	4	5	6	7	8
Treated response	4	6	6	5	5	5	3	3
Control response	1	2	6	2	7	3	3	4
Difference	3	4	0	3	−2	2	0	−1

The sign test is based on the number of positive, or equivalently the number
of negative, differences.   In the present case we do not know whether
pairs 3 and 7 are positive or negative, although this might have been
known had the responses been observed with greater precision.   As things
stand, these two pairs are of no help in deciding which of the treatments
is better, and we shall therefore disregard them.   (A high proportion of
zeros would of course be informative in another way: it would suggest
that the treatment effect, whatever its direction, is small in relation to the
precision with which the comparisons are being made.)

Under the null hypothesis, the remaining six pairs are independent and
each is equally likely to be positive or negative.   As in Section 12.5, we
can use the number B of positive signs among the six as test statistic, with
large values providing significant evidence of a favorable treatment effect.
The null distribution of B is the binomial distribution ($n = 6$, $p = \frac{1}{2}$), and
Table C shows the significance probability to be $P(B \geq 4) = P(B \leq 2) =$
.3438.

If we wish to use the Wilcoxon paired-comparison test instead of the
sign test, we are faced with the same difficulty of not knowing whether
the differences of pairs 3 and 7 are positive or negative.   These pairs are
again uninformative, and we shall disregard them as before.   In view of
the distribution of the signs of the remaining six pairs discussed above,
the $2^6 = 64$ possible assignments of $+$ or $-$ to the six differences are
equally likely, and this provides the probabilistic basis for the test.

The ties encountered when we attempt to rank the six pairs by their
absolute differences can be handled by the mean-rank method as follows:

Pair	8	5	6	1	4	2
Difference	−1	−2	2	3	3	4
Rank	1	2.5	2.5	4.5	4.5	6

The rank sum V_- of the negative differences is $1 + 2.5 = 3.5$. Its significance probability can be found by enumeration. Among the 64 equally likely assignments of $+$ or $-$ to the differences, V_- is seen to be 3.5 or less in just six cases, indicated below by listing for each case the set of negative pairs: the empty set (no negatives); pair 8; pair 5; pair 6; pairs 8 and 5; pairs 8 and 6. Thus, $P(V_- \leqq 3.5) = 6/64 = .0938$. That the Wilcoxon test attaches greater significance to the data than the sign test reflects the fact that the two negative signs are associated with differences of low absolute rank.

When enumeration becomes too cumbersome because there are too many cases to be listed, we can again fall back on the normal approximation. The expectation of V_- is given by (6.1) even in the presence of ties, where n is now the number of nonzero differences. The previous value $n(n + 1)(2n + 1)/24$ of the variance is reduced by

$$(2) \qquad\qquad d(d^2 - 1)/48$$

for each set of d tied ranks among the (nonzero) absolute differences (Problem 14). In the example computed by enumeration above, the number of nonzero differences is $n = 6$, so that $E(V_-) = 6 \cdot 7/4 = 10.5$. If there were no ties, formula (6.2) would have shown the variance to be $6 \cdot 7 \cdot 13/24 = 22.75$. From this value, however, we must subtract $2(4 - 1)/48 = .125$ for each of the pairs of tied ranks, getting $\mathrm{Var}(V_-) = 22.5$. The normal approximation to $P(V_- \leqq 3.5)$ is thus

$$\Phi\left[\frac{3.75 - 10.5}{\sqrt{22.5}}\right] = \Phi(-1.4230) = .0774,$$

compared with the correct value .0938. The number $n = 6$ is too small for the normal approximation to be very accurate.

To conclude the section, let us show how Table H can be used to compute the significance probability of the Wilcoxon two-sample test when all the ties occur at one end.

EXAMPLE 1. *Breaking strength.* In an experiment to compare the breaking strengths of $t = 9$ steel rods given a special tempering, with the strengths of $s = 8$ control rods, three of the treated rods and one control rod failed to break even when the testing machine was exerting its maximum stress. Each of these four rods is accordingly assigned the mean rank $(14 + 15 + 16 + 17)/4 = 15.5$. The other 13 rods broke at stresses measured sufficiently precisely to avoid ties; the ranks of the control rods were 1, 2, 4, 5, 7, 9, and 12. The value of W_C was thus 55.5, with small values being significant.

To calculate $P(W_C \leqq 55.5)$, note from Table A that there are $\binom{17}{8} = 24{,}310$ ways of choosing 8 of the (mean) ranks 1, 2, 3, . . . , 13, 15.5, 15.5, 15.5, 15.5, and that all these choices are equally likely under the null hypothesis. How many of these cases are "favorable," in the sense of giving a total $W_C \leqq 55.5$? Let us classify the favorable cases according to the number of mean ranks 15.5 included in the set of 8. Consider first the cases where none is included. The 8 numbers whose sum is W_C must then be chosen from the set $1, 2, \ldots, 13$. The number of ways of choosing 8 of these 13 integers so that their sum is not greater than 55.5 (or, since this sum is an integer, not greater than 55) can be found in row (5, 8) of Table H, in the column corresponding to $u = 55 - \frac{1}{2} \cdot 8 \cdot 9 = 19$, and is seen to be 607. Next consider the cases where just one of the mean ranks 15.5 is included in the set of 8. There are 4 of these mean ranks, any of which may be chosen. For each such choice, the set is completed by choosing 7 of the first 13 integers to produce a total of $55.5 - 15.5 = 40$ or less. From the (6, 7) row of Table H at $u = 40 - \frac{1}{2} \cdot 7 \cdot 8 = 12$, we see that this can be done in 201 ways. Similarly, two of the mean ranks 15.5 can be chosen in $\binom{4}{2} = 6$ ways, and for each choice the set of 8 can be completed in 7 ways, as shown by the entry at $u = 3$ in row (6, 7) of Table H. Since the inclusion of three or more of the mean ranks 15.5 in the set of 8 would lead to a total above 55.5, this completes the count. The total number of favorable cases is therefore

$$607 + 4 \times 201 + 6 \times 7 = 1453,$$

and the significance probability is

$$P(W_C \leqq 55.5) = 1453/24310 = .0598.$$

For comparison, let us compute the normal approximation. From our earlier results we see (Problem 17) that $E(W_C) = 72$ and $\mathrm{Var}(W_C) = 106.676$. The normal approximation to $P(W_C \leqq 55.5)$ is therefore

$$\Phi\left[\frac{55.75 - 72}{10.328}\right] = \Phi(-1.573) = .0579.$$

PROBLEMS

1. Suppose that $s = t = 4$, the treated observations are .7, .8, 1.8, 1.1 and the controls .1, .5, .7, .8.
   (i) Find the mean ranks of the 8 observations.
   (ii) If small values of W_C are significant, enumerate cases to find the significance probability of the data.

2. Solve the preceding problem if $s = 4$, $t = 5$, the treated observations are .6, .9, .9, 1.0, 1.1 and the controls .3, .6, .8, .9.

3. Suppose that $s = t = 4$, the treated observations are .6, .8, 1.0, 1.4 and the controls .1, .4, .4, .7. The argument is sometimes given that the value of W_C is independent of how the tie at .4 is broken, and that the usual null distribution of W_C (given in Table H) is therefore applicable. Do you find this convincing?

4. Suppose that $s = 3$, $t = 5$, and that there are ties between ranks 1 and 2, and also between ranks 5, 6 and 7. By listing all 56 cases, find the distribution of W_C, and compare each possible value of $P(W_C \leqq w)$ with its normal approximation.

5. Let $u_1, \ldots, u_d$ be any d numbers and $\bar{u}$ their arithmetic mean. If each of $u_1, \ldots, u_d$ is replaced by $\bar{u}$, show that
  (i) the sum of the numbers is unchanged;
  (ii) the sum of the squares of the numbers is reduced by

$$(3) \qquad (u_1 - \bar{u})^2 + \cdots + (u_d - \bar{u})^2.$$

[Hint for (ii): Use (5) of Section 5.7 with $u_1, \ldots, u_d$ in place of $v_1, \ldots, v_N$.]

6. In the preceding problem, let $u_1 = a + 1$, $u = a + 2, \ldots, u_d = a + d$ be d successive integers. Show that
  (i) $\bar{u} = a + \frac{1}{2}(d + 1)$;
  (ii) if the u's are replaced by $\bar{u}$, the sum of squares of the u's is reduced by $d(d^2 - 1)/12$.

[Hint for (ii): Use part (ii) of the preceding problem; note that (3) is unchanged if the u's are replaced by $1, \ldots, d$ (see for example Problem 9.1.3); apply the formula for τ^2 at the bottom of p. 000 with d in place of N.]

7. Use the results of Problems 5 and 6 to show that if W_C is the sum of the mean ranks of the controls,
  (i) the expectation of W_C is still given by (12.3.1);
  (ii) the variance of W_C is obtained from (3.2) by subtracting the quantity (1) for any d values that are tied.

8. Verify the values of the expectation and variance required in the normal approximation to $P(W_C \leq 18.5)$ in the text.

9. Consider the extreme case of ties, in which only two different response values are observed, say 0 and 1. Suppose that the total number of responses equal to 0 is a and those equal to 1 is $b = N - a$; and that the number of control responses equal to 1 is Z and those equal to 0 is $s - Z$.
  (i) Express W_C in terms of Z.
  (ii) How does the Wilcoxon test in this case compare with the Fisher-Irwin test?

10. Headache sufferers are asked to say whether a remedy gives no relief, a little relief, or substantial relief. Discuss how the Wilcoxon two-sample test with mean ranks can be used to analyze comparative data in such a situation.

11. In a paired-comparison experiment the responses are $(4, 3)$, $(4, 4)$, $(3, 6)$, $(4, 5)$, $(5, 5)$, $(5, 5)$, $(8, 10)$, $(5, 4)$, with the first member of each pair being the control and the second the treatment observation.
  (i) Find the mean ranks of the absolute values of the nonzero differences.
  (ii) Find the significance probability of V_- if small values are significant.

12. Solve the preceding problem if the observations are $(4, 3)$, $(4, 4)$, $(3, 6)$, $(4, 5)$, $(5, 5)$, $(7, 5)$, $(8, 11)$, $(4, 5)$.

13. Obtain the normal approximation to the significance probability of the two preceding problems, and compare it with the exact value.

14. Prove that the expectation and variance of V_- (or V_+) in the presence of ties is as stated in the text.

[Hint: Use the representation (3) of Problem 6.12 and the results of Problems 5 and 6.]

The remaining problems relate to the material in small print.

15. In Example 1 find the significance probability if the ranks of the untied controls are 1, 2, 4, 6, 7, 9, 12 instead of 1, 2, 4, 5, 7, 9, 12.

16. Under the assumptions of Example 1, find the significance probability if 4 of the treated rods and one of the controls fail to break under maximum stress, and if the ranks of the remaining control rods are 1, 2, 4, 6, 8, 9, 11.

17. Check the expectation and variance of W_C given at the end of Example 1.

18. Compute the normal approximation to the significance probability obtained in (i) Problem 15; (ii) Problem 16.

CHAPTER 13

THE CONCEPT OF POWER

13.1 THE TWO KINDS OF ERROR

In Section 11.1 we introduced one of the desiderata of a test of a statistical hypothesis. Suppose that the test consists in rejecting the hypothesis whenever a test statistic T is greater than or equal to a critical value c. Because it is usually a serious mistake to reject the hypothesis when it is correct, we are anxious that the probability

$$\alpha = P_H(T \geq c)$$

of making this mistake should be small, and we have spent some time seeing in a variety of problems how to determine c so that α will have an acceptably small value.

A little reflection will show, however, that it is not enough to achieve satisfactory control of α. If we had to worry only about the error of false rejection, we could simply decide always to accept the hypothesis. This would reduce α to 0, since if the hypothesis is never rejected at all, it can never be rejected falsely. Furthermore, such a policy would save us the trouble and expense of experimentation, for there is no need to collect data if we have decided in advance to ignore them!

The difficulty is of course that the hypothesis may in fact be wrong, in which case the policy of universal acceptance will result in a different kind of mistake. If the hypothesis is wrong and is mistakenly accepted, we say that the error of *false acceptance* has been committed. In most circumstances this will also be a serious mistake, though quite different in its nature and consequences from the error of false rejection, as the consideration of a few examples will show.

EXAMPLE 1. *Changing the sex ratio.* Recall Example 11.1.1 where we considered an experiment to test a biochemist's claim of being able to increase the chance that a calf will be female. The null hypothesis asserts that the claim is without substance. Here, false rejection would mean that the dairy association, which is conducting the test, recommends the method as effective though in fact it is not. As a result the dairy industry might waste a great deal of money on an ineffective treatment. False acceptance, on the other hand, would occur if the association decides the method is valueless when in fact it works. This error would be a source of regret to the biochemist, whose discovery is denied recognition, and would deprive the industry of a technique of great commercial value. The seriousness of this error will of course depend on how effective the method really is: if it serves to increase the chance of a female only from .5 to .53 the consequences of accepting the null hypothesis would be much less serious than if the chance is increased to .7.

EXAMPLE 2. *Chemotherapy.* The U.S. Government is supporting a large-scale screening program for cancer therapy. New chemicals by the tens of thousands are tried out on animal cancers, in the hope of finding a few that will at least retard the speed of growth of human cancer. We may think of the test of each such chemical as a test of the null hypothesis that it is useless. False rejection of this null hypothesis means that the experimental facilities of the program (cancer patients who volunteer for the new treatment, doctors who are willing to apply it under controlled conditions, etc.) are wasted on a worthless chemical. False acceptance means that society loses, perhaps forever, a treatment that is in fact helpful.

Only one of the two errors is possible in any given situation, since false rejection can occur only if the null hypothesis H is correct, while false acceptance can arise only if H is wrong. Unfortunately, however, it is not known whether H is correct or wrong, and we must therefore try to protect ourselves against both errors. To measure the risk of false rejection one uses the probability $\alpha = P_H(T \geq c)$ of rejecting H when it is true. Analogously, the risk of committing the other error is measured by the probability, which we shall denote by β, of accepting the hypothesis when it is false. Since a null hypothesis may be false in many different ways and to varying extents, β will be precisely specified only if one specifies an alternative, say A, to the null hypothesis. If P_A denotes a probability computed under the assumption that A is true, β is given by

$$\beta = P_A(T < c)$$

since the null hypothesis is accepted when $T < c$. Let us now illustrate the computation of β.

EXAMPLE 1. *Changing the sex ratio (continued).* Suppose the dairy association decides to recommend the biochemist's technique to its members if the number F of females among 20 calves is 15 or more; that is, the null hypothesis (which says the treatment is worthless) is rejected if $F \geq 15$, so that the critical value is $c = 15$. The significance level of this test was seen in Section 11.1 to be $\alpha = P_H(F \geq 15) = .021$; this was computed by giving to F the binomial distribution ($n = 20$, $p = \frac{1}{2}$). Let us now consider the alternative hypothesis A that the biochemist's technique raises the proportion of female calves to 70% on the average, this being a degree of success that would make the technique economically very attractive. According to A, the test statistic F still has a binomial distribution with $n = 20$, but now with success probability $p = .7$. The probability $\beta = P_A(F < 15) = P_A(F \leq 14)$ can therefore be computed by summing the relevant binomial probabilities, giving

$$\beta = P_A(F = 14) + P_A(F = 13) + \ldots$$

$$= \binom{20}{14} (.7)^{14}(.3)^6 + \binom{20}{13} (.7)^{13}(.3)^7 + \ldots$$

$$= .192 + .199 + \ldots = .584.$$

(This value can also be read directly from a table of the binomial distribution, or can be obtained approximately from a normal table.)

The biochemist will undoubtedly be very dissatisfied with a test which has so great a probability of deciding against his technique if in fact it increases the chance of a female calf to 70%, and the association would not want to take that large a risk of overlooking so valuable a discovery. Thus, while the test gives satisfactory control of false rejection ($\alpha = .021$), the value $\beta = .584$ shows that the risk of false acceptance is not controlled sufficiently well.

Can anything be done to reduce this large value of β to a more satisfactory level? We had previously chosen the critical value $c = 15$, so that the null hypothesis of the biochemist's treatment being worthless was rejected when the number F of female calves is 15 or more. Let us now see whether it is possible to lower β, without increasing α to an unacceptable level, by changing c. Since the test statistic F has the binomial distribution ($n = 20, p$) with $p = .5$ under H and $p = .7$ under A, we can compute $\alpha = P_H(F \geq c)$ and $\beta = P_A(F < c)$, obtaining the following values:

c	11	12	13	14	15	16
α	.412	.252	.132	.058	.021	.006
β	.048	.113	.228	.392	.583	.762

An inspection of the bottom row shows that the use of $c = 11$ gives β a reasonably small value .048 — but now α is too large. The dairy association certainly would not be willing to run a 41% chance of recommending the technique if it is worthless. Perhaps the most reasonable compromise

would be $c = 13$, giving a 13% chance of rejecting H when $p = .5$ and a
23% chance of accepting H when $p = .7$. Actually, in this case both α
and β are undesirably large. There is no satisfactory choice; the experiment is too small to permit us to distinguish with reasonable reliability
between $p = .5$ and $p = .7$.

Two main considerations are involved in striking a balance between the
conflicting requirements of low α and low β. We saw earlier in this section
that the consequences of false rejection and false acceptance are usually
unpleasant in different ways. If the consequences are about equally
serious, we would have an approximately equal interest in α and β, and we
would then choose c so as to get these two probabilities as nearly equal as
possible. If on the other hand the consequences of one error are much
more serious than those of the other, it would be reasonable to reduce the
probability of the more serious error at the expense of increasing that of the
less serious one.

The following example may help to illustrate how the seriousness of the
errors depends on the circumstances of the problem. Suppose a botanist,
who is mapping the flora of a region, is uncertain whether to record a plant
as species A or B. The two errors he may make (listing A if it is actually
B, or listing B if it is A) are presumably about equally serious—in either
case he has made an erroneous report that will mislead other scholars and
damage his reputation if it is later found out. He might accordingly want
α and β about equal. Let us now change the circumstances of the problem:
suppose we have found what are either A edible or B poisonous mushrooms;
shall we eat them for supper? If the null hypothesis is species A, false
rejection means foregoing a treat, while the consequences of false acceptance will be considerably more unpleasant. It is now desirable to have β
extremely small, even if this forces α to be rather large. In fact, many
people refuse to eat field mushrooms at all, thereby reducing β to 0 at
the expense of having $\alpha = 1$. Let us finally change the circumstances
once more: a castaway on a desert island finds mushrooms as the only
source of food. Now false rejection means death from starvation, and
the balance will shift to favor small α.

The example illustrates the fact that the reasonable compromise in choosing the critical value will depend on the consequences of the two errors.
However, it also depends on the circumstances of the problem in another
way. If the null hypothesis is very firmly believed, on the basis of much
past experience or of a well-verified theory, one would not lightly reject it
and hence would tend to use a very small α. On the other hand, a large
α would be appropriate for testing a null hypothesis about which one is
highly doubtful prior to the experiment. For example, most psychologists
are firmly convinced that extrasensory powers do not exist. They would

therefore demand overwhelmingly convincing evidence (that is, an extremely small α) before publishing a rejection of the hypothesis that a supposed clairvoyant is just guessing. On the other hand in Example 1, if the biochemist can present a convincing theory explaining why his treatment should work, the dairy association might be prepared to believe the null hypothesis to be false, and accordingly be willing to use a rather large α.

The concepts and computations relevant to testing a null hypothesis have more general applications. It is frequently necessary on the basis of chance data to choose between two actions or decisions. In making such a choice there are two possible errors and it is helpful to know their probabilities. Suppose that the data are summarized by a random variable T, large values of which indicate that one decision is appropriate while small values indicate the other decision. It is then necessary to choose a critical value c and to take one or the other decision as $T \geqq c$ or $T < c$. The choice of c should be made in the light of the resulting error probabilities.

EXAMPLE 3. *College admission.* A college bases its admission of a student on his score T in an entrance examination, requiring a score of c or more ($T \geqq c$) for admission. The two possible errors here consist in admitting students who will prove to be unsuccessful in their studies, or conversely in excluding students who would have been successful. The minimum score for admission (the critical value c) should be chosen to balance the risks of these two errors.

EXAMPLE 4. *Lot-sampling inspection.* A manufacturer is negotiating with a consumer to whom he will send regular shipments of a lot of certain mass-produced items. As a safeguard against poor quality, it is agreed that the customer will inspect a random sample of s items from each lot, and will return the lot to the manufacturer if too many of the inspected items prove to be defective. It is necessary to specify in the contract how many of the s inspected items must be defective before the lot is to be returned.

If N denotes the total number of items and r the number of defectives in the lot, then r is a measure of lot quality. The number D of defective items found in the sample of size s has a hypergeometric distribution (Section 6.2). Suppose the lot is rejected if $D \geqq c$, and accepted if $D < c$. There are again two possible errors: returning the lot although its quality (i.e., the number r of defectives) is satisfactory, or failing to return an unsatisfactory lot. The probabilities of these two errors depend on the choice of c.

The situation is quite analogous to that of testing a statistical hypothesis.

The role of the test statistic is played by D, with c serving as the critical value. Rejecting the lot corresponds to rejecting the hypothesis that the lot is of good quality. Suppose that r_H is the largest value of r that would correspond to good lot quality. Perhaps the manufacturer guarantees that r will not exceed r_H. Then the probability that $D \geq c$ when $r = r_H$ is the probability that a lot will be rejected, assuming that it just barely meets the manufacturer's guarantee. This probability, say $P_H(D \geq c)$, then corresponds to the level of significance α. In industrial practice it is called the "producer's risk," as it is the risk the producer runs of having rejected a lot which does (just barely) conform to his guarantee. Of course, if r is smaller than r_H, the chance of finding $D \geq c$ would be even smaller than α, so that $\alpha = P_H(D \geq c)$ is the maximum chance of rejecting a good lot.

Now let us look at the inspection scheme from the customer's point of view. His main interest is to protect himself against accepting a lot of really poor quality. Let $r_A > r_H$ be a number of defective items so large that the customer would definitely want to reject the lot if r were as great as r_A. The customer will then want the probability of accepting such a lot, say $P_A(D < c)$, to be small. This probability is known as the "consumer's risk." It represents the maximum chance of accepting a lot with r_A or more defectives. The consumer's risk corresponds to the probability β of falsely accepting the hypothesis.

Now consider how $\alpha = P_H(D \geq c)$ and $\beta = P_A(D < c)$ will vary as c is changed. When c is made smaller, it becomes more likely that the lot will be rejected, so that α goes up and β comes down. This will favor the customer at the expense of the manufacturer. The value of c is often negotiated between the parties and written into the contract.

To illustrate these ideas, suppose $N = 1000$, $r_H = 20$, and $r_A = 120$. That is, the lot of 1000 would be considered of good quality if there are 20 or fewer defective items in it (2% defective), and to be of bad quality if there are 120 or more defective items (12% defective).

Suppose that $s = 100$ items are inspected. Then the following values of α and β for $c = 5$, 6, and 7, are obtained from formula (6.2.1) for the hypergeometric distribution.

c	5	6	7
α	.0415	.0104	.00211
β	.00375	.0117	.0308

If the two kinds of error are about equally important, it might be desirable to have the producer's risk α and the consumer's risk β approximately equal, and this is achieved by taking c equal to 6. Actually, the two parties might be satisfied with somewhat larger values of α and β, say both α and

β not exceeding .05, if something were gained by this. Permitting larger error probabilities does of course have a compensating advantage: it permits a smaller sample. How to determine the smallest size giving a stated protection for both error probabilities will be discussed in the next section.

PROBLEMS

1. Describe the consequences of false rejection and false acceptance in the following cases.

    (i) In a court trial the hypothesis is that the accused is innocent; the alternative that he is guilty.

    (ii) The hypothesis is that a contemplated operation will cure a patient from a severe chronic backache; the alternative is that it will paralyze him.

2. In Example 1, use the normal approximation to

    (i) find β against the alternative $p = .8$ when H is rejected for $F \geq 15$;

    (ii) discuss the choice of c so as to achieve a balance between α and β for the alternative $p = .8$.

3. In Example 1, suppose that the treatment is applied in $n = 60$ cases. Use the normal approximation

    (i) to find the critical values c giving $\alpha = .01$, $\alpha = .05$, $\alpha = .1$;

    (ii) for each case of part (i) to find the value of β when $p = .7$.

4. Solve the preceding problem if the alternative of interest is $p = .6$ rather than $p = .7$.

5. In Problem 3, use the normal approximation to find β when $p = .4$ and (i) $\alpha = .05$, (ii) $\alpha = .1$.

Do you find this value of β alarming or reassuring?

6. The standard technique for a certain heart operation has shown a mortality of 30%. A new technique, hoped to be safer, is tried on ten patients. Consider the null hypothesis H that the new technique is just as dangerous as the standard one, and suppose that the hypothesis is to be rejected if all ten patients survive the new operation.

    (i) What is the significance level of the test?

    (ii) What is the value of β for the alternatives that the new technique has long-run mortality (a) 20%, (b) 10%, (c) 5%, (d) 0%?

    (iii) Work parts (i) and (ii) if the test rejects H when either nine or ten patients survive the new operation.

7. Work parts (i) and (ii) of the preceding problem under the assumptions that the number of patients is 60 and that H is rejected if 48 or more patients survive the operation.

8. Suppose that in Problem 6 the new operation is performed on 80 patients rather than ten, and that B denotes the number of deaths.

(i) To test H against the alternative that the new operation is safer, should H be rejected for small B, large B, or both?

(ii) Using the normal approximation, find a critical value that will give α near .05.

(iii) With this critical value, find the approximate value of β for the alternative that the new technique has a 20% mortality rate.

(iv) Against what mortality rate does the test give approximately 10% chance of false acceptance?

9. In the preceding problem, find the probability of rejecting H when the mortality rate of the new technique is in fact 35%. Discuss the consequences of this finding.

10. In Problem 11.2.6(ii), find β if the critical value of the test is $c = 11$, and if the probability of heads is (i) .55, (ii) .60, (iii) .65.

11. Under the assumptions of Example 11.2.2, find the value of β when (i) $p = .36$, (ii) $p = .4$.

12. In Problem 11.1.10, suppose that the hypothesis of no extrasensory powers (and hence random selection) is rejected if the spiritualist correctly picks all four red cards.

(i) What is the value of α?

(ii) If the spiritualist manages to see one of the red cards as it is put down, and selects the other three at random from the remaining seven, what is the value of β?

13. In Example 4, with $r_A = 80$ instead of 120,

(i) find β when $c = 6$,

(ii) by trial and error find a value of c which makes α and β nearly equal. (Use the normal approximation.)

14. In Example 1, discuss the choice of c so as to achieve a balance between α and β for the alternative $p = .7$ when $n = 50$. [Hint: Use the normal approximation for α and β, and equate the two to get an equation for c.]

13.2 DETERMINATION OF SAMPLE SIZE. POWER OF A TEST

In the preceding section we have seen that it may not be possible, with a given sample size, to reduce α and β simultaneously to a satisfactorily low level. When this happens, it is natural to ask whether the desired values could be achieved by increasing the sample size. More observations should provide more information and permit a reduction in the probabilities of erroneous decisions.

EXAMPLE 1. *Changing the sex ratio.* Let us investigate this possibility for Example 11.1, in which the results of 20 binomial trials were used to test the hypothesis H: $p = .5$ against the alternative A: $p = .7$. Suppose we wish to reduce α and β from the values .132 and .228, which we were able to obtain with $n = 20$, to the values $\alpha = .05$ and $\beta = .10$. To try to achieve this, let us increase the number of trials from $n = 20$ to $n = 100$. For any critical value c, the error probabilities α and β can then be com-

puted, or obtained from a large binomial table. The results, for several
values of c are shown below.

c	59	60	61	62	63	64
α	.0555	.0364	.0230	.0140	.00825	.00467
β	.00877	.0150	.0249	.0397	.0611	.0907

With 100 trials (that is, in the specific context of the example, with 100
calves), it is thus possible to obtain much more satisfactory control of the
error probabilities than with 20 trials. For example, the critical value
$c = 62$ gives error probability $\alpha = .0140$, which is well below the specified
objective of $\alpha = .05$, and at the same time gives a β (.0397) well below the
specified $\beta = .10$. Is it desirable to have α and β so much smaller than
required? One might say, "the smaller the better," but the cost of the
experiment must also be considered. Presumably $n = 100$ is larger than
necessary. How large must n be taken so that α will be about .05 and β
about .10?

This question can be answered by inspection of a set of binomial tables,
but if these are unavailable or if the required values lie outside the tabu-
lated range, the normal approximation can be used. If as before F denotes
the number of successes (female calves) in the n trials, then

$$E(F) = np \quad \text{and} \quad \text{Var}(F) = np(1 - p).$$

Putting $p = \frac{1}{2}$ as specified by the null hypothesis, we find $1 - \alpha =
P_{\mathrm{H}}(F < c)$ to be approximately $\Phi[(c - \frac{1}{2} - \frac{1}{2}n)/\sqrt{.25n}]$. Since α is to be
.05, we must have $1 - \alpha = .95$ and hence get the approximate equation

$$\Phi[(c - \tfrac{1}{2} - \tfrac{1}{2}n)/\sqrt{.25n}] = .95.$$

The normal table gives

$$\Phi(1.645) = .95,$$

and by combining these two equations it follows that $(c - \frac{1}{2} - \frac{1}{2}n)/\sqrt{.25n}$
$= 1.645$, which may be rewritten as

$$c - \tfrac{1}{2} = 0.5n + 0.8225 \sqrt{n}.$$

Similarly, putting $p = .7$, $\beta = P_{\mathrm{A}}(F < c)$ is seen to be approximately
$\Phi[(c - \frac{1}{2} - .7n)/\sqrt{.21n}]$. Since β is to be .10, c and n must also satisfy

$$\Phi[(c - \tfrac{1}{2} - .7n)/\sqrt{.21n}] = .10.$$

The normal table shows

$$\Phi(-1.282) = .10$$

so that $(c - \frac{1}{2} - .7n)/\sqrt{.21n} = -1.282$, or equivalently

$$c - \tfrac{1}{2} = 0.7n - 0.5875 \sqrt{n}.$$

Equating the two expressions for $c - \frac{1}{2}$ and dividing by $\sqrt{n}$ leads to the equation

$$.5 \sqrt{n} + .8225 = .7 \sqrt{n} - .5875.$$

This may be solved for $\sqrt{n}$ to give $\sqrt{n} = 7.05$ or $n = 49.7$. Substitution of this value in either of the equations for $c - \frac{1}{2}$ gives $c = 31.1$.

These values are of course not exact, because the actual binomial probabilities were replaced by normal approximations, and because n and c must be integers. It is natural to try the integers closest to the values obtained: $n = 50$ and $c = 31$. The corresponding error probabilities α and β can be computed from the binomial formula (or read from a suitable table) as $\alpha = .0595$, $\beta = .0848$. The first of these is still somewhat too large. If we insist on $\alpha \leq .05$ and $\beta \leq .10$, binomial tables show the solution to be $n = 53$, $c = 33$, giving $\alpha = .0492$ and $\beta = .0862$.

The choice of a sample size n and critical value c involves a balancing of three conflicting objectives: to keep down α, β and n. If cost were no object, one could make both α and β arbitrarily small by taking n sufficiently large. Our intuitive notion of probability suggests that F/n will be very close to p if n is very large. In fact, with n sufficiently large, it is possible to distinguish between any two values of p, such as $p = .5$ and $p = .7$, with negligible risk of error. Unfortunately, the cost of an experiment can never be ignored, and an experimenter must settle for a degree of control of the error probabilities that he can afford.

A careful consideration of the levels at which the error probabilities are to be controlled, the sample size necessary to achieve these levels and, if this exceeds the resources available for the experiment, a reasonable compromise between the various requirements, is one of the most important aspects of planning an experiment. Unless such a prior analysis is made, the experiment may either achieve quite unnecessarily high precision (as in the above example with $n = 100$), and thereby waste resources which could be used more profitably for other experimental purposes, or not permit satisfactory control of both error probabilities (as in the above example with $n = 20$). In the latter case, the conclusions to be drawn from the experiment may be so unreliable as to defeat its purpose.

EXAMPLE 2. *Sampling inspection.* We shall now illustrate the methods and principles of sample-size determination on a lot sampling problem similar to that of Example 4 of the preceding section. In writing a contract for $N = 500$ electric lamps, the manufacturer asserts that not more than 5% of his lamps will fail under a specified degree of overload, while the customer is anxious to be at least sure that not more than 15% of the lamps will do so. The parties agree that a random sample of s of the lamps should be overloaded, and that the customer need not accept the order if the number D which fail is as large as c. We may say that the lot is "good" if 5% of the lamps would fail and "bad" if this proportion is instead 15%. What values must be given to s and c so that the manufacturer's risk of having a good lot rejected is only $\alpha = .10$ (and hence even less if the lot is better than "good"), while the customer's risk of accepting a bad lot is only $\beta = .05$ (and hence even less if the lot is worse than "bad")?

According to the null hypothesis H that 5% of the lamps would fail, the number of defective lamps in the lot is $r = 25$, and D has a hypergeometric distribution with $E_H(D) = rs/N = .05 s$ and

$$\mathrm{Var}_H(D) = \frac{N - s}{N - 1} \cdot s \cdot \frac{r}{N} \cdot \frac{N - r}{N} = .00009519(500 - s)s$$

or $\mathrm{SD}_H(D) = .009757 \sqrt{(500 - s)s}$. Combining the normal approximation

$$P_H(D < c) = \Phi[(c - \tfrac{1}{2} - .05s)/.009757 \sqrt{(500 - s)s}]$$

with the condition

$$P_H(D < c) = 1 - \alpha = .9 = \Phi(1.282)$$

gives

$$c - \tfrac{1}{2} = .05s + .01251 \sqrt{(500 - s)s}.$$

Similarly, if 15% of the lamps would fail, $r/N = .15$, so that

$$E_A(D) = .15s \quad \text{and} \quad \mathrm{SD}_A(D) = .01598 \sqrt{(500 - s)s}.$$

We therefore get the approximation

$$P_A(D < c) = \Phi[(c - \tfrac{1}{2} - .15s)/.01598 \sqrt{(500 - s)s}],$$

which we combine with the condition

$$P_A(D < c) = \beta = .05 = \Phi(-1.645)$$

to obtain the second equation

$$c - \tfrac{1}{2} = .15s - .02629 \sqrt{(500 - s)s}.$$

Equating these two expressions for $c - \tfrac{1}{2}$, we find

$$.1s = .03880 \sqrt{(500 - s)s}.$$

Squaring both sides and cancelling s gives a linear equation for s, which has the solution $s = 65.4$. Substitution of this value in either of the equations for $c - \tfrac{1}{2}$ gives $c = 5.88$. As before, these values are only approximations since s and c have to be integers. If we put $s = 65$ and $c = 6$, the normal approximation gives $\alpha = .0851$ and $\beta = .0569$. (For comparison the correct values, computed from the hypergeometric formula, are $\alpha = .0911$ and $\beta = .0498$.)

PROBLEMS

1. In Example 1, find the sample size required if
  (i) $\alpha = .01, \beta = .1$;　　(ii) $\alpha = .05, \beta = .05$;　　(iii) $\alpha = .01, \beta = .05$.

2. Solve the three parts of the preceding problem if the stated value is to be achieved against the alternative A $= p = .6$ instead of $p = .7$.

3. In Problem 11.1.4, find the sample size required if $\alpha = .03$ and if β is to be .05 when the mortality under the improved technique is (i) .25, (ii) .2, (iii) .15.

4. Assume that S has the binomial distribution (n, p). We wish to test the null hypothesis H: $p = .3$ against·the alternative A: $p = .8$, and insist on having $\alpha \leqq .1$ and $\beta \leqq .2$. By inspection of Table B, determine the smallest possible value of n.

5. The Army will accept a shipment of rockets only if all of those that are test-fired function properly. Regulations require that $\beta \leqq .1$ whenever $r/N = 10\%$.
  (i) What is the value of c?
  (ii) How many of a shipment of 20 rockets must be test-fired?
  (iii) Is the sampling plan realistic? How could it be made more so?
[Hint: (ii) Use formula (6.2.1) and Table A.]

6. A geneticist is told that in a certain community the fraction p of males is about .55, instead of the customary .514. He plans to study the recorded births at the community hospital to see if there are indeed unusually many males. He is willing to take only one chance in a hundred of publishing the statement that the community is exceptional if it is not. On the other hand, he will tolerate one chance in twenty of failing to publish if the community in fact has $p = .55$. How many birth records need he examine?

7. In Example 2, find the sample size required if
  (i) $\alpha = .1, \beta = .01$;　　(ii) $\alpha = .05, \beta = .05$;　　(iii) $\alpha = .05, \beta = .01$.

8. Solve the three parts of the preceding problem if the stated value of β is to be achieved when the proportion of failing lamps is 10% (instead of 15%).

9. Design a sampling plan for inspecting a lot of 25 items, so that $\alpha \leqq .1$ when there are 5 defectives and $\beta \leqq .2$ when there are 10 defectives. [Hint: First use the normal approximation to determine trial values of s and c, and then use Table A to correct the trial values.]

10. In Example 11.2.2, find the sample size required if $\alpha = .05$ and if it is desired to have $\beta = .08$ when (i) $p = \frac{7}{20}$; (ii) $p = \frac{9}{25}$.

13.3 THE POWER CURVE

When discussing the probabilities of the two kinds of error, it is necessary to specify not only the null hypothesis but also a particular alternative A, such as $p = .7$ in Example 1.1, or $r/N = 15\%$ in Example 2.2. The reader may have felt that the choice of such an alternative is rather arbitrary, and a clearer picture of the performance of a test is in fact obtained by computing the probability of false acceptance for a number of different alternatives. A certain unification is then possible, as will be seen below, by considering, instead of β itself, the quantity $1 - \beta$.

For a particular alternative A, $1 - \beta$ is known as the *power* of the test against the alternative A, and we shall denote it by π. By the law of complementation,

$$\pi = 1 - \beta$$

is the probability of rejecting the null hypothesis when the alternative A is true, and hence of correctly detecting the hypothesis to be false. Thus in Example 1.1, we found for the alternative $p = .7$ that $\beta = .584$, so that $\pi = .416$. This is the probability that the test will correctly detect the effectiveness of the biochemist's treatment when actually $p = .7$. Similarly, in Example 1.4 the probability β was seen to have the value .0038 for the alternative $r = 120$, if critical value $c = 5$ is used. Hence $\pi = .9962$, which is the probability of returning the shipment when it contains 120 defective motors, using $c = 5$.

Let us now consider the quantity π for varying alternatives. In Example 1.1, for instance, π depends on the true probability p of a female calf, and we may write it as $\pi(p)$ to emphasize this dependence: thus $\pi(p)$ is the probability of recommending the biochemist's treatment when it actually increases the probability of a female calf to the value p. Let us generalize our earlier notations P_H and P_A and write P_p to mean a probability computed using p as the probability of a female calf. Then

$$\pi(p) = P_p(\text{rejection}) = P_p(F \geq c).$$

Given values of n and c, for example our earlier values $n = 20$ and $c = 15$, we can compute this probability for a number of values of p, as β was computed for $p = .7$ earlier, and then graph these values to obtain a *power curve*.

The unification alluded to above is obtained by extending the power curve also to values $p \leq \frac{1}{2}$. If in particular we put $p = \frac{1}{2}$, we are assuming the null hypothesis to be correct, so that $\pi(\frac{1}{2})$ — the value of the curve at $p = \frac{1}{2}$ — equals $P_H(F \geq c) = \alpha$. For values of $p < \frac{1}{2}$, $\pi(p)$ is the probability of recommending the technique when it actually reduces the chance of a female. It is intuitively clear that the smaller p is, the smaller is the probability that F will be c or greater and hence that the null hypothesis will be rejected, so that $\pi(p)$ decreases as p decreases. It follows that for $p < \frac{1}{2}$, $\pi(p) < \pi(\frac{1}{2}) = \alpha$.

Table 1 shows $\pi(p)$ for $c = 15$ and for several different values of p.

TABLE 1

p	0	.4	.5	.6	.7	.8	.9	1
$\pi(p)$	0	.00161	.0207	.126	.416	.804	.9888	1

From the entries in this table, we can plot the points shown in Figure 1 and then sketch the entire power curve. This curve shows just how the

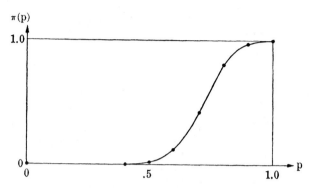

FIGURE 1. POWER CURVE FOR EXAMPLE 2.1

chance of our test discovering that the treatment works varies with the effectiveness of the treatment. At one extreme, if the treatment raises the probability of a female to $p = 1$, we are certain to decide that it works: $\pi(1) = 1$. At the other extreme, if in fact the treatment reduces the chance of a female much below $\frac{1}{2}$, we are very unlikely to recommend it as useful. As we expected on intuitive grounds, the graph shows $\pi(p)$ to be increasing as p increases.

The performance of the test of Example 1.4 can be studied quite analogously. In this case, the probabilities π and β depend on the (discrete) number r of defective motors in the shipment. Some values of $\pi(r) = P_r(D \geqq 5)$ are given in Table 2 (the binomial approximation was used),

TABLE 2

r	0	10	20	30	40	50	60	70	80	90	100
$\pi(r)$	0	.003	.051	.182	.371	.564	.723	.837	.910	.953	.976

and permit sketching the power curve shown in Figure 2. This curve summarizes the entire story of how well the inspection plan is performing

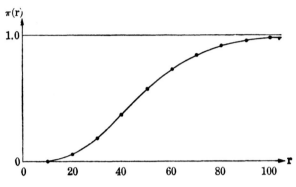

FIGURE 2. POWER CURVE FOR EXAMPLE 1.4

its tasks. The probability that the shipment will be returned as unacceptable increases with r, rising from negligible values for small r to near certainty when r is sufficiently large. The value for $r = 20$ is of course just α by definition. (The discrepancy between the value .051 given in Table 2 and the correct value $\alpha = .0415$ given in Section 1 is explained by the use here of the binomial approximation.)

We conclude by giving two other illustrations of power computations.

EXAMPLE 1. *The Wilcoxon two-sample test.* A farmer wishes to know whether pigs will grow faster when an antibiotic is added to their feed. He has a litter of six pigs, and selects three of these at random to receive the antibiotic. If these three turn out to gain more weight than the others, he will decide to use the antibiotic regularly. Since $\binom{6}{3} = 20$, and since he will reject the null hypothesis only if the rank-sum W of the controls has its smallest value $1 + 2 + 3 = 6$, the significance level of his test is $\alpha = \frac{1}{20} = .05$. What is its power? This depends on how effective the antibiotic is, and also on how variable the pigs are in normal growth.

To illustrate, let us suppose that without the treatment the six pigs would gain 23, 27, 32, 35, 41, and 48 pounds, and let Δ denote the additional gain caused by giving the antibiotic, which for simplicity we assume to be the same for all pigs. The null hypothesis states that $\Delta = 0$. The power π of the test depends on Δ. If in fact $\Delta = 10$, what is the chance that the farmer will decide to use the antibiotic? We can by inspection list the different treatment groups that will lead to rejection of the null hypothesis. For example, if the treatment is given to the pigs with natural weight gains 32, 35, and 48 pounds, they will actually gain 42, 45, and 58 pounds compared with gains of 23, 27, and 41 pounds for the controls, and the hypothesis will be rejected. On the other hand, the hypothesis will be accepted if the treated pigs happen to be those with natural gains 27, 35, and 48, since in this case the observed treatment gains 37, 45, and 58 are not all above the control gains 23, 32, and 41. The treatment groups leading to rejection are:

$$(35, 41, 48), \quad (27, 41, 48), \quad (32, 41, 48), \quad (32, 35, 48).$$

Thus $\pi(10) = \frac{4}{20} = .2$. If a weight gain of 10 pounds is economically important, this power is unsatisfactorily low, and a better test or more informative experiment is needed.

EXAMPLE 2. *A poetry quiz.* An instructor in English wishes to test his students' knowledge of English poetry. He gives them the titles of eight poems, and (in random order) quotations from these poems. The students are asked to match the quotations with the titles. How many correct matchings should constitute a passing grade?

Let M be the number of titles matched correctly by a student, and suppose that a value of $M \geq c$ is passing. The instructor wants c to be high enough so that a student who knows nothing and merely matches at random is very unlikely to pass. Referring to Table 6.8.1, we see that $P_H(M \geq 4) = \frac{771}{40320} = .019$ where H denotes the null hypothesis of complete ignorance. Thus with a passing grade of 4, on the average only about one out of every 50 completely ignorant students will pass.

A student who definitely recognizes four or more of the poems is sure to pass this test, but even recognizing one considerably improves his chances. If he matches the remaining seven at random and M' denotes the number among these that he gets right by chance, he will pass if $M' \geq 3$. The probability of his passing if he knows a single poem is therefore $P_H(M' \geq 3) = \frac{407}{5040} = .081$, four times as large as when he knows none of the poems. The power of the test here is the probability of the student passing, and this depends on the number k of poems he knows. Denoting it by $\pi(k)$, we have seen that $\pi(1) = .081$ and that $\pi(k) = 1$ when $k \geq 4$. By the same method that led to $\pi(1)$ we find $\pi(2) = .265$ and $\pi(3) = .633$ (Problem 9). (These calculations are for simplicity made under the unrealistic assumption that the student either knows the title for sure, or else matches at random.)

Figure 2 showed that $\pi(r) = P_r(D \geq 5)$ increases as the number r of defective items in the lot is increased. The following argument proves quite generally that if D has the hypergeometric distribution (N, r, s) then $P(D \geq c)$ increases as r is increased. To see this, let $r < r' \leq N$ and consider a box containing N marbles, of which r are red, $r' - r$ pink and $N - r'$ white. Then the number D of red marbles in a sample of s has the hypergeometric distribution (N, r, s), while the number D' of colored marbles in the sample has the hypergeometric distribution (N, r', s). Since the number of colored marbles in a sample is at least as large as the number of red marbles, it is always true that $D' \geq D$ and for some samples D'

is actually greater than D. Thus there are more samples with $D' \geq c$ than with $D \geq c$ and hence $P(D' \geq c) > P(D \geq c)$. A similar argument for binomial tests is given in Problem 9.

PROBLEMS

1. Sketch the power curve of the test whose β-values were obtained in Problem 1.6(ii).

2. In Problem 1.6 suppose that the mortality of the standard operation is 40%, and that the hypothesis is to be rejected if at least nine of the ten patients survive the new operation. Sketch the power curve of the test.

3. Use the normal approximation to sketch the power curve of the test whose critical value was found in Problem 11.2.3(ii).

4. Use the normal approximation to sketch the power curve of the test of Problem 11.2.6(ii).

5. (i) Is it reasonable to use the binomial approximation in computing Table 2?
(ii) Compute by the normal approximation the value of $\pi(90)$ for comparison with the entry in Table 2.

6. (i) In Example 1, find $\pi(20)$.
(ii) How large must Δ be to give $\pi \geq .7$?

7. In Example 1, find the power of the test against the alternative that the additional gain caused by giving the antibiotic, instead of being constant, is 40% of the amount the pigs would gain without the treatment.

8. In Example 1, suppose that the weight gains without treatment would be 10, 12, 17, 18, 26 and 42 pounds. Find (i) $\pi(3)$, (ii) $\pi(4)$, (iii) $\pi(10)$, (iv) $\pi(13)$.

9. Verify the values of $\pi(2)$ and $\pi(3)$ given in Example 2, and plot the power "curve."

10. In Example 12.6.1, suppose that there are only four subjects and that H is rejected if $V_- \leq 1$. Suppose that the new lotion has the effect of reducing the amount of burn by Δ from the amount the same skin area would receive with the old lotion. Find $\pi(\Delta)$ for (i) $\Delta = 0$, (ii) $\Delta = .1$, (iii) $\Delta = .5$, (iv) $\Delta = .7$ if the burns on the subjects with both sides receiving the old lotion would have been (38, 38.35), (46.1, 46.51), (51.2, 51.73), (41.9, 42.2), (38.3, 38.7), (43.5, 44.1).

11. Solve the preceding problem if H is rejected when $V_- \leq 2$.

12. Solve the preceding problem if the sign test is used instead of the Wilcoxon test, but at the same significance level.

13. If B has the binomial distribution (n, p), prove that $P(B \geq c)$ is increased by increasing p. [Hint: Let $p < p' \leq 1$ and consider a sequence of n trinomial trials (Section 7.3) whose three possible outcomes have probabilities p, $p' - p$, $1 - p'$, with numbers of occurrences B_1, B_2, B_3 ($B_1 + B_2 + B_3 = n$).   As pointed out in Section 7.3, B_1 has the binomial distribution (n, p) while by Problem 7.3.3, $B_1 + B_2$ has the binomial distribution (n, p').]

13.4   ONE- AND TWO-SIDED TESTS

Once the test statistic T has been chosen, it is necessary to decide what range of its values will be regarded as significant.   In the preceding sections, we considered this choice as rather self-evident.   Thus in Example 11.1.1 it clearly seemed appropriate to reject the null hypothesis for large values of F (the number of female calves), while in Example 12.2.1 it seemed equally clear that small values of D (the number of treated patients who did not survive five years) should be regarded as significant.   In general, values t of the test statistic T were considered as offering significant evidence against the null hypothesis H and in favor of the alternative A, if they were relatively likely to occur according to A but relatively unlikely according to H.   We shall in this section examine more carefully the issues involved in deciding which values of a test statistic should be regarded as significant.

EXAMPLE 1.   *Changing the sex ratio.*   To fix the ideas, let us return to the sex-ratio example considered in several previous sections.   To test the null hypothesis that the probability p of a female calf is $\frac{1}{2}$, we observed the number F of females among $n = 20$ calves, and rejected the hypothesis if F is large, say $F \geq 15$.   The reader may wonder why we do not also reject the hypothesis when $F \leq 5$.   After all, if $p = \frac{1}{2}$, the distribution of F is symmetric about its expected value of 10, and $F = 5$ is just as great a departure from what is expected, and is just as unlikely to occur, as $F = 15$.

There could be two rather different reasons for using the asymmetric test that we have adopted.   Perhaps the biochemist has given convincing reasons for believing that his treatment, if it has any effect at all, can only serve to increase p.   If the reasons are sufficiently convincing, we might want to assume that $p \geq \frac{1}{2}$.   In this case, while $F \leq 5$ is unlikely to occur when $p = \frac{1}{2}$, it is still less likely according to the only permissible alternatives, namely that $p > \frac{1}{2}$.   The observation $F \leq 5$ should therefore not be considered evidence against H in favor of the permissible alternatives.

There is, however, in the present case a second, and stronger, reason for using the asymmetric test.   If p were known to be $\frac{1}{2}$, the dairy association would want to recommend against adoption of the treatment.   If instead p were known to be less than $\frac{1}{2}$, there would be even more reason for a negative recommendation.   Thus, the same action (negative recommendation) is appropriate when $p < \frac{1}{2}$ as when $p = \frac{1}{2}$.   Since small values of F

indicate small values of p, it is therefore reasonable to recommend negatively whenever F is sufficiently small. Thus, a test which "rejects the hypothesis" (i.e., which leads to a favorable recommendation) when F is large, is appropriate even when one cannot exclude the possibility that $p < \frac{1}{2}$. (In this case one could reformulate the problem, and take the hypothesis to be $p \leq \frac{1}{2}$ rather than $p = \frac{1}{2}$.)

Let us now change the conditions of the problem. Suppose the biochemist believes that a certain treatment might affect the sex ratio, but if so does not know whether it would favor males or females. He applies the treatment to 20 laboratory animals, which naturally have sex ratio near $\frac{1}{2}$, to see whether the method is sufficiently promising to warrant further study. Let us suppose that for simplicity he formulates the null hypothesis as $p = \frac{1}{2}$. Neither of the two earlier reasons for an asymmetric test now applies. In the new circumstances there is no reason for excluding the possibility $p < \frac{1}{2}$. Furthermore, the same action (continued investigation of the treatment) would be appropriate both when p is, say, .3 and when it is .7. It is therefore reasonable to reject the null hypothesis (i.e., to continue investigation of the treatment) if F departs significantly from 10 in either direction. One might, for example, reject $p = \frac{1}{2}$ if either $F \leq 5$ or $F \geq 15$.

Since rejection is now desirable when p departs materially from $\frac{1}{2}$ in either direction, we would wish the power of the test to be large against both these types of alternatives. In the present example, where the null hypothesis is rejected if either $F \leq 5$ or $F \geq 15$, the power against any alternative p is given by

$$\pi(p) = P_p(F \leq 5) + P_p(F \geq 15).$$

The following is a table of $\pi(p)$ for a number of different values of p:

p	0	.1	.2	.3	.4	.5	.6	.7	.8	.9	1.0
$\pi(p)$	1.000	.989	.804	.416	.127	.041	.127	.416	.804	.989	1.000

From these values, the power curve of the test is sketched in Figure 1. The value at $p = \frac{1}{2}$, $\pi(\frac{1}{2}) = .041$, is of course the significance level of the test.

In comparing the power curve of the present test with Figure 3.1, the power curve of the test that rejects only when $F \geq 15$, a striking feature is the different shape. This corresponds to the fact, already mentioned, that in the earlier test it was desired to have a low probability of rejection when $p < \frac{1}{2}$, while for the present test we wish this probability to be high. There is however another important difference: the significance level of the earlier test was .021, only half of what it is for the present test. This is of course clear without computation since previously $\alpha = P_{1/2}(F \geq 15)$ while now

$$\alpha = P_{1/2}(F \leqq 5) + P_{1/2}(F \geqq 15),$$

and the binomial distribution for $p = \frac{1}{2}$ is symmetric about $\frac{1}{2}n$.

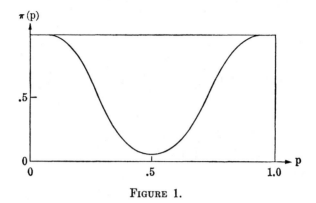

FIGURE 1.

It is important to realize this difference in α since it is a common mistake in hypothesis testing to pretend one is using a one-sided test when in fact a two-sided test is being employed. To illustrate, let us consider the work of John Graunt, who in 1662 published a collection of figures of birth and death for London. One finding that particularly interested him was the excess of men over women. Suppose that in 4000 births he had found the numbers to be 2050 and 1950 respectively. How greatly should he have been surprised if he expected the probability to be $\frac{1}{2}$? Applying the normal approximation we find that

$$P_{1/2}(F \leqq 1950) = \Phi\left(\frac{1950.5 - 2000}{\frac{1}{2}\sqrt{4000}}\right) = \Phi(-1.565) = .059,$$

and hence we might be tempted to attach to his result the significance probability .059. However, Graunt would clearly have been equally surprised if the numbers had been reversed. The relevant significance probability is therefore

$$P_{1/2}(F \leqq 1950) + P_{1/2}(F \geqq 2050) = .118.$$

Actually, Graunt's data were much more extensive. He writes that "there have been buried from the year 1628 to the year 1662 exclusive, 209436 Males and but 190474 Females," and draws from this the blunt conclusion "that there must be more Males than Females." The probability of observing so large a deviation from what would be expected if p were $\frac{1}{2}$, is indeed negligible.

Let us consider somewhat more generally the hypothesis that a parameter θ (such as the quantities p or r specifying a binomial or hypergeometric distribution), has the value θ_H. What issues are involved in deciding between the *two-sided* test, which rejects when the test statistic T satisfies

$$T \leq c \quad \text{or} \quad T \geq d,$$

or one of the two *one-sided* tests which reject the hypothesis only for $T \geq d$, or only for $T \leq c$, but not for both? In this discussion, we shall suppose that large values of T are likely to occur when θ is large, small values of T are likely to occur when θ is small, while for θ near θ_H intermediate values of T are most likely. (The reader may check that these suppositions hold in our binomial and hypergeometric examples.)

Our earlier discussion suggests the following three possibilities.

(i) If it is assumed that the permissible alternatives to the null hypothesis all lie in the same direction from θ_H, either to its right ($\theta > \theta_H$), or to its left ($\theta < \theta_H$), clearly the one-sided test appropriate to this direction should be used.

(ii) Suppose there is a choice between two decisions, one of which is appropriate if the null hypothesis is true or if it is false in one direction (for example $\theta \leq \theta_H$) while the other decision is appropriate if the null hypothesis is false in the other direction ($\theta > \theta_H$). One is then really testing the hypothesis $\theta \leq \theta_H$ against the alternative $\theta > \theta_H$, and again a one-sided test is appropriate.

(iii) If alternatives θ on both sides of θ_H are assumed possible, and if the same decision or action would be called for against both, a two-sided test is appropriate.

We shall now give one example of each of these possibilities.

EXAMPLE 2. *Triangular taste tests.* Recall Example 11.2.2, in which a manufacturer of powdered coffee finds a method of producing his product more cheaply, and wonders if it causes a detectable alteration of taste. He prepares the product by both methods and presents to each of 500 consumers a tray with three cups, identical in appearance and randomly ordered, one of which contains the new product while the other two contain the old. The consumers are asked to identify the cup that is different from the other two.

If there is no difference in taste, each consumer has one chance in three of correctly guessing which cup contains the new product. Under the null hypothesis of no difference, the number B of consumers who guess correctly therefore has the binomial distribution ($n = 500$, $p = \frac{1}{3}$). What are the permissible alternatives? If there is a difference in taste, the chance of a consumer correctly identifying the cup that differs from the other two is increased. It is hard to see how a change of taste could lead to a decrease in the ability of guessing correctly unless the customers are willfully uncooperative. The manufacturer may therefore wish to assume $p \geq \frac{1}{3}$, so that we are in case (i). The appropriate test would be one-sided and reject if $B \geqq c$, where c is the critical value chosen so as to provide the desired compromise between α and β.

EXAMPLE 3. *A heart treatment.* In Example 12.2.1 we formulated the null hypothesis that a certain treatment for victims of coronary attack would have no influence on whether the patient survives five years. The test proposed was based on the number D of early deaths among treated patients, with small values considered significant. Clearly, a very large value of D would also make one doubt that the treatment is without effect, and lead one to believe on the contrary that it is harmful. However, although alternatives on both sides of the null hypothesis of no effect are reasonable here, a one-sided test is appropriate. We have here an example of case (ii), since the physician would come to the same decision—against the use of the treatment—whether he knew it to be merely useless or actually harmful.

EXAMPLE 4. *The Mendelian theory.* With an inherited trait which is controlled by a single pair of genes, each offspring of a cross of two hybrids has probability $\frac{1}{4}$ of being recessive according to the simplest assumption of the Mendelian theory. (We have given a discussion of this theory in Section 4.5, but an understanding of it is not necessary to follow this example.) A geneticist wonders whether a certain trait is inherited in this way. He collects 30 cases that are offspring of hybrid crosses, and observes the number T possessing the trait, in order to test the null (Mendelian) hypothesis. If the 30 trials are assumed to be unrelated, then according to the hypothesis T will have the the binomial distribution ($n = 30$, $p = \frac{1}{4}$). If the geneticist has no particular alternative value of p in mind, he would want to reject the hypothesis if p departed materially from $\frac{1}{4}$ in either direction. He would accordingly use a two-sided test, rejecting the hypothesis H: $p = \frac{1}{4}$ if $T \leqq c$ or if $T \geqq d$. The value of the significance level will depend on the values used for c and d. For $c = 2$ and $d = 13$, for example we may calculate from the binomial formula that $P_{\mathrm{H}}(T \leqq c) = .0106$ and $P_{\mathrm{H}}(T \geqq d) = .0216$, so that $\alpha = .0106 + .0216 = .0322$. With this test, the power $\pi(p) = 1 - P_p(3 \leqq T \leqq 12)$ may also be calculated from the binomial formula. A few values are shown below.

p	0	.1	.2	.25	.3	.4	.5
$\pi(p)$	1	.411	.0473	.0322	.0866	.422	.819

Unfortunately, the test is not very sensitive to small departures of p from $\frac{1}{4}$. More than 30 cases would therefore be needed before acceptance of the hypothesis would be very convincing.

For two-sided tests, the problem arises of choosing two critical values c and d. The most common practice is to choose them so that the two one-sided probabilities $P_H(T \leq c)$ and $P_H(T \geq d)$ are equal, or at least approximately so. The resulting test is said to have *equal tails*. Actually, an equal-tailed test is reasonable only if the interest in the alternatives on the two sides is about equal. If in a two-sided problem one is nearly (but not quite) certain that $\theta < \theta_H$ may be ruled out, it would be reasonable to take $P_H(T \leq c)$ smaller than $P_H(T \geq d)$, so that the test will have greater power against the alternatives $\theta > \theta_H$, while still providing some power against the implausible alternatives $\theta < \theta_H$.

For a test to be equal-tailed when the null distribution of T is symmetric about a point μ, the critical value d must be just as far above μ as c is below it, that is, $d - \mu = \mu - c$. Furthermore, since the probabilities in the two tails are equal, the value of α will be just twice what it is for the one-sided test which rejects for $T \leq c$. Therefore, a table of the significance levels of the one-sided test can also be used for the two-sided test: it is only necessary to double the tabled value of α. For example, from Table H of the Wilcoxon distribution when $s = 4$ and $t = 7$, we see that the test which rejects for $W \leq 13$ has $\alpha = P_H(W \leq 13) = \frac{7}{330} = .0212$. Since W is symmetric about $\mu = \frac{1}{2} \cdot 4 \cdot 12 = 24$, the corresponding two-sided test rejects when $W \leq 13$ or $W \geq 35$, and this test has significance level $\alpha = 2P_H(W \leq 13) = 2(.0212) = .0424$.

Two-sided tests are sometimes used for a class of problems, which are actually more complicated. Suppose one action would be appropriate if $\theta = \theta_H$, a second one if $\theta < \theta_H$, and a third one if $\theta > \theta_H$. The choice then lies not between two actions but three: we are dealing with a *three-decision problem*. These problems are more complicated than testing statistical hypotheses since, for example, six different errors are possible instead of only two. However, such problems can be handled by the machinery of a two-sided test. If the null hypothesis $\theta = \theta_H$ is accepted, the first action is taken; if it is rejected, either the second or third action is taken according as $T \leq c$ or $T \geq d$.

PROBLEMS

1. Suppose in Example 1 that there are only 10 experimental animals instead of 20.
   (i) Find the equal-tailed, two-sided test for $H: p = \frac{1}{2}$ which has significance level as near .1 as possible.
   (ii) Compute and sketch the power curve of this test.

2. Suppose S has the binomial distribution $(400, p)$. To test $H: p = .5$ an experimenter proposes to use the statistic $T = |S - 200|$.
   (i) Is this a one-sided or a two-sided test?

(ii) If he wants α to be approximately .1, what critical value should be used for T?

(iii) What is the power against the alternative $p = .55$?

(iv) If the experimenter were willing to assume $p \geq .5$, what test should he use?

(v) What would be the power of his test against $p = .55$ at level $\alpha = .1$?

3. In Example 2 the manufacturer wants α to be near .05.

(i) What critical value should be used for B?

(ii) What is the power of this test against the alternative $p = .4$?

4. Of 15 experimental subjects, 7 are selected at random for treatment, the other 8 serving as controls. A two-sided Wilcoxon test is to be used to test the hypothesis of no treatment effect against the two-sided alternatives $\Delta \neq 0$ (positive or negative) approximately at the 10% significance level.

(i) Find critical values for which the test is approximately equal-tailed.

(ii) Suppose you are fairly sure (but not certain) that $\Delta \geq 0$. Determine critical values so that the 10% rejection probability under H is distributed approximately in the ratio 3:1 between the two tails.

5. In Example 4, suppose the geneticist is fairly sure (but not certain) that $p \geq \frac{1}{4}$. If he collects 200 cases and wishes to use a two-sided test at level .1, determine approximate critical values so that the 10% rejection probability is distributed in the ratio (i) 9:1; (ii) 4:1; (iii) 2:1.

6. In the three parts of the preceding problem, use the normal approximation to sketch the power curve of the test.

13.5 CHOICE OF TEST STATISTIC

In the examples considered so far, we have presented on intuitive grounds a function of the experimental observations to be used as the test statistic. In most examples there actually appeared to be only one reasonable statistic. In order to test the quality of a lot, for example, after having observed the number D of defectives in a random sample from this lot, it is difficult to imagine any reasonable test not based on the test statistic D.

In more complicated testing problems, however, there will frequently exist a variety of apparently reasonable and essentially different test statistics, and it then becomes necessary to choose one of them. The main criterion of choice is the power of the resulting test: if one statistic provides greater power than another when both are applied at the same level of significance, we prefer the more powerful statistic since its use results in a smaller probability of false acceptance.

That the first statistic which comes to mind is not necessarily the best may be illustrated by the following example.

EXAMPLE 1. *The loaded die.* A gambler produces a die, and offers to bet that it will fall with the ace on top. His colleagues not unnaturally suspect that it may be loaded so that $P(\text{Ace}) > \frac{1}{3}$. They decide to throw it

200 times to see whether Ace does tend to occur more often than one time in six. If this is not the case, they will accept his bet, and otherwise will take appropriate action.

Perhaps the most natural test statistic is the number X of times Ace appears in the 200 throws, with large values of X significant evidence against the null hypothesis H that $P(\text{Ace}) = \frac{1}{6}$, in favor of the alternative that $P(\text{Ace}) > \frac{1}{6}$. According to H, the random variable X has the binomial distribution ($n = 200$, $p = \frac{1}{6}$). If the critical value is taken as $c = 43$, the level of significance is $\alpha = P_{\text{H}}(X \geq 43) = .0444$. The power of the test depends on how much $P(\text{Ace})$ exceeds $\frac{1}{6}$, that is, on how heavily the die is loaded. Against the alternative A that $P(\text{Ace}) = .20$, for example, the power is

$$\pi(.20) = P_{\text{A}}(X \geq 43) = .324.$$

This power is not at all satisfactory, since the other gamblers would not want to take a chance as large as 68% of playing with a die so heavily loaded against them. One possible method of increasing the power, without raising the significance level, would be to increase the sample size by throwing the die more than 200 times. However, some improvement can be obtained simply by using the available data to better advantage. If a die is loaded to favor Ace, it will at the same time decrease the probability of the opposite side (Six) turning up. Let us therefore consider the number Y of times that Six appears. Small values of Y will also constitute evidence against the null hypothesis, and this evidence is ignored when X is used as the test statistic.

Whether the use of X or Y will lead to the more powerful test will depend on just how the die is loaded. For purposes of illustration, suppose that under the alternative A the six faces have the following probabilities:

(1)

Face	Ace	Two	Three	Four	Five	Six
Probability	.20	.17	.17	.17	.17	.12

In testing H with the aid of Y, we should want to reject when Y is small. If H is rejected when $Y \leq 24$, we have

$$\alpha = P_{\text{H}}(Y \leq 24) = .0426 \quad \text{and} \quad \pi(.12) = P_{\text{A}}(Y \leq 24) = .554.$$

Against this particular alternative, the test based on Y is therefore better than the test based on X, since it has a slightly smaller probability of false rejection and a considerably larger power.

However, Y is still not the best available test statistic. Both X and Y contain information on whether or not H is true, and a better test can be constructed by utilizing all of this information. If the die is loaded in the manner indicated, X will tend to be large and Y small. Hence the statistic

$T = X - Y$ will be large for two reasons, and large values of T will reflect both sources of information.

The exact distribution of T is complicated. However, the distribution of T can be approximated by the normal curve. We obtain this approximation by extending the idea of an indicator random variable, defining J_1 as

$$J_1 = \begin{cases} 1 \\ -1 \text{ if the first toss gives} \\ 0 \end{cases} \begin{cases} \text{Ace} \\ \text{Six} \\ \text{Two, Three, Four, or Five,} \end{cases}$$

and analogously $J_2, \ldots, J_{200}$ as $1, -1, 0$ as the 2nd, $\ldots$, 200th throw gives Ace, Six, or one of the other results. The sum of all the J's is then equal to the number of J's that are equal to 1, minus the number that are equal to -1, and hence

$$(2) \qquad T = J_1 + J_2 + \ldots + J_{200}.$$

If the 200 throws are unrelated, so that a product model is appropriate, the random variables $J_1, \ldots, J_{200}$ are independent (Section 5.2) and hence according to the central limit theorem (Section 6.4) the normal approximation may be applied to T. This approximation requires the expectation and variance of T, which we shall now compute.

Let us write $P(\text{Ace}) = p_1$, $P(\text{Two}) = p_2, \ldots, P(\text{Six}) = p_6$. Note that

$$E(J_1) = \ldots = E(J_{200}) = p_1 - p_6.$$

Since J_1^2 is an indicator,

$$E(J_1^2) = P(J_1^2 = 1) = p_1 + p_6$$

so that

$$\text{Var}(J_1) = \ldots = \text{Var}(J_{200}) = p_1 + p_6 - (p_1 - p_6)^2.$$

From (2) it follows that

$$E(T) = 200(p_1 - p_6),$$

and, from the addition law for variance (5.7.15),

$$\text{Var}(T) = 200[p_1 + p_6 - (p_1 - p_6)^2].$$

When H is true, $P(\text{Ace}) = P(\text{Six}) = \frac{1}{6}$, so that

$$E_H(T) = 0 \quad \text{and} \quad \text{Var}_H(T) = \tfrac{200}{3}.$$

By the normal approximation, the test which rejects H when $T \geq c$, has for $c = 15$ the significance level $\alpha = .0379$. Under the alternative A, $P(\text{Ace}) = .20$ and $P(\text{Six}) = .12$. Hence

$$E_A(T) = 200(.20 - .12) = 16,$$
$$\text{Var}_A(T) = 200[.32 - (.08)^2] = 62.72,$$

and the normal approximation shows the power of the test to be $\pi = P_A(T \geq 15) = .575$. Comparison with the test based on Y alone, shows the new test to have slightly higher power and a considerably smaller significance level. The test based on T is therefore better than either of the two previously considered. We shall see below that even this test is not the best possible.

The example shows that it is not always obvious what test statistic should be used, and that the choice can be of importance. We shall now give a method for constructing the best possible test.

A hypothesis or alternative is called *simple* if it completely specifies the distribution of the observed random variables. (For example $p = .5$ and $r = 20$ are simple, while $p > .5$ and $r \leqq 20$ are not.) Let us consider the problem of testing a simple hypothesis H against a simple alternative A. We have suggested in Section 11.1 that H should be rejected for those values t of a test statistic T, for which $P_A(T = t)$ is relatively large compared with $P_H(T = t)$. Intuitively, an event which is, relatively speaking, unlikely to occur if H is true, but likely to occur if A is true, would seem to provide a good basis for rejecting H in favor of A. This idea suggests classifying the simple events e in the event set $\mathcal{E}$ according to how many times more likely they are under A than under H; that is, according to the value of the *probability ratio*

$$\lambda(e) = \frac{P_A(e)}{P_H(e)}.$$

The larger this ratio is, the stronger is the evidence against H and in favor of A. Accordingly, we may choose a critical value c and decide to reject H whenever $\lambda(e) \geqq c$. The resulting tests are called *probability ratio tests*. A basic theorem of the theory of testing hypotheses, known as the *Neyman-Pearson Fundamental Lemma*, states that a probability ratio test is more powerful (against A) than any other test with the same or a smaller level of significance. When it is applicable (i.e., when we have a particular alternative in mind), this theorem answers two questions that we have discussed before: what test statistic should be used, and what range of values of the statistic should be regarded as significant. Before giving a proof of the theorem, we shall illustrate its use on a number of examples.

EXAMPLE 2. *Testing binomial p.* Suppose that B is a binomial random variable corresponding to n trials with success probability p, and that it is desired to test the hypothesis $p = p_H$ against the alternative $p = p_A$, where p_H and p_A are given numbers with $p_A > p_H$. Here the simple events are the possible values $b = 0, 1, \ldots, n$ of B. By (6.1.2), the probabilities of these values under A and H are

$$P_A(b) = \binom{n}{b} p_A^b (1 - p_A)^{n-b}$$

and

$$P_H(b) = \binom{n}{b} p_H^b (1 - p_H)^{n-b}.$$

By definition, $\lambda(b)$ is the ratio $P_A(b)/P_H(b)$, which may be written

(3)
$$\lambda(b) = \left(\frac{1 - p_A}{1 - p_H}\right)^n \left(\frac{p_A}{p_H} \cdot \frac{1 - p_H}{1 - p_A}\right)^b.$$

Since $p_A > p_H$, both of the ratios p_A/p_H and $(1 - p_H)/(1 - p_A)$ are greater than 1, and therefore so is their product. The first factor on the right side of (3) does not change with b, while the second factor, as the bth power of a number greater than 1, increases as b is increased. It follows that

(4)
$$\lambda(0) < \lambda(1) < \ldots < \lambda(n).$$

A probability ratio test always rejects for large values of $\lambda(e)$. In the present example, this means rejection for large values of $\lambda(b)$, which by (4) means rejection for large observed values b of B.

It follows from the Neyman-Pearson lemma that a test which rejects when $B \geq c$, is most powerful at its level of significance for testing H: $p = p_H$ against the alternative $p = p_A$, provided $p_A > p_H$. For the application of the lemma it was necessary to specify a simple alternative value p_A. It is seen however that the same test ($B \geq c$) is most powerful regardless of which alternative $p_A > p_H$ we selected. The test is therefore simultaneously (or *uniformly*) most powerful for testing H: $p = p_H$ against all alternatives $p > p_H$. This conclusion justifies our intuitive choice of test in the sex-ratio problem (Example 11.1).

That the same test is the most powerful solution for all alternatives of interest in Example 2 is a happy accident, which one should not expect to happen in general. Suppose for example that we are testing H: $p = \frac{1}{2}$ against the alternatives that p may be either less or greater than $\frac{1}{2}$. In Example 4.1 we have suggested for this problem the two-sided test that rejects when $F \leq c$ or $F \geq d$. The most powerful test of H against the alternatives $p > \frac{1}{2}$ has just been seen to reject when F is too large; analogously (Problem 3), the most powerful test against the alternatives $p < \frac{1}{2}$ rejects when F is too small. Since these tests do not agree, there exists in this case no test which is simultaneously most powerful against all alternatives of interest. The two-sided test is a compromise solution which, by sacrificing some power on each side of $p = \frac{1}{2}$, achieves reasonably good power simultaneously against all alternatives.

EXAMPLE 3. *Combination of two binomials.* Suppose that in the preceding example, while our experiment is in progress, we hear of another experiment conducted to test the same hypothesis. In order to obtain a more powerful test, it is agreed to pool the observations from the two experiments. Suppose that in the other experiment, the number B' of successes in n' trials is being observed, so that the total observational material consists of the pair of numbers (B, B'). The simple events in this case consist of the nn' points (b, b') with $b = 0, 1, \ldots, n$ and $b' = 0, 1, \ldots, n'$. If the two experiments are unrelated, a product model is appropriate, and the probability of the simple event (b, b') under $A: p = p_A$ is

$$P_A(b, b') = \binom{n}{b} p_A^b (1 - p_A)^{n-b} \binom{n'}{b'} p_A''(1 - p_A)^{n'-b'}$$

and under $H: p = p_H$ is

$$P_H(b, b') = \binom{n}{b} p_H^b (1 - p_H)^{n-b} \binom{n'}{b'} p_H''(1 - p_H)^{n'-b'}.$$

The ratio of these two probabilities is, after some simplification,

$$\lambda(b, b') = \left(\frac{1 - p_A}{1 - p_H}\right)^{n+n'} \left(\frac{p_A}{p_H} \cdot \frac{1 - p_H}{1 - p_A}\right)^{b+b'}$$

This is exactly the expression obtained for the corresponding ratio in the preceding example, with $n + n'$ in place of n and $b + b'$ in place of b. From the discussion given there, it therefore follows that the probability ratio test in the present case rejects when $B + B' \geqq c$. Since $B + B'$ is the number of successes among the total $n + n'$ trials in the two experiments, the test is the same as if we were observing the number of successes in a single experiment of $n + n'$ trials.

EXAMPLE 1. *The loaded die (continued)*. In Example 1, we proposed $T = X - Y$ as a reasonable test statistic for deciding whether a die is loaded to favor Ace, but stated that further improvement is still possible. The best test against the simple alternative A defined by (1) is, according to the Neyman-Pearson lemma, given by the probability ratio test, which we shall now derive. When the die is thrown n times, one can observe the numbers of times each face appears. We shall label these as follows:

Face	Ace	Six	Two	Three	Four	Five
Number of appearances	X	Y	U	V	W	Z

These six random variables have a multinomial distribution (Section 7.3), so that according to the alternative A,

$P(X = x$ and $Y = y$ and $U = u$ and $V = v$ and $W = w$ and $Z = z) =$
$$K(.2)^x(.12)^y(.17)^{u+v+w+z}$$

where K is an integer whose value we shall not need. According to the hypothesis H that the die is fair, the probability of the same event is

$$K(\tfrac{1}{6})^{x+y+u+v+w+z}.$$

Since $x + y + u + v + w + z = n$, the probability ratio is therefore
$$6^n(.2)^x(.12)^y(.17)^{n-x-y},$$

which (except for a constant factor) is

(5) $$\left(\frac{.2}{.17}\right)^x \left(\frac{.12}{.17}\right)^y = (1.176)^x(.706)^y.$$

The desirability of rejecting H for various observed pairs (x,y) is measured by the size of the quantity (5). The calculations are simplified by using logarithms. Since the logarithm of (5) is

(6) $x \log 1.176 + y \log .706 = .0704\, x - .1512\, y = (.0704)(x - 2.15\, y)$,

and since the probability ratio test calls for rejecting for large values of (5), which means large values of (6), we see that the probability ratio test rejects for large values of $X - 2.15\, Y$. The best statistic is therefore not $X - Y$ but rather $X - 2.15\, Y$. Intuitively, Y contains more information than X, and should be more heavily weighted when the two are combined. However, the particular weighting $X - 2.15\, Y$ depends on the particular alternative A that we selected. For certain alternatives (Problem 4) greater weight should be given to X than to Y. Since the precise probabilities that result from loading the die are in practice unknown, the statistic $T = X - Y$ may be a reasonable one to use.

We shall now give a proof of the Neyman-Pearson lemma. Let R denote the set of simple events for which the probability ratio test rejects the hypothesis H. Then R consists of the simple events e for which $\lambda(e) \geqq c$ or equivalently

(7) $P_A(e) \geqq cP_H(e)$,

where c is a positive number. If α denotes the significance level of this test, $\alpha = P_H(R)$. Consider any other test whose significance level α' satisfies $\alpha' \leqq \alpha$, and let S denote the set of simple events for which this test rejects the hypothesis. The set R may be partitioned into the two exclusive parts (R and S) and (R and $\bar{S}$), so that

$$\alpha = P_H(R \text{ and } S) + P_H(R \text{ and } \bar{S})$$

by the addition law. Similarly,

$$\alpha' = P_H(R \text{ and } S) + P_H(\bar{R} \text{ and } S).$$

Since $\alpha \geqq \alpha'$, we must therefore have

(8) $P_H(R \text{ and } \bar{S}) \geqq P_H(\bar{R} \text{ and } S)$.

By (7),

$$P_A(e)/c \geqq P_H(e)$$

for every simple event e in R, and therefore for every simple event in (R and $\bar{S}$), so that

(9) $P_A(R \text{ and } \bar{S})/c \geqq P_H(R \text{ and } \bar{S})$.

Similarly, $P_H(e) > P_A(e)/c$ for every simple event e in $\bar{R}$, and therefore for every simple event e in ($\bar{R}$ and S), so that

(10) $P_H(\bar{R} \text{ and } S) > P_A(\bar{R} \text{ and } S)/c$.

By combining (9), (8) and (10), we see that

$$P_A(R \text{ and } \bar{S})/c \underset{(9)}{\geqq} P_H(R \text{ and } \bar{S}) \underset{(8)}{\geqq} P_H(\bar{R} \text{ and } S) \underset{(10)}{>} P_A(\bar{R} \text{ and } S)/c,$$

and it follows that the leftmost member of this string of inequalities is larger than the rightmost member. Cancelling the common factor $1/c$, we find

$$P_A(R \text{ and } \bar{S}) > P_A(\bar{R} \text{ and } S).$$

Finally, by adding $P_A(R \text{ and } S)$ to both sides, it follows that

$$P_A(R) > P_A(S).$$

This proves that the probability ratio test R is more powerful than any other test S whose significance level does not exceed that of R.

PROBLEMS

1. In Example 1, use the normal approximation to determine the number n of tosses required to distinguish between H and A with $\alpha = \beta = .01$, with the test statistic (i) X, (ii) Y, (iii) T, (iv) $X - 2Y$.

2. In Example 1, calculate the correlation of X and Y (i) under H, (ii) under A. [Hint: Recall $\text{Var}(X - Y) = \text{Var}(X) + \text{Var}(Y) - 2 \text{Cov}(X,Y)$.]

3. Show that in Example 2, the most powerful test for testing the hypothesis $p = p_H$ against any alternative value $p_A < p_H$ rejects when B is less than or equal to some constant c.

4. Find an alternative loading for the loaded die of Example 1, such that the most powerful test statistic for testing the hypothesis of no loading gives more weight to X than to Y.

5. A box contains N items numbered from 1 to N. One item is selected at random and its number X is observed. We wish to test the hypothesis H that $N = 5$ at significance level $\alpha = .2$. Since an observation $X > 5$ definitely disproves H, the hypothesis should of course be rejected whenever $X > 5$. The following five tests $T_1, \ldots, T_5$ are thus available at the desired level: T_1 rejects H if $X = 1$ or $X > 5$; T_2 rejects H if $X = 2$ or $X > 5$; etc.
  (i) Calculate and sketch the power curve of each of these tests against the alternatives $N = 1, 2, 3, 4, 6, 7, \ldots$.
  (ii) Is there a uniformly most powerful test among $T_1, T_2, \ldots, T_5$?
  (iii) Suppose that $N \geq 2$, that two items are drawn at random from the box without replacement, and are found to be numbered X and Y. Suggest a good test for H at level .1, and sketch its power function against $N = 2, 3, 4, 6, 7, \ldots$.

6. Consider a model in which the event set $\mathcal{E}$ has four points whose probabilities according to H and A are as follows:

e	e_1	e_2	e_3	e_4
P_H	.05	.05	.10	.80
P_A	.10	.20	.36	.34

  (i) List the possible probability ratio tests, and give the significance level α and power π of each.
  (ii) Suppose that α is required to be .1 or less. List all tests available at this level (not necessarily probability ratio tests) and give the power of each.
  (iii) Consider the following rejection rule: H is rejected if e_2 is observed and is accepted if e_1 or e_4 is observed; if e_3 is observed, a fair coin is tossed, and H is rejected if the coin falls Tails and accepted if it falls Heads. Compute α and π for this *randomized* test and compare with the tests of part (ii).

13.6 THE $\overline{X}$, t AND WILCOXON ONE-SAMPLE TESTS

Consider a process for obtaining a sequence X_1, X_2, $\cdots$ of observations or measurements of the same kind.   For example, the X's may be the heights of a sequence of male freshmen enrolling at a college, the yields of a variety of wheat grown on a sequence of test plots, or quality judgements on a sample of cans taken from the production line of a packing plant.   The long-run average value of X is represented by the expectation $\mu = E(X)$.

The problem of estimating μ was treated in Section 9.2.   We shall now take up the problem of testing the hypothesis that μ is equal to some specified value.   The following examples will illustrate how this problem may arise.

EXAMPLE 1.   *Life testing.*   A manufacturer guarantees that his light bulbs will on the average burn for 1000 hours.   A testing laboratory buys 20 of the bulbs and burns them until they fail, observing the 20 times to failure.   The 20 observed life times are to be used to test the manufacturer's guarantee.

EXAMPLE 2.   *Quality improvement.*   The quality of a mass-produced item is measured on a numerical scale, for example, by length of life, brightness, hardness, and so on.   The average value of quality is known for the currently used production process.   It is believed that the quality can be raised by making a certain change in the process.   A number of items is produced by the new method and their quality is measured.   Do they justify the belief that the new method constitutes an improvement?

EXAMPLE 3.   *Testing for adulteration.*   The purchaser of a gold ring suspects that the ring is adulterated with silver.   He measures the specific gravity of the ring several times to find out whether the ring has the known specific gravity of gold, or a lower value.

In all three examples, the problem is that of testing the hypothesis H that μ has a specified value μ_H.   Thus, in Example 1, μ_H is 1000 hours; in Example 2, it is the average quality with the current process; in Example 3, it is 19.3, the specific gravity of gold.   The alternative in some cases is that $\mu > \mu_H$ (as in Example 2), in others, that $\mu < \mu_H$ (as in Examples 1 and 3).   To be specific, suppose that the alternative under consideration is $\mu > \mu_H$.

To obtain a test of $\mu = \mu_H$ against $\mu > \mu_H$, one may take a number, say n, of observations and compute their arithmetic mean

$$\overline{X} = (X_1 + \cdots + X_n)/n,$$

which in Section 9.2 was proposed as a reasonable estimate of μ.   If $\overline{X}$ is sufficiently much larger than μ_H, it would seem reasonable to conclude that the expectation μ of the X's is greater than μ_H and hence to reject H.   The proposed test, which we shall call the $\bar{X}$-test, therefore rejects H if

(1) $\bar{X} - \mu_H \geqq c,$

where c is an appropriate critical value.

In order to determine c, we shall employ the model for repeated measurements developed in Section 9.2.   This model represents the observations $X_1, \ldots, X_n$ as independent random variables, all having the same distribution.   Their common expectation is denoted by μ and their common standard deviation by σ.

The critical value c in (1) is determined so that the test has the prescribed significance level α.   In principle, this critical value or the significance probability could be calculated once the precise form of the common distribution had been specified.   In practice, this calculation would often be difficult; in addition, one is frequently uncertain about the form of the distribution of the observations X.   It is therefore doubly fortunate that the central limit theorem (Section 6.4) provides a simple approximation, which does not require any assumption about the form of the distribution of X.

To apply the central limit theorem, we need to express $\bar{X}$ in standard units, as in Section 6.5.   This standardization requires knowledge only of the expectation and standard deviation of $\bar{X}$, which were found in Section 9.2 to be

(2) $E(\bar{X}) = \mu,$ $\mathrm{SD}(\bar{X}) = \sigma/\sqrt{n}.$

If the null hypothesis $\mu = \mu_H$ is correct, the standardized test statistic is therefore

(3) $\dfrac{\bar{X} - \mu_H}{\sigma/\sqrt{n}} = \dfrac{\sqrt{n}(\bar{X} - \mu_H)}{\sigma}.$

The hypothesis is rejected if $\bar{X} - \mu_H$ is too large, or equivalently if $\sqrt{n}(\bar{X} - \mu_H)/\sigma$ is too large, say, if

(4) $\sqrt{n}(\bar{X} - \mu_H)/\sigma \geqq u.$

In this form of the $\bar{X}$-test, which is equivalent but more convenient than the earlier (1), the critical value u is determined by

(5) $P_H\left[\dfrac{\sqrt{n}(\bar{X} - \mu_H)}{\sigma} \geqq u \right] = \alpha.$

By the central limit theorem, the null distribution of the statistic (3) is approximated by the appropriate area under the normal curve, and in particular the left-hand side of (5) is approximately equal to the area $1 - \Phi(u)$ to the right of u.   The critical value u of (4) can therefore be obtained approximately from the equation

$$(6) \qquad 1 - \Phi(u) = \alpha.$$

This approximation typically proves to be adequate even for samples as small as 5 or 10, provided the distribution of X is not too strongly asymmetric.

An exactly analogous argument applies if the hypothesis H is to be tested against the alternatives $\mu < \mu_H$. The hypothesis is then rejected when

$$(7) \qquad \sqrt{n}(\overline{X} - \mu_H)/\sigma \leqq v,$$

where v is determined by the equation (Problem 3)

$$(8) \qquad \Phi(v) = \alpha.$$

EXAMPLE 3. *Testing for adulteration (continued).* Suppose a method for measuring specific gravity has standard deviation $\sigma = .2$. (This value represents the precision of the method as revealed by past experience.) Suppose that 10 readings are to be taken on the suspected gold ring. It is desired to have $\alpha = .01$; that is, the customer wishes to run only a 1% risk of mistakenly accusing the seller of fraud if in fact the ring is unadulterated. Since Table E shows that $\Phi(-2.327) = .01$, it is seen that the critical value v in (8) is $v = -2.327$.

If the readings are

18.83, 19.03, 18.61, 19.46, 18.80, 18.96, 19.37, 19.20, 18.88, 19.34

we have $\overline{X} = 19.048$ or $\sqrt{n}(\overline{X} - \mu_H)/\sigma = -3.984$, which is less than v. With these measurements, the customer would therefore allege fraud.

The same method can be used to find (approximately) the power curve of the test (1). The power depends on the value of μ, and it will be convenient to express it in terms of the difference $\Delta = \mu - \mu_H$. The event (4) of rejecting H is the same as the events

$$\frac{\sqrt{n}(\overline{X} - \mu)}{\sigma} + \frac{\sqrt{n}\Delta}{\sigma} \geqq u \qquad \text{or} \qquad \frac{\sqrt{n}(\overline{X} - \mu)}{\sigma} \geqq u - \frac{\sqrt{n}\Delta}{\sigma}.$$

Since $\sqrt{n}(\overline{X} - \mu)/\sigma$ is just the random variable $\overline{X}$ expressed in standard units, the probability of this event is approximately equal to the area under the normal curve to the right of $u - (\sqrt{n}\Delta/\sigma)$. The power $\pi(\Delta)$ against an alternative Δ is the probability of rejection when this particular value of Δ is true, and is therefore approximately equal to

$$(9) \qquad \pi(\Delta) = 1 - \Phi\left(u - \frac{\sqrt{n}\Delta}{\sigma}\right).$$

When H is true, $\mu = \mu_H$ and hence $\Delta = 0$, and (9) reduces to $1 - \Phi(u) = \alpha$, as it should.

Analogously, the power of the test (7) is approximated by

$$(10) \qquad \pi(\Delta) = \Phi\left(v - \frac{\sqrt{n}\Delta}{\sigma}\right).$$

As an illustration of (10), let us compute the power of the test in Example 3 (continued) against the alternative $\mu = 19.1$; that is, the probability of detecting such a degree of adulteration with 10 observations. Since $\Delta = -.2$, we have $\sqrt{n}\Delta/\sigma = -\sqrt{10} = -3.162$ and $\pi(\Delta) = \Phi(-2.328 + 3.162) = \Phi(.834) = .798$.

Suppose that this power is too low. Can we find a larger sample size n for which the power against the alternative $\mu = 19.1$ is equal to .9? Formula (10) shows that this aim will be achieved (approximately) provided

$$\Phi\left(-2.328 - \frac{\sqrt{n}\Delta}{\sigma}\right) = .9.$$

From Table E we see that $\Phi(1.282) = .9$, so that n must satisfy the equation

$$-2.328 - \frac{\sqrt{n}\Delta}{\sigma} = 1.282.$$

Since in our case $\Delta/\sigma = -1$, it follows that $\sqrt{n} = 3.61$ and hence $n = 13$.

In this way, formulas (9) and (10) can be used quite generally to determine the sample size required, at a given significance level, to achieve a given power against a specific alternative.

Let us finally consider the shape of the power curve $\pi(\Delta)$. When testing $\mu = \mu_H$ against $\mu > \mu_H$, and hence $\Delta = 0$ against $\Delta > 0$, one would intuitively expect the power to increase as Δ increases. This is in fact true and is shown approximately by formula (9) as follows. As Δ increases, $u - (\sqrt{n}\Delta/\sigma)$ decreases, and the area under the normal curve to the right of $u - (\sqrt{n}\Delta/\sigma)$ therefore increases; but this is just the right-hand side of (9). Also, as Δ increases indefinitely, so does $\sqrt{n}\Delta/\sigma$, so that $u - (\sqrt{n}\Delta/\sigma)$ takes on increasingly large negative values; the area to the right of this value then increases toward 1. Similarly, $\pi(\Delta)$ decreases toward zero as Δ decreases indefinitely, and the power curve is thus seen to have the same general shape as those depicted in Figures 3.1 and 3.2.

In the foregoing, we have assumed that the standard deviation σ is known, as it might be, for example, from extensive past observation on X. In practice however, it is usually unwise to assume that σ still has the same value it used to have. The new factors, whose effect on the expectation of X is being investigated, may also have changed the variability of the observations. To avoid the serious misinterpretation of the data that would follow from the use of an assumed false value of σ in (4) (see for example Problem 23), it has been the general practice to use instead a value for σ that is estimated from the current observations themselves. In Section 9.5, we developed in formula (2) and Problem 5 the unbiased estimate

$$S^2 = [(X_1 - \overline{X})^2 + (X_2 - \overline{X})^2 + \cdots + (X_n - \overline{X})^2]/(n - 1)$$

for σ^2. Replacement of σ by S in (3) yields the test statistic known as *Student's t:*

(11)
$$T = \frac{\sqrt{n}(\overline{X} - \mu_{\mathrm{H}})}{S}.$$

An extension of the central limit theorem shows that the distribution of T (and hence the significance probability of the associated test) can be approximated by the same area under the normal curve as the distribution of the statistic (3). This is not surprising, since for very large n, S will be a very accurate estimate of σ, so that T and the statistic (3) will have nearly the same distribution. However, in the present case, much larger values of n are required for the approximation to be reliable, particularly if the distribution of X may not be symmetric. Unfortunately, it is not known just how large n has to be for the approximation to be safe for the kinds of distribution often encountered in practice.

When n is small, the distribution of T depends more heavily than that of $\overline{X}$ on the precise form of the distribution of X. For any given form, it is in principle possible to work out the null distribution of T, and hence to obtain the significance probability of the test. Historical and theoretical interest attaches to one such result. If the histogram of the distribution of an observation X looks like the normal curve (see Figure 6.4.1), the observations are said to be *normally distributed*. The null distribution of T for such observations was conjectured by W. S. Gosset ("Student") in 1908, and verified by R. A. Fisher in 1923. The critical values associated with this Student's t-distribution have been extensively published and the test is widely used with these values.

We shall here give only an approximation to these published values. Consider once more the test against the alternatives $\mu > \mu_H$ so that large values of T are significant. If n is not too small (say, $n > 10$), the critical value corresponding to significance level α is approximately equal to

(12)
$$u + \frac{u(u^2 + 1)}{4n}.$$

(Note that as n increases, (12) approaches u.) Furthermore, for an observed value t of T near such a critical value, the significance probability is about

(13)
$$1 - \Phi\left[t - \frac{t(t^2 + 1)}{4n}\right].$$

We stated earlier that the critical value of the t-test can be approximated by u regardless of the distribution of X, but that at least a moderately large value of n

is required before this approximation becomes reliable. The modification of u given by (12) or by the critical value found in a table of the t-distribution tends to provide an improvement over this approximation not only when the distribution of X is normal, but fairly generally. However, even with this improvement, the remarks made in connection with the normal approximation continue to apply.

A reasonable rank test is available for testing the hypothesis $\mu = \mu_H$ against the alternative $\mu > \mu_H$ if we can assume that (under the null hypothesis) the observations are symmetrically distributed about μ_H. By (5.4.3), this assumption means that when an observation X falls at a given distance from μ_H, it is as likely to fall to the left as to the right. As a result, the sign attached to the absolute difference $|X - \mu_H|$ is equally likely to be $+$ or $-$. Consider now n independent observations $X_1, \ldots,$ X_n, each having the same distribution which under the hypothesis is symmetric about μ_H. Then the sign of each of the absolute differences

(14) $|X_1 - \mu_H|, |X_2 - \mu_H|, \ldots, |X_n - \mu_H|$

is as likely to be $+$ or $-$. The 2^n possible ways of assigning $+$ or $-$ signs to these n absolute values are therefore equally likely, so that each has probability $1/2^n$.

The reader will find the situation reminiscent of that encountered in Section 12.6 in connection with the problem of paired comparisons, and we can in fact use the null distribution developed there. Let us rank the absolute differences (14), and let V_- denote the sum of the ranks of those differences to which a negative sign is attached. Then small values of V_- are significant, with the significance probability obtainable from Tables I, J and K. The test based on this null distribution is known as the Wilcoxon one-sample test.

EXAMPLE 4. *The effect of a growth hormone.* An experiment station wishes to test whether a growth hormone will increase the yield of wheat above the average value of 100 units per plot produced under the currently standard conditions. Twelve plots treated with the hormone give these yields:

141, 102, 73, 171, 137, 91, 81, 157, 146, 69, 121, 134.

We subtract the hypothetical value of 100 from each yield, and arrange the differences in order of increasing absolute value, with the following results:

2, −9, −19, 21, −27, −31, 34, 37, 41, 46, 57, 71.

The negative differences have ranks 2, 3, 5 and 6, so that the sum of their ranks is $V_- = 16$. Reference to Tables I and K shows the significance probability of these data to be $158/4096 = 0.0386$ by the Wilcoxon one-sample test. The validity of this result requires that the yields are symmetrically distributed, but no further distributional assumption is needed.

If in addition we are willing to assume that the yields are normally distributed, we may use the Student t-test. The values of $\bar{X}$ and S are (Problem 20) $\bar{X} = 118.583$ and $S = 34.437$, so that

$$T = \sqrt{12}(118.583 - 100)/34.437 = 1.8693.$$

Using (12), the approximate significance probability is $1 - \Phi(1.694) = .045$. (The value given by a table of Student's t-distribution is .044; the approximation (12) is not usually quite this good.)

If the reader will work through the details of these calculations, he will see how much easier the Wilcoxon test is to apply than is the t-test. In addition, the significance probability of the t-test is exactly valid only for normal data, while the Wilcoxon test requires only the weaker assumption of symmetry. (However, if the sample sizes are very large, the t-test has an advantage: its significance probabilities are then approximately correct even if the sampled population is not symmetric.) From the standpoint of power, the two tests are about equally good. The Wilcoxon test is slightly inferior to the t-test for normally distributed observations, but it is superior for certain types of non-normal data fairly common in practice.

PROBLEMS

1. In Example 3 (continued), find the approximate significance probability of the observations.

2. Find the relationship between the quantities u and v defined by (6) and (8), respectively.

3. Show that the critical value v of (7) is approximately determined by (8).

4. Find the approximate significance probability of n observations $X_1, \ldots, X_n$ if large values of $\bar{X}$ are significant and

  (i) $\mu_H = -2.5$, $\sigma = 1.3$, $\bar{X} = -1.7$, $n = 12$;

  (ii) $\mu_H = 4.01$, $\sigma = .2$, $\bar{X} = 4.26$, $n = 4$;

  (iii) $\mu_H = .352$, $\sigma = .08$, $\bar{X} = .369$, $n = 20$.

5. For each part of the preceding problems determine whether the data are significant at significance level (a) $\alpha = .01$, (b) $\alpha = .05$, (c) $\alpha = .1$.

6. Find the approximate significance probability of n observations $X_1, \ldots, X_n$ if small values of $\bar{X}$ are significant and

  (i) $\mu_H = 0$, $\sigma = 3$, $\bar{X} = 3$, $n = 10$;

  (ii) $\mu_H = 2.13$, $\sigma = .03$, $\bar{X} = 2.12$, $n = 30$;

  (iii) $\mu_H = .5$, $\sigma = 1$, $\bar{X} = .7$, $n = 20$.

7. The (symmetrical) two-sided test of the hypothesis $\mu = \mu_H$ against the alternatives $\mu \neq \mu_H$ rejects if

$$\sqrt{n}|\bar{X} - \mu_H|/\sigma \geq w. \tag{15}$$

Find the equation, corresponding to (6) and (8), that determines w for a given value of α.

8. Find the significance probability of the data of (i) Problem 4(i); (ii) Problem 6(i); (iii) Problem 6(iii) if the two-sided test of the preceding problem is used.

9. Based on the results of the preceding problem, make a general conjecture concerning the relation of the significance probabilities of a value of X based on the one-sided tests (4) or (7) and the two-sided test (15).

10. Find the approximate power of the test (4) at level $\alpha = .01$ when $n = 25$, $\mu_H = 1$, $\sigma = 2$ against the alternatives (i) $\mu = 1.5$, (ii) $\mu = 2.0$, (iii) $\mu = 2.5$.

11. Find the approximate power of the test (4) at level $\alpha = .02$ against the alternative $\mu = 1$ when $\mu_H = .5$, $\sigma = 1$ and (i) $n = 9$, (ii) $n = 25$, (iii) $n = 36$.

12. How large does n have to be in the test (4) if $\alpha = .05$, $\mu_H = -.8$, $\sigma = 3$ and if it is desired to have power .95 against the alternative (i) $\mu = -.5$, (ii) $\mu = -.6$, (iii) $\mu = -.7$?

13. Discuss the shape of the power function $\pi(\Delta)$ given by (10).

14. Find the approximate power of the two-sided test (15) at level $\alpha = .01$ when $n = 25$, $\mu_H = 2$, $\sigma = 2$ against the alternatives (i) $\mu = 2.5$, (ii) $\mu = -2.5$, (iii) $\mu = 3.0$, (iv) $\mu = 3.5$.

15. Solve Problem 12 if the two-sided test (15) is used rather than the one-sided test (4).

16. Discuss the shape of the power function $\pi(\Delta)$ of the test given by (15).

17. Use the test based on (11) and (13) to find the approximate significance probability of the data of Problem 9.2.5 for testing that the physical quantity being measured is 6.4 against the alternatives that it is greater than 6.4.

18. Let $X_1, \ldots, X_n$ be n measurements of a quantity μ and $X'_1, \ldots, X'_k$ k measurements of a quantity μ', and denote their averages by X and X' respectively.
   (i) Express $X' - X$ in standard units.
   (ii) Letting $\Delta = \mu' - \mu$, use the normal approximation to find an approximate test of the hypothesis $\Delta = 0$ against the alternatives $\Delta > 0$ which is an analogue of (4).
[Note: By an extension of the central limit theorem, the distribution of $X' - X$ in standard units can be approximated by the appropriate area under the normal curve. See Example 9.3.1.]

19. If $n = k = 20$, $\sigma = 1$, $\sigma' = 2$, determine the approximate significance probability under the test of the preceding problem if (i) $X = .3$, $X' = 1.4$; (ii) $X = -2$, $X' = -1.5$; (iii) $X = 1.2$, $X' = 2.8$.

20. Verify the values of X and S given in Example 4.

21. How must the Wilcoxon one-sample test be modified to test $\mu = \mu_H$ against the alternative $\mu < \mu_H$? Against alternatives on both sides?

22. Find the significance probability for the data of Example 3 (continued) when (i) the Student t-test is used with formula (13); (ii) the Wilcoxon one-sample test is used.   What use is made by these tests of the fact that σ is known to be .2?

23.   Compare the estimate S for σ in Problem 22(i) with the value assumed for σ in Example 3 (continued).   Use this comparison to discuss the contrast between the significance probabilities of the $\overline{X}$-test and t-test for this data.

TABLES

TABLE A. NUMBER OF COMBINATIONS $\binom{N}{s}$ OF N THINGS TAKEN s AT A TIME

N \ s	2	3	4	5	6	7	8	9	10	11	12	13
2	1											
3	3	1										
4	6	4	1									
5	10	10	5	1								
6	15	20	15	6	1							
7	21	35	35	21	7	1						
8	28	56	70	56	28	8	1					
9	36	84	126	126	84	36	9	1				
10	45	120	210	252	210	120	45	10	1			
11	55	165	330	462	462	330	165	55	11	1		
12	66	220	495	792	924	792	495	220	66	12	1	
13	78	286	715	1,287	1,716	1,716	1,287	715	286	78	13	1
14	91	364	1,001	2,002	3,003	3,432	3,003	2,002	1,001	364	91	14
15	105	455	1,365	3,003	5,005	6,435	6,435	5,005	3,003	1,365	455	105
16	120	560	1,820	4,368	8,008	11,440	12,870	11,440	8,008	4,368	1,820	560
17	136	680	2,380	6,188	12,376	19,448	24,310	24,310	19,448	12,376	6,188	2,380
18	153	816	3,060	8,568	18,564	31,824	43,758	48,620	43,758	31,824	18,564	8,568
19	171	969	3,876	11,628	27,132	50,388	75,582	92,378	92,378	75,582	50,388	27,132
20	190	1,140	4,845	15,504	38,760	77,520	125,970	167,960	184,756	167,960	125,970	77,520
21	210	1,330	5,985	20,349	54,264	116,280	203,490	293,930	352,716	352,716	293,930	203,490
22	231	1,540	7,315	26,334	74,613	170,544	319,770	497,420	646,646	705,432	646,646	497,420
23	253	1,771	8,855	33,649	100,947	245,157	490,314	817,190	1,144,066	1,352,078	1,352,078	1,144,066
24	276	2,024	10,626	42,504	134,596	346,104	735,471	1,307,504	1,961,256	2,496,144	2,704,156	2,496,144
25	300	2,300	12,650	53,130	177,100	480,700	1,081,575	2,042,975	3,268,760	4,457,400	5,200,300	5,200,300
26	325	2,600	14,950	65,780	230,230	657,800	1,562,275	3,124,550	5,311,735	7,726,160	9,657,700	10,400,600

TABLE B. $P(B = b)$ FOR THE BINOMIAL DISTRIBUTION (n, p)

n	b	p = .05	p = .1	p = .2	p = .3	p = .4
2	0	.9025	.8100	.6400	.4900	.3600
	1	.0950	.1800	.3200	.4200	.4800
	2	.0025	.0100	.0400	.0900	.1600
3	0	.8574	.7290	.5120	.3430	.2160
	1	.1354	.2430	.3840	.4410	.4320
	2	.0071	.0270	.0960	.1890	.2880
	3	.0001	.0010	.0080	.0270	.0640
4	0	.8145	.6561	.4096	.2401	.1296
	1	.1715	.2916	.4096	.4116	.3456
	2	.0135	.0486	.1536	.2646	.3456
	3	.0005	.0036	.0256	.0756	.1536
	4		.0001	.0016	.0081	.0256
5	0	.7738	.5905	.3277	.1681	.0778
	1	.2036	.3280	.4096	.3602	.2592
	2	.0214	.0729	.2048	.3087	.3456
	3	.0011	.0081	.0512	.1323	.2304
	4		.0005	.0064	.0284	.0768
	5			.0003	.0024	.0102
6	0	.7351	.5314	.2621	.1176	.0467
	1	.2321	.3543	.3932	.3025	.1866
	2	.0305	.0984	.2458	.3241	.3110
	3	.0021	.0146	.0819	.1852	.2765
	4	.0001	.0012	.0154	.0595	.1382
	5		.0001	.0015	.0102	.0369
	6			.0001	.0007	.0041
7	0	.6983	.4783	.2097	.0824	.0280
	1	.2573	.3720	.3670	.2471	.1306
	2	.0406	.1240	.2753	.3176	.2613
	3	.0036	.0230	.1147	.2269	.2903
	4	.0002	.0026	.0287	.0972	.1935
	5		.0002	.0043	.0250	.0774
	6			.0004	.0036	.0172
	7				.0002	.0016
8	0	.6634	.4305	.1678	.0576	.0168
	1	.2793	.3826	.3355	.1977	.0896
	2	.0515	.1488	.2936	.2965	.2090
	3	.0054	.0331	.1468	.2541	.2787
	4	.0004	.0046	.0459	.1361	.2322
	5		.0004	.0092	.0467	.1239
	6			.0011	.0100	.0413
	7			.0001	.0012	.0079
	8				.0001	.0007
9	0	.6302	.3874	.1342	.0404	.0101
	1	.2985	.3874	.3020	.1556	.0605
	2	.0629	.1722	.3020	.2668	.1612
	3	.0077	.0446	.1762	.2668	.2508
	4	.0006	.0074	.0661	.1715	.2508
	5		.0008	.0165	.0735	.1672
	6		.0001	.0028	.0210	.0743
	7			.0003	.0039	.0212
	8				.0004	.0035
	9					.0003
10	0	.5987	.3487	.1074	.0282	.0060
	1	.3151	.3874	.2684	.1211	.0403
	2	.0746	.1937	.3020	.2335	.1209
	3	.0105	.0574	.2013	.2668	.2150
	4	.0010	.0112	.0881	.2001	.2508
	5	.0001	.0015	.0264	.1029	.2007
	6		.0001	.0055	.0368	.1115
	7			.0008	.0090	.0425
	8			.0001	.0014	.0106
	9				.0001	.0016
	10					.0001

n	b	p = .05	p = .1	p = .2	p = .3	p = .4
11	0	.5688	.3138	.0859	.0198	.0036
	1	.3293	.3835	.2362	.0932	.0266
	2	.0867	.2131	.2953	.1998	.0887
	3	.0137	.0710	.2215	.2568	.1774
	4	.0014	.0158	.1107	.2201	.2365
	5	.0001	.0025	.0388	.1321	.2207
	6		.0003	.0097	.0566	.1471
	7			.0017	.0173	.0701
	8			.0002	.0037	.0234
	9				.0005	.0052
	10					.0007
12	0	.5404	.2824	.0687	.0138	.0022
	1	.3413	.3766	.2062	.0712	.0174
	2	.0988	.2301	.2835	.1678	.0639
	3	.0173	.0852	.2362	.2397	.1419
	4	.0021	.0213	.1329	.2311	.2128
	5	.0002	.0038	.0532	.1585	.2270
	6		.0005	.0155	.0792	.1766
	7			.0033	.0291	.1009
	8			.0005	.0078	.0420
	9			.0001	.0015	.0125
	10				.0002	.0025
	11					.0003
13	0	.5133	.2542	.0550	.0097	.0013
	1	.3512	.3672	.1787	.0540	.0113
	2	.1109	.2448	.2680	.1388	.0453
	3	.0214	.0997	.2457	.2181	.1107
	4	.0028	.0277	.1535	.2337	.1845
	5	.0003	.0055	.0691	.1803	.2214
	6		.0008	.0230	.1030	.1968
	7		.0001	.0058	.0442	.1312
	8			.0011	.0142	.0656
	9			.0001	.0034	.0243
	10				.0006	.0065
	11				.0001	.0012
	12					.0001
14	0	.4877	.2288	.0440	.0068	.0008
	1	.3593	.3559	.1539	.0407	.0073
	2	.1229	.2570	.2501	.1134	.0317
	3	.0259	.1142	.2501	.1943	.0845
	4	.0037	.0349	.1720	.2290	.1549
	5	.0004	.0078	.0860	.1963	.2066
	6		.0013	.0322	.1262	.2066
	7		.0002	.0092	.0618	.1574
	8			.0020	.0232	.0918
	9			.0003	.0066	.0408
	10				.0014	.0136
	11				.0002	.0033
	12					.0005
	13					.0001
15	0	.4633	.2059	.0352	.0047	.0005
	1	.3658	.3432	.1319	.0305	.0047
	2	.1348	.2669	.2309	.0916	.0219
	3	.0307	.1285	.2501	.1700	.0634
	4	.0049	.0428	.1876	.2186	.1268
	5	.0006	.0105	.1032	.2061	.1859
	6		.0019	.0430	.1472	.2066
	7		.0003	.0138	.0811	.1771
	8			.0035	.0348	.1181
	9			.0007	.0116	.0612
	10			.0001	.0030	.0245
	11				.0006	.0074
	12				.0001	.0016
	13					.0003

TABLE C. $P(B = b)$ FOR THE BINOMIAL DISTRIBUTION $(n, .5)$

n	b	p = .5	n	b	p = .5	n	b	p = .5	n	b	p = .5	n	b	p = .5
2	0	.2500	13	0	.0001	18	0	.0000	23	2	.0000	27	3	.0000
	1	.5000		1	.0016		1	.0001		3	.0002		4	.0001
3	0	.1250		2	.0095		2	.0006		4	.0011		5	.0006
	1	.3750		3	.0349		3	.0031		5	.0040		6	.0022
4	0	.0625		4	.0873		4	.0117		6	.0120		7	.0066
	1	.2500		5	.1571		5	.0327		7	.0292		8	.0165
	2	.3750		6	.2095		6	.0708		8	.0584		9	.0349
5	0	.0312	14	0	.0001		7	.1214		9	.0974		10	.0629
	1	.1562		1	.0009		8	.1669		10	.1364		11	.0971
	2	.3125		2	.0056		9	.1855		11	.1612		12	.1295
6	0	.0156		3	.0222	19	1	.0000	24	2	.0000		13	.1494
	1	.0938		4	.0611		2	.0003		3	.0001	28	3	.0000
	2	.2344		5	.1222		3	.0018		4	.0006		4	.0001
	3	.3125		6	.1833		4	.0074		5	.0025		5	.0004
7	0	.0078		7	.2095		5	.0222		6	.0080		6	.0014
	1	.0547	15	0	.0000		6	.0518		7	.0206		7	.0044
	2	.1641		1	.0005		7	.0961		8	.0438		8	.0116
	3	.2734		2	.0032		8	.1442		9	.0779		9	.0257
8	0	.0039		3	.0139		9	.1762		10	.1169		10	.0489
	1	.0312		4	.0417	20	1	.0000		11	.1488		11	.0800
	2	.1094		5	.0916		2	.0002		12	.1612		12	.1133
	3	.2188		6	.1527		3	.0011	25	2	.0000		13	.1395
	4	.2734		7	.1964		4	.0046		3	.0001		14	.1494
9	0	.0020	16	0	.0000		5	.0148		4	.0004	29	4	.0000
	1	.0176		1	.0002		6	.0370		5	.0016		5	.0002
	2	.0703		2	.0018		7	.0739		6	.0053		6	.0009
	3	.1641		3	.0085		8	.1201		7	.0143		7	.0029
	4	.2461		4	.0278		9	.1602		8	.0322		8	.0030
10	0	.0010		5	.0667		10	.1762		9	.0609		9	.0187
	1	.0098		6	.1222	21	1	.0000		10	.0974		10	.0373
	2	.0439		7	.1746		2	.0001		11	.1328		11	.0644
	3	.1172		8	.1964		3	.0006		12	.1550		12	.0967
	4	.2051	17	0	.0000		4	.0029	26	3	.0000		13	.1264
	5	.2461		1	.0001		5	.0097		4	.0002		14	.1445
11	0	.0005		2	.0010		6	.0259		5	.0010	30	4	.0000
	1	.0054		3	.0052		7	.0554		6	.0034		5	.0001
	2	.0269		4	.0182		8	.0970		7	.0098		6	.0006
	3	.0806		5	.0472		9	.1402		8	.0233		7	.0019
	4	.1611		6	.0944		10	.1682		9	.0466		8	.0055
	5	.2256		7	.1484	22	1	.0000		10	.0792		9	.0133
12	0	.0002		8	.1855		2	.0001		11	.1151		10	.0280
	1	.0029					3	.0004		12	.1439		11	.0509
	2	.0161					4	.0017		13	.1550		12	.0806
	3	.0537					5	.0063					13	.1115
	4	.1208					6	.0178					14	.1354
	5	.1934					7	.0407					15	.1445
	6	.2256					8	.0762						
							9	.1186						
							10	.1542						
							11	.1682						

TABLE D. SQUARE ROOTS

n	$\sqrt{n}$	$\sqrt{10n}$	n	$\sqrt{n}$	$\sqrt{10n}$	n	$\sqrt{n}$	$\sqrt{10n}$
1	1.0000	3.1623	34	5.8310	18.439	67	8.1854	25.884
2	1.4142	4.4721	35	5.9161	18.708	68	8.2462	26.077
3	1.7321	5.4772	36	6.0000	18.974	69	8.3066	26.268
4	2.0000	6.3246	37	6.0828	19.235	70	8.3666	26.458
5	2.2361	7.0711	38	6.1644	19.494	71	8.4261	26.646
6	2.4495	7.7460	39	6.2450	19.748	72	8.4853	26.833
7	2.6458	8.3666	40	6.3246	20.000	73	8.5440	27.019
8	2.8284	8.9443	41	6.4031	20.248	74	8.6023	27.203
9	3.0000	9.4868	42	6.4807	20.494	75	8.6603	27.386
10	3.1623	10.000	43	6.5574	20.736	76	8.7178	27.568
11	3.3166	10.488	44	6.6332	20.976	77	8.7750	27.749
12	3.4641	10.954	45	6.7082	21.213	78	8.8318	27.928
13	3.6056	11.402	46	6.7823	21.448	79	8.8882	28.107
14	3.7417	11.832	47	6.8557	21.679	80	8.9443	28.284
15	3.8730	12.247	48	6.9282	21.909	81	9.0000	28.460
16	4.0000	12.649	49	7.0000	22.136	82	9.0554	28.636
17	4.1231	13.038	50	7.0711	22.361	83	9.1104	28.810
18	4.2426	13.416	51	7.1414	22.583	84	9.1652	28.983
19	4.3589	13.784	52	7.2111	22.804	85	9.2195	29.155
20	4.4721	14.142	53	7.2801	23.022	86	9.2736	29.326
21	4.5826	14.491	54	7.3485	23.238	87	9.3274	29.496
22	4.6904	14.832	55	7.4162	23.452	88	9.3808	29.665
23	4.7958	15.166	56	7.4833	23.664	89	9.4340	29.833
24	4.8990	15.492	57	7.5498	23.875	90	9.4868	30.000
25	5.0000	15.811	58	7.6158	24.083	91	9.5394	30.166
26	5.0990	16.125	59	7.6811	24.290	92	9.5917	30.332
27	5.1962	16.432	60	7.7460	24.495	93	9.6437	30.496
28	5.2915	16.733	61	7.8102	24.698	94	9.6954	30.659
29	5.3852	17.029	62	7.8740	24.900	95	9.7468	30.822
30	5.4772	17.321	63	7.9373	25.100	96	9.7980	30.984
31	5.5678	17.607	64	8.0000	25.298	97	9.8489	31.145
32	5.6569	17.889	65	8.0623	25.495	98	9.8995	31.305
33	5.7446	18.166	66	8.1240	25.690	99	9.9499	31.464

TABLE E. AREA $\Phi(z)$ UNDER THE NORMAL CURVE TO THE LEFT OF z

z	.00	.01	.02	.03	.04	.05	.06	.07	.08	.09
.0	.5000	.5040	.5080	.5120	.5160	.5199	.5239	.5279	.5319	.5359
.1	.5398	.5438	.5478	.5517	.5557	.5596	.5636	.5675	.5714	.5753
.2	.5793	.5832	.5871	.5910	.5948	.5987	.6026	.6064	.6103	.6141
.3	.6179	.6217	.6255	.6293	.6331	.6368	.6406	.6443	.6480	.6517
.4	.6554	.6591	.6628	.6664	.6700	.6736	.6772	.6808	.6844	.6879
.5	.6915	.6950	.6985	.7019	.7054	.7088	.7123	.7157	.7190	.7224
.6	.7257	.7291	.7324	.7357	.7389	.7422	.7454	.7486	.7517	.7549
.7	.7580	.7611	.7642	.7673	.7704	.7734	.7764	.7794	.7823	.7852
.8	.7881	.7910	.7939	.7967	.7995	.8023	.8051	.8078	.8106	.8133
.9	.8159	.8186	.8212	.8238	.8264	.8289	.8315	.8340	.8365	.8389
1.0	.8413	.8438	.8461	.8485	.8508	.8531	.8554	.8577	.8599	.8621
1.1	.8643	.8665	.8686	.8708	.8729	.8749	.8770	.8790	.8810	.8830
1.2	.8849	.8869	.8888	.8907	.8925	.8944	.8962	.8980	.8997	.9015
1.3	.9032	.9049	.9066	.9082	.9099	.9115	.9131	.9147	.9162	.9177
1.4	.9192	.9207	.9222	.9236	.9251	.9265	.9279	.9292	.9306	.9319
1.5	.9332	.9345	.9357	.9370	.9382	.9394	.9406	.9418	.9429	.9441
1.6	.9452	.9463	.9474	.9484	.9495	.9505	.9515	.9525	.9535	.9545
1.7	.9554	.9564	.9573	.9582	.9591	.9599	.9608	.9616	.9625	.9633
1.8	.9641	.9649	.9656	.9664	.9671	.9678	.9686	.9693	.9699	.9706
1.9	.9713	.9719	.9726	.9732	.9738	.9744	.9750	.9756	.9761	.9767
2.0	.9772	.9778	.9783	.9788	.9793	.9798	.9803	.9808	.9812	.9817
2.1	.9821	.9826	.9830	.9834	.9838	.9842	.9846	.9850	.9854	.9857
2.2	.9861	.9864	.9868	.9871	.9875	.9878	.9881	.9884	.9887	.9890
2.3	.9893	.9896	.9898	.9901	.9904	.9906	.9909	.9911	.9913	.9916
2.4	.9918	.9920	.9922	.9925	.9927	.9929	.9931	.9932	.9934	.9936
2.5	.9938	.9940	.9941	.9943	.9945	.9946	.9948	.9949	.9951	.9952
2.6	.9953	.9955	.9956	.9957	.9959	.9960	.9961	.9962	.9963	.9964
2.7	.9965	.9966	.9967	.9968	.9969	.9970	.9971	.9972	.9973	.9974
2.8	.9974	.9975	.9976	.9977	.9977	.9978	.9979	.9979	.9980	.9981
2.9	.9981	.9982	.9982	.9983	.9984	.9984	.9985	.9985	.9986	.9986
3.0	.9987	.9987	.9987	.9988	.9988	.9989	.9989	.9989	.9990	.9990
3.1	.9990	.9991	.9991	.9991	.9992	.9992	.9992	.9992	.9993	.9993
3.2	.9993	.9993	.9994	.9994	.9994	.9994	.9994	.9995	.9995	.9995
3.3	.9995	.9995	.9995	.9996	.9996	.9996	.9996	.9996	.9996	.9997
3.4	.9997	.9997	.9997	.9997	.9997	.9997	.9997	.9997	.9997	.9998

AUXILIARY TABLE OF z IN TERMS OF $\Phi(z)$

$\Phi(z)$	z	$\Phi(z)$	z	$\Phi(z)$	z
.50	0	.91	1.341	.995	2.576
.55	.126	.92	1.405	.999	3.090
.60	.253	.93	1.476	.9995	3.291
.65	.385	.94	1.555	.9999	3.719
.70	.524	.95	1.645	.99995	3.891
.75	.674	.96	1.751	.99999	4.265
.80	.842	.97	1.881	.999995	4.417
.85	1.036	.98	2.054	.9999999	4.753
.90	1.282	.99	2.326	.9999999	5.199

TABLE F. THE POISSON APPROXIMATION $P(T = t)$

t	$E(T) = \lambda$ .1	.2	.3	.4	.5	.6	.7	.8	.9	1.0
0	.9048	.8187	.7408	.6703	.6065	.5488	.4966	.4493	.4066	.3679
1	.0905	.1637	.2222	.2681	.3033	.3293	.3476	.3595	.3659	.3679
2	.0045	.0164	.0333	.0536	.0758	.0988	.1217	.1438	.1647	.1839
3	.0002	.0011	.0033	.0072	.0126	.0198	.0284	.0383	.0494	.0613
4		.0001	.0003	.0007	.0016	.0030	.0050	.0077	.0111	.0153
5				.0001	.0002	.0004	.0007	.0012	.0020	.0031
6							.0001	.0002	.0003	.0005
7										.0001

t	$E(T) = \lambda$ 1	2	3	4	5	6	7	8	9	10
0	.3679	.1353	.0498	.0183	.0067	.0025	.0009	.0003	.0001	.0000
1	.3679	.2707	.1494	.0733	.0337	.0149	.0064	.0027	.0011	.0005
2	.1839	.2707	.2240	.1465	.0842	.0446	.0223	.0107	.0050	.0023
3	.0613	.1804	.2240	.1954	.1404	.0892	.0521	.0286	.0150	.0076
4	.0153	.0902	.1680	.1954	.1755	.1339	.0912	.0572	.0337	.0189
5	.0031	.0361	.1008	.1563	.1755	.1606	.1277	.0916	.0607	.0378
6	.0005	.0120	.0504	.1042	.1462	.1606	.1490	.1221	.0911	.0631
7	.0001	.0034	.0216	.0595	.1044	.1377	.1490	.1396	.1171	.0901
8		.0009	.0081	.0298	.0653	.1033	.1304	.1396	.1318	.1126
9		.0002	.0027	.0132	.0363	.0688	.1014	.1241	.1318	.1251
10			.0008	.0053	.0181	.0413	.0710	.0993	.1186	.1251
11			.0002	.0019	.0082	.0225	.0452	.0722	.0970	.1137
12			.0001	.0006	.0034	.0113	.0264	.0481	.0728	.0948
13				.0002	.0013	.0052	.0142	.0296	.0504	.0729
14				.0001	.0005	.0022	.0071	.0169	.0324	.0521
15					.0002	.0009	.0033	.0090	.0194	.0347
16						.0003	.0014	.0045	.0109	.0217
17						.0001	.0006	.0021	.0058	.0128
18							.0002	.0009	.0029	.0071
19							.0001	.0004	.0014	.0037
20								.0002	.0006	.0019
21								.0001	.0003	.0009
22									.0001	.0004
23										.0002
24										.0001

TABLE G. APPROXIMATE SIGNIFICANCE PROBABILITIES $P(Q \geq q)$ OF THE CHI-SQUARE TEST

ν \ $q-\nu$	0	.4	.8	1.2	1.6	2.0	2.5	3.0	3.5	4.0	4.5	5.0	5.5	6	7	8	9	10	12	14	16	18
2	.368	.301	.246	.202	.165	.135	.105	.082	.064	.050	.039	.030	.024	.018	.011	.007	.004	.002	.001			
3	.392	.334	.284	.241	.204	.172	.139	.112	.090	.072	.058	.046	.037	.029	.019	.012	.007	.005	.002	.001		
4	.406	.355	.308	.267	.231	.199	.165	.136	.112	.092	.075	.061	.050	.040	.027	.017	.011	.007	.003	.001		
5	.416	.369	.326	.287	.252	.221	.186	.156	.131	.109	.091	.075	.062	.051	.035	.023	.016	.010	.004	.002	.001	
6	.423	.380	.340	.303	.269	.238	.204	.174	.147	.125	.105	.088	.074	.062	.043	.030	.020	.014	.006	.003	.001	.001
7	.429	.388	.351	.315	.283	.253	.219	.189	.162	.139	.118	.101	.085	.072	.051	.036	.025	.017	.008	.004	.002	.001

ν \ $q-\nu$	0	.5	1.0	1.5	2.0	2.5	3.0	3.5	4.0	4.5	5.0	5.5	6	7	8	9	10	12	14	16	20	24
8	.433	.386	.342	.302	.265	.231	.202	.175	.151	.130	.112	.096	.082	.059	.042	.030	.021	.010	.005	.002		
10	.440	.398	.358	.320	.285	.253	.224	.197	.173	.151	.132	.115	.100	.074	.055	.040	.029	.015	.008	.004	.001	
12	.446	.406	.369	.334	.301	.270	.241	.215	.191	.169	.150	.132	.116	.089	.067	.050	.038	.020	.011	.006	.002	.001
14	.450	.413	.378	.345	.313	.284	.256	.231	.207	.185	.165	.147	.130	.102	.079	.060	.046	.026	.014	.008	.003	.001
16	.452	.419	.386	.354	.324	.295	.269	.244	.220	.199	.179	.160	.143	.114	.090	.070	.054	.032	.018	.010	.003	.001
18	.456	.423	.392	.362	.333	.305	.279	.255	.232	.211	.191	.172	.155	.125	.100	.079	.062	.037	.022	.013	.004	.001
20	.458	.427	.397	.368	.341	.314	.289	.265	.242	.221	.201	.183	.166	.135	.109	.088	.070	.043	.026	.015	.005	.002

ν \ $q-\nu$	0	1	2	3	4	5	6	7	8	9	10	11	12	13	14	16	18	20	24	28	32	36
20	.458	.397	.341	.289	.242	.201	.166	.135	.109	.088	.070	.055	.043	.034	.026	.015	.009	.005	.002	.001		
25	.462	.408	.356	.308	.264	.224	.189	.158	.131	.108	.088	.072	.058	.046	.037	.023	.014	.008	.003	.002	.001	
30	.466	.415	.368	.323	.281	.243	.208	.177	.150	.126	.105	.087	.072	.059	.048	.031	.020	.012	.005	.003	.001	
40	.470	.426	.384	.344	.306	.271	.238	.208	.180	.156	.134	.114	.097	.082	.069	.048	.040	.033	.009	.004	.002	.001
50	.473	.434	.396	.359	.324	.291	.260	.231	.204	.180	.157	.137	.119	.103	.088	.064	.046	.032	.015	.007	.003	.001

TABLE H. WILCOXON TWO-SAMPLE DISTRIBUTION

$u = w -$ (minimum value of W)

Group sizes	$\binom{N}{s}$	0	1	2	3	4	5	6	7	8	9	10	11	12	13	14	15	16	17	18	19	20
3,3	20	1	2	4	7	10	13	16	18	19	20											
3,4	35	1	2	4	7	11	15	20	24	28	31	33	34	35								
4,4	70	1	2	4	7	12	17	24	31	39	46	53	58	63	66	68	69	70				
3,5	56	1	2	4	7	11	16	22	28	34	40	45	49	52	54	55	56					
4,5	126	1	2	4	7	12	18	26	35	46	57	69	80	91	100	108	114	119	122	124	125	126
5,5	252	1	2	4	7	12	19	28	39	53	69	87	106	126	146	165	183	199	213	224	233	240
3,6	84	1	2	4	7	11	16	23	30	38	46	54	61	68	73	77	80	82	83	84		
4,6	210	1	2	4	7	12	18	27	37	50	64	80	96	114	130	146	160	173	183	192	198	203
5,6	462	1	2	4	7	12	19	29	41	57	76	99	124	153	183	215	247	279	309	338	363	386
6,6	924	1	2	4	7	12	19	30	43	61	83	111	143	182	224	272	323	378	433	491	546	601
3,7	120	1	2	4	7	11	16	23	31	40	50	60	70	80	89	97	104	109	113	116	118	119
4,7	330	1	2	4	7	12	18	27	38	52	68	87	107	130	153	177	200	223	243	262	278	292
5,7	792	1	2	4	7	12	19	29	42	59	80	106	136	171	210	253	299	347	396	445	493	539
6,7	1716	1	2	4	7	12	19	30	44	63	87	118	155	201	253	314	382	458	539	627	717	811
7,7	3432	1	2	4	7	12	19	30	45	65	91	125	167	220	283	358	445	545	657	782	918	1064
3,8	165	1	2	4	7	11	16	23	31	41	52	64	76	89	101	113	124	134	142	149	154	158
4,8	495	1	2	4	7	12	18	27	38	53	70	91	114	141	169	200	231	264	295	326	354	381
5,8	1287	1	2	4	7	12	19	29	42	60	82	110	143	183	228	280	337	400	466	536	607	680
6,8	3003	1	2	4	7	12	19	30	44	64	89	122	162	213	272	343	424	518	621	737	860	994
7,8	6435	1	2	4	7	12	19	30	45	66	93	129	174	232	302	388	489	609	746	904	1080	1277
8,8	12870	1	2	4	7	12	19	30	45	67	95	133	181	244	321	418	534	675	839	1033	1254	1509
$s \geq u, t \geq u$		1	2	4	7	12	19	30	45	67	97	139	195	272	373	508	684	915	1212	1597	2089	2714

TABLE I.   AUXILIARY TABLE FOR THE WILCOXON PAIRED-COMPARISON TEST

n	2^n	n	2^n	$E(V)$	$SD(V)$	n	$E(V)$	$SD(V)$
1	2	11	2,048	33.0	11.25	21	115.5	28.77
2	4	12	4,096	39.0	12.75	22	126.5	30.80
3	8	13	8,192	45.0	14.31	23	138.0	32.88
4	16	14	16,384	52.0	15.93	24	150.0	35.00
5	32	15	32,768	60.0	17.61	25	162.5	37.17
6	64	16	65,536	68.0	19.34	26	175.5	39.37
7	128	17	131,072	76.5	21.12	27	189.0	41.62
8	256	18	262,144	85.5	22.96	28	203.0	43.91
9	512	19	524,288	95.0	24.85	29	217.5	46.25
10	1,024	20	1,048,576	105.0	26.79	30	232.5	48.62

TABLE J. $\#(V \leqq v)$ FOR $v \leqq n$ IN THE WILCOXON PAIRED-COMPARISON TEST

v	$\#(V \leqq v)$	v	$\#(V \leqq v)$	v	$\#(V \leqq v)$	v	$\#(V \leqq v)$
0	1	6	14	11	55	16	169
1	2	7	19	12	70	17	207
2	3	8	25	13	88	18	253
3	5	9	33	14	110	19	307
4	7	10	43	15	137	20	371
5	10						

TABLE K. #(V ≦ v) FOR v > n IN THE WILCOXON PAIRED-COMPARISON TEST

v − n \ n	3	4	5	6	7	8	9	10	11	12	13	14	15	16	17	18	19	20
1	6	9	13	18	24	32	42	54	69	87	109	136	168	206	252	306	370	446
2	7	11	16	22	30	40	52	67	85	107	134	166	204	250	304	368	444	533
3	8	13	19	27	37	49	64	82	104	131	163	201	247	301	365	441	530	634
4		14	22	32	44	59	77	99	126	158	196	242	296	360	436	525	629	751
5		15	25	37	52	70	92	119	151	189	235	289	353	429	518	622	744	886
6		16	27	42	60	82	109	141	179	225	279	343	419	508	612	734	876	1,041
7			29	46	68	95	127	165	211	265	329	405	494	598	720	862	1,027	1,219
8			30	50	76	108	146	192	246	310	386	475	579	701	843	1,008	1,200	1,422
9			31	54	84	121	167	221	285	361	450	554	676	818	983	1,175	1,397	1,653
10			32	57	91	135	188	252	328	417	521	643	785	950	1,142	1,364	1,620	1,916
11				59	98	148	210	285	374	478	600	742	907	1,099	1,321	1,577	1,873	2,213
12				61	104	161	253	320	423	545	687	852	1,044	1,266	1,522	1,818	2,158	2,548
13				62	109	174	256	356	476	617	782	974	1,196	1,452	1,748	2,088	2,478	2,926
14				63	114	186	279	394	532	695	886	1,108	1,364	1,660	2,000	2,390	2,838	3,350
15				64	118	197	302	433	591	779	999	1,254	1,550	1,890	2,280	2,728	3,240	3,825
16					121	207	324	472	653	868	1,120	1,414	1,753	2,143	2,591	3,103	3,688	4,356
17					123	216	345	512	717	962	1,251	1,587	1,975	2,422	2,934	3,519	4,187	4,947
18					125	224	366	552	783	1,062	1,391	1,774	2,218	2,728	3,312	3,980	4,740	5,604
19					126	231	385	591	851	1,166	1,539	1,976	2,481	3,062	3,728	4,487	5,351	6,333
20					127	237	403	630	920	1,274	1,697	2,192	2,766	3,427	4,183	5,045	6,026	7,139
21					128	242	420	668	989	1,387	1,863	2,423	3,074	3,823	4,680	5,658	6,769	8,028
22						246	435	704	1,059	1,502	2,037	2,669	3,404	4,251	5,222	6,328	7,584	9,008
23						249	448	739	1,128	1,620	2,219	2,929	3,757	4,714	5,810	7,059	8,478	10,084
24						251	460	772	1,197	1,741	2,408	3,203	4,135	5,212	6,447	7,856	9,455	11,264
25						253	470	803	1,265	1,863	2,603	3,492	4,536	5,746	7,136	8,721	10,520	12,557
26						254	479	832	1,331	1,986	2,805	3,794	4,961	6,318	7,878	9,658	11,681	13,968
27						255	487	859	1,395	2,110	3,012	4,109	5,411	6,928	8,675	10,673	12,941	15,506
28						256	493	883	1,457	2,233	3,223	4,437	5,884	7,576	9,531	11,766	14,306	17,180
29							498	905	1,516	2,355	3,438	4,776	6,380	8,265	10,445	12,942	15,783	18,997
30							502	925	1,572	2,476	3,656	5,126	6,901	8,993	11,420	14,206	17,377	20,966

SELECTED ANSWERS TO PROBLEMS

CHAPTER 1

1.3.1. (i) no; (ii) yes; (iii) no
3.7. 36
3.13. (i) number of white marbles: 0, 1, 2
3.14. $P(ABC) = .182$, $P(ABD) = P(BCD) = P(CAD) = .2727$
1.4.1. (i) no; (iii) yes
4.2. (ii) largest .93, smallest 0
4.4. (ii) $\frac{1}{3}$
4.5. (iii) .27
4.7. (ii) .34
4.8. (iii) .10; (v) .22
4.9. (iii) .85; (v) .21
4.10. (iii) .55; (v) .37
1.5.1. (i) $\frac{1}{2}$; (iii) $\frac{1}{26}$
5.2. (ii) $P(W) = \frac{1}{5}$
5.3. (iii) $\frac{1}{2}$; (vi) $\frac{7}{12}$; (ix) $\frac{1}{9}$
5.5. $\frac{3}{4}$
5.6. (iii) $\frac{3}{8}$; (vi) $\frac{1}{2}$; (ix) $\frac{1}{4}$
5.7. (iii) $\frac{1}{6}$
5.8. (i) $\frac{9}{100}$; (iii) $\frac{34}{100}$
5.9. (iii) $\frac{7}{15}$; (v) $\frac{1}{3}$
5.11. $P(T = 9) = \frac{25}{216}$, $P(T = 10) = \frac{27}{216}$

1.6.1. (i) not more than five heads
6.7. (iii) no; (vi) yes
6.9. (iii) no; (vi) no; (ix) no
6.13. (ii) yes
6.15. yes
1.7.1. (iv) $\frac{1}{4}$; (viii) $\frac{4}{9}$
7.2. (ii) $\frac{1}{2}$
7.3. (ii) $\frac{7}{15}$; (iv) $\frac{1}{15}$
7.4. (ii) $\frac{3}{4}$
7.5. (iii) $\frac{1}{3}$
7.6.

M	1	2	3	4	5	6
$P(M)$	$\frac{1}{36}$	$\frac{3}{36}$	$\frac{5}{36}$	$\frac{7}{36}$	$\frac{9}{36}$	$\frac{11}{36}$

7.9. (ii) $\frac{1}{8}$
7.11. (ii) $\frac{11}{16}$
7.13. (ii) $\frac{1}{2}$
7.14. (ii)

V	0	1	2	3	4
$P(V = v)$	.04	.13	.44	.27	.12

7.15. (ii) .44
7.16. (ii) .93
7.17. (iii) .65
7.18. (ii) true

CHAPTER 2

2.1.1. (ii) $\frac{18}{35}$; (iv) $\frac{3}{7}$
1.2. (ii) $\frac{5}{126}$; (iv) $1\frac{21}{126}$
1.3. (ii) $\frac{1}{21}$; (iv) $\frac{16}{21}$
1.5. (ii) $\frac{2}{3}$; (iv) $\frac{4}{15}$
1.6. (ii) N
1.7. (ii) $\frac{4}{35}$
1.8. (ii) $\frac{2}{7}$
1.9. (ii) $\frac{1}{3}$
1.11. $\frac{35}{210}$
1.14. (ii) $\frac{9}{28}$
1.15. (ii) $\frac{1}{7}$
1.16. $\frac{56}{2024}$
1.17. (ii) $\frac{5}{8}$
2.2.1. (ii) 319,770
2.2. (ii) 8,436,285
2.3. (ii) 40,116,600; (iv) 4,292,145
2.4. (ii) $\frac{29}{57}$
2.5. (ii) $\frac{229}{230}$
2.6. (ii) .998
2.7. (ii) .0919
2.8. .975

2.10. $\frac{33}{16.660}$
2.12. (ii) 7
2.13. (ii) 2
2.14. (ii) 40
2.21. (ii) $.43747/.43758 = .999$
2.3.2. (iii) yes
3.6. (ii) 720
3.7. (iii) $\frac{1}{210}$
3.8. (ii) $\frac{1}{10}$
3.13. (ii) $\frac{1}{6}$; (iv) $\frac{5}{18}$
3.14. (iii) $\frac{2}{3}$
3.15. (ii) $\frac{81}{253}$
3.16. (ii) $\frac{1}{4}$; (iv) $\frac{1}{3}$
3.17. $\frac{7}{24}$
3.19. s/N
2.4.4. (ii) $\frac{1}{13}$
4.7. $0! = 1$
4.9. (ii) $\frac{4}{9}$
4.10. $\frac{5}{324}$
4.11. .027
4.12. (ii) 792; (iv) 1

CHAPTER 3

3.1.5. $P(e_1 \text{ and } f_1) = .02$, $P(e_2 \text{ and } f_3)$
    $= .10$

1.7. (ii) $P(e_1) = .1$, $P(f_2) = .6$

1.9. (ii) $P(e_3) = .60$

1.10. $P(e_i \text{ and } f_1) = P(e_i \text{ and } f_2)$ $i = 1$,
    $\ldots, m$

1.11. .08

1.12. (ii) .12

1.14. $P(e_2 \text{ and } f_1) = .12$, $P(e_2 \text{ and } f_2)$
    $= .28$, $P(e_2 \text{ and } f_2) = .40$

1.15. .1003

1.18. (ii) .86

1.19. (i) .24

3.2.1. (ii) $\frac{8}{25}$

2.2. (i) $\frac{2}{7}$

2.3. (ii) $\frac{59}{200}$; (iv) $\frac{1}{20}$

2.6. related

2.10. (i) yes

3.3.1. (ii) $\frac{26}{27}$

3.2. (iii) $\frac{5}{72}$

3.6. (iii) $4pq^3$; (v) $4p^3q$; (vii) $q^4 + 4pq^3 + 6p^2q^2$

3.7. $\frac{1}{512}$

3.8. (ii) $\frac{3}{16}$

3.9. (i) $\frac{1}{243}$

3.10. (ii) $\frac{3}{256}$

3.11. (ii) 5

3.13. $\frac{1}{27}$

3.14. (ii) $\frac{1}{12}$

3.15. (ii) .011; (iv) .812

3.16. (ii) .000; (v) .004

CHAPTER 4

4.1.2. $P(e_3|E) = .4$

1.3. $\frac{9}{13}$

1.4. (ii) .292

1.5. (ii) $\frac{2}{3}$

1.8. (i) $1/\binom{26}{5}$

1.10. (i) $P(e_2|E)/P(e_1|E) = \frac{1}{2}$

1.11. (ii) $\frac{1}{4}$

1.13. (ii) $\frac{1}{4}$

4.2.2. (ii) $\frac{42}{80}$

2.4. (ii) $\frac{63}{143}$

2.5. (i) $\frac{25}{126}$

2.6. (ii) $\frac{7}{102}$

2.7. (i) $\frac{13}{216}$

2.8. (iii) $\frac{17}{27}$

2.9. 28%

2.14. 9

2.15. (ii) $\frac{64}{729}$

2.17. (i) .001

2.19. $\frac{91}{216}$

2.20. (ii) $\frac{1}{243}$

4.3.1. (ii) $\frac{1}{8}$

3.2. $\frac{1}{3}$

3.3. (ii) $\frac{1}{4}$, $\frac{1}{10}$

3.4. (ii) $\frac{1}{30}$

3.6. (ii) .208

3.8. (ii) $\frac{3}{5}$

3.11. (ii) $\frac{1}{126}$

3.13. (ii) .55

3.15. (ii) $\frac{1}{2}$

4.4.1. $\frac{9}{20}$

4.4. $\frac{17}{50}$

4.5. (ii) $\frac{1}{2}$

4.8. (i) .101

4.9. $\frac{3}{20}$

4.11. (iii) $\frac{2}{3}$

4.12. .00224

4.5.3. $\frac{14}{69}$

5.4. (ii) $\frac{1}{2}$

5.6. (ii) $9\lambda/(16 - 7\lambda)$

5.9. (ii) $\frac{1}{8}\lambda(4 - 3\lambda)$

5.11. $\frac{1}{48}$

4.6.3. (ii) $\lambda/8$

6.4. $\frac{136}{1835} = .074$

6.6. (iii) safer for $\lambda \geq 2 - \sqrt{2} = .59$

4.7.1. (i) (b) $\frac{4}{13}$

CHAPTER 5

5.1.2. (ii) $\{0, 1, 2, \ldots, 8\}$;
(iv) $\{5, 6, 7, 8\}$
1.3. (iii) $\{4, 5, \ldots, 12\}$
1.5. (ii) $\{0, 1, 2, 3, 5\}$
1.11. (iv) $\frac{7}{8}$
1.12. (iv) $\frac{27}{100}$
1.14. (ii) $\frac{1}{9}$
1.15. (ii) $\frac{1}{6}$
1.16. (ii) .0833

5.2.1.
B	1
$P(B = b)$	$\frac{3}{8}$

2.3.
D	-3	2
$P(D = d)$	$\frac{3}{36}$	$\frac{4}{36}$

2.5. (ii) $\frac{21}{40}$
2.6. (i) $\frac{3}{4}$
2.10. .2025
2.11. (ii) $\frac{8}{15}$

2.18. (i)
j	11
$P(Z + W = j)$	$\frac{1}{2}$
;
(iii)
j	18
$P(WZ = j)$	$\frac{1}{4}$

5.3.2. 4.5
3.4. (i) 3.25
3.5. 2.07
3.7. (i) 4.1985; (ii) 0
3.10. (ii) $\frac{45}{8}$; (iv) $\frac{228}{8}$

3.14. 1.75
5.4.1. (ii) $\frac{2}{3}$ dollar
4.3. 6.5
4.4. (ii) 7
4.6. (ii) 4.47
4.7. (i)
m	3
$P(\max = m)$	.1581

4.8. (ii) 10.5625
4.9. (ii) 2.5
4.10. (ii) 2.30; (iv) $-.26$
4.14. (i) $\frac{17}{25}$
5.5.3. (i) $\frac{21}{2}$
5.8. (ii) (b) .81; (iii) (a) .9
5.11. $-.26$
5.13. $\frac{5}{13}$
5.15. $\frac{6}{5}$
5.16. (ii) 10
5.6.2. 1.9715
6.3. (ii) .75
6.4. (i) 1.025
6.5. (ii) $\frac{35}{6}$
6.7. (i) .9264
6.8. (ii) .75
6.10. (b) $11\frac{7}{4}$
6.11. (ii) $432\frac{8}{9}$
6.15. (ii) .7924
5.7.7. $-\frac{3}{4}$
7.10. .467

CHAPTER 6

6.1.2. (ii) .2048
1.3. (ii) .9294
1.5. .0439
1.7. .1094
1.9. (i) $p < \frac{1}{6}$
1.12. .3222
1.13. (ii) .6563
1.17. (ii) 1.44

1.24.
S	1
Exact	.177
Approx.	.175

6.2.2.
d	2
$P(D = d)$	.325

2.4. .2487
2.6. .103
2.11. .0862
2.13. .3222
2.16. (ii) 2.64
2.17. $\text{Var}(A) = 1.468$
2.19. .049
2.21. (i) $\frac{26}{21}$
2.22. (ii) $\frac{20}{9}$
2.23. (i) $\frac{109}{72}$

6.3.2. (ii) $-.7, -.5$
3.4. (i) $-.988, -.329$
3.5. (ii) $-1.449, -1.035$
3.6. (iii) .335
3.8. (ii) 216.79
6.4.1. (ii) .0721
4.2. (iii) .3361
4.3. (ii) .5860
4.4. (ii) .7114; (v) .5892
4.5. (ii) $-.55$
4.6. (ii) 2.45
4.7. (ii) .6745
4.8. (ii) $\pm.345$ or ±1.11
6.5.1. (ii) .2659 (iv) .1215
5.2. (ii) .0414
5.3. (ii) 23.87
5.4. (ii) 235.12
5.5. (i) 212.16
5.6. (ii) 4.63
5.7. 564
5.9. (ii) .9988
5.10. (ii) .1746; (iv) .4791
5.11. (ii) 14.5
5.12. (ii) 36.58
5.13. (iii) .5849
5.14. (ii) .6186
5.15. (ii) .5841
5.16. (ii) 4094

5.17. (iii) 4171, 6676, 12,304
5.20. (ii)

b	1
$P(B = b)$	.2464
Normal Approx.	.2414

6.6.3. (ii) .0797
6.4. .0037
6.7. .2642
6.9. (ii) .2707
6.10. (ii) .1494
6.7.4. (ii) .0008
7.5. (iii) 14
7.6. (ii) 13
7.8. 798
7.10. (ii) .3033
7.12.

T	1
Exact	.102
Poisson	.099

7.15. $E(T^2) = \lambda + \lambda^2$
6.8.8. .0191
6.9.3. 1.19
9.10. (i) $c = .1$, $n = 200$, $\frac{2}{3}nc^2 = .1111$, normal approx. $= .0035$

CHAPTER 7

7.1.2. $P(S = 4 \text{ and } D = 2) = \frac{2}{36}$
1.7. (i) $P(F = 4|G = 1) = \frac{2}{9}$
1.8. (i) $P(D_2 = 2|D_1 = 1) = \frac{260}{1632}$
1.9. (ii) $P(V = 1|U = 1) = \frac{5}{12}$
1.10. (iii) $P(W = 3|U = 2) = \frac{4}{36}$
1.11. (i) $E(F|G = 1) = 4\frac{1}{9}$
1.13. (ii) $E(D|U = 2) = -\frac{1}{36}$
7.2.1. $-\frac{1}{4}$
2.4. $\mathrm{Cov}(F, G) = \frac{1}{4}$

2.6. $\mathrm{Cov}(U, V) = .642$
2.9. (ii) $-\sqrt{2}/2$
2.12. $(1 - 101p)/(1 + 99p)$
7.3.2. (iii) .1184
3.6. 2520
3.14. hypergeometric $(N - r_1, s - d_1, r_2)$
3.16. $\left(\dfrac{N - s}{N - 1}\right)s \cdot \dfrac{1}{N^2} \cdot (-r_1 r_2)$

CHAPTER 8

8.1.2. (ii) .6563
1.3. (iii) .746
1.5. (ii) $\frac{1}{3}(2 - T)$
1.7. (iii) $(T_1 + 2T_2)/3$
1.14. (ii) 20

1.17. (ii) $\dfrac{(N - s)D(s - D)}{N(s - 1)}$

1.21. 3333
8.2.2. (ii) 9 Var(T)
2.3. (ii) $[\text{Var}(T_1) + \text{Var}(T_2)]/4$
2.4. (ii) $(T_1 + 2T_2)/3$
2.7. $(T_1 + T_2 + T_3)/3$
2.8. (ii) $(T_1 + 2T_2 + T_3)/4$
2.9. (ii) Var(T_1) = 87.9, Var(T_2) = 85.9
2.13. (ii) 1.316
2.14. .25

2.15. $d/\text{SD}(T) = 1.75$, $\pi = .9198$
2.18. (ii) .8974
2.19. (a) (ii) .6552, (b) (iii) .5704, (c) (iii) .5616
2.21. $b = 1.16$
8.3.1. (ii) 105
3.2. (ii) .736
3.4. (i) 75
3.6. (ii) 20%
3.7. (ii) 1667; (iv) 2381
3.8. (ii) 556; (v) 625
3.9. (ii) 1429; (iv) 2174
3.11. (ii) 22; (iv) 62
3.12. (a) (ii) 28, (iv) 80; (b) (ii) 12, (iv) 35
3.16. (ii) 2537
3.17. (ii) 3112

CHAPTER 9

9.1.2. $\bar{v} = 1.27$, $\tau^2 = .356$
1.4. $\bar{v} = 874.4$, $\tau^2 = 8.04$
1.6. $\bar{v} = 1012.45$, $\tau^2 = .45$
1.8. (i) $P(\bar{Y} = \frac{5}{2}) = .2$

1.11. (ii) $\dfrac{N^2(N - s)}{(N - 1)} \cdot \tau^2/s$

1.14. 139
1.15. (ii) 843
9.2.6. (ii) 400
2.8. (ii) $a = \frac{1}{2}$, Var(T) = $(\frac{1}{2})\sigma^2$
2.12. (ii) 576
2.13. (ii) $p + r - (p - r)^2$
9.3.2. (ii) $n = 3$, $k = 6$

3.3. (ii) 500
3.5. (ii) $-.06$

3.7. (iii) $\left(\dfrac{N - s}{N - 1}\right) \cdot \tau^2/s +$

$\left(\dfrac{k - t}{k - 1}\right) \cdot \tau'^2/t$

3.9. (ii) 94
3.10. $s = 60$, $t = 90$
9.4.1. (ii) $P(\bar{D} = 3) = \frac{1}{5}$
4.2. $\omega^2 = 84.8$
4.5. 9.44
9.5.1. (iii) .038
5.4. (ii) 121

CHAPTER 10

10.1.2. (ii) $\frac{1}{2}\bar{X} + \frac{1}{2}\bar{Y}$

1.4. (ii) $\dfrac{2\bar{X} + \bar{Y} + 12\bar{Z}}{15}$

1.9. $k > 3$
1.11. $k = 3 + 2\sqrt{2}$
1.14. (i) $\frac{1}{15}(1 + \Delta^2)$; (ii) $\Delta = 1.5$, $P(\text{med} = 1) = .3$

10.2.1. (ii) $n = 15$
2.4. $k = 6$, $n = 8$
2.6. $n = 4$, $k = 8$
2.12. (iii)
$\frac{1}{4}[Z_1 + Z_2 + Z_3 + Z_4]$ for α
$\frac{1}{4}[Z_1 - Z_2 + Z_3 - Z_4]$ for β
etc.
Variances all $\sigma^2/4$
10.3.2. (ii) 5

CHAPTER 11

11.1.2. (ii) $S \geq 8$
  1.9. (ii) $D = 5$
  1.11. (ii) .0083, .0529
  1.12. (ii) .0548
  1.18. $\alpha = .05$ Reject if $M = 0, 1, 2,$
             12, 13, 14
  1.20. .0328
  1.21. $N = 5$
11.2.1. (ii) .1493; (iv) .1667
  2.2. (ii) $S \geq 19$; (v) $S \geq 30$
  2.3. (ii) $S \geq 97$

  2.4. (iii) .0780
  2.5. (ii) 61; (iv) 57
  2.6. (ii) 10
  2.9. .049
  2.10. (ii) .0176
  2.11. .21
  2.13. (ii) .0004
  2.14. .0213
11.3.4. .202
  3.9. (ii) .057
  3.12. (iv) .1950

CHAPTER 12

12.1.2. (ii) .103
  1.3. (ii) .1875
  1.4. (ii) 924, 64
12.2.1. (ii) .083
  2.3. (ii) .5758
  2.5. .006, significant
  2.7. .0803, not significant
  2.9. (ii) .096; (iv) .026
  2.10. 68
12.3.2. $(s/2)[N + t + 1]$
  3.6. (i) $s(s + 1)/2 + 1$
  3.7. .095
  3.9. .004
  3.11. (ii) .175
  3.12. (ii) .007; (iv) .074
  3.13. .018
12.4.1. (ii) Exact .0974, approx. .0946
  4.2. (ii) .0576
  4.4. .089
  4.5. (ii) .0154; (iv) .0023
  4.8. (ii) .0212
  4.9. (ii) .0326
  4.12. (ii) .399, .415, .426

12.5.1. .0482
  5.3. (i) reject if $B \geq 58$
  5.4. (ii) .0577
  5.5. (iii) .0539
  5.7. (ii) .377
  5.8. (ii) .0287
  5.10. (ii) .4602, .4160, .3863
  5.12. 8
12.6.2. $(\frac{1}{2})^n$
  6.4. (ii) .1445
  6.5. (ii) .3125
  6.8. (iii) Exact .2106, approx. .2051
  6.12. (ii) .0068
  6.15. .1698
12.7.2. (ii) .0555
  7.4. $w = 10.5$ Exact .2321, approx. .1990
  7.9. (i) $W_C = \dfrac{Z_N + s(a + 1)}{2}$
  7.12. (ii) .219
  7.16. .0667
  7.18. (ii) .0655

CHAPTER 13

13.1.4. (ii) .8218, .5524, .4476
  1.5. (ii) .9988
  1.6. (ii) (b) .6513; (iii) (a) .1493
  1.8. (iv) .155

  1.10. (ii) .6203
  1.11. (ii) .0786
  1.14. 31
13.2.1. (ii) 63

2.2. (ii) 266

2.3. (iii) 94

2.6. 3030

2.7. (ii) 76

2.8. (iii) 237

2.9. $s = 14, c = 5$

2.10. (ii) 2957

13.3.5. (ii) .951

3.8. (ii) .1; (iv) .25

3.10. (ii) $\frac{1}{4}$; (iv) 1

3.11. (ii) $\frac{1}{4}$; (iii) $\frac{3}{4}$

3.12. (ii) $\frac{1}{16}$; (iii) $\frac{1}{4}$

13.4.3. (ii) .9214

4.4. (ii) accept if $39 < W_T < 69$

4.5. (ii) accept if $37 < T < 60$

13.5.1. (ii) 1054; (iv) 953

5.2. (ii) $-.1846$

5.6. (iii) $\alpha = .1, \pi = .38$

13.6.4. (ii) .0062

6.6. (ii) .0339

6.8. (ii) .0016

6.10. (ii) .5688

6.11. (ii) .6722

6.12. (ii) 2436

6.15. (ii) 2925

6.17. .0034

6.19. (ii) .1587

INDEX*

Acceptance of a null hypothesis 311; false 370, 373

Accuracy (of an unbiased estimate) 250, 252, 269, 283. *See also* Laplace's measure of dispersion; Precision; Sample size; Standard deviation; Variance

Addition law: for exclusive events 34, 35; for general events 35, 38; of expectation 151; of variance 162, 163, 227; of covariance 225

Additivity (of treatment effects) 296

Algebra of events 28, 31

Allocation (of observations): equal 293; optimum 293, 298, 303, 307; in estimating a difference 293, 307; in estimating a treatment effect 294; optimum 293, 298; in estimating the effects of two treatments 294; in stratified sampling 301, 303, 308; proportional 301, 303

Alternative (to the null hypothesis) 372, 381

Arithmetic mean 145, 164, 261; coded computation of 263. *See also* Average of n random variables; Population mean

Assessment bias 336, 355. *See also* Bias

Assignment bias 353, 363. *See also* Bias

Average of n random variables: expectation of 155; variance of 164; law of large numbers for 211; as estimate 261, 267; as best weighted average 288; as test statistic 401. *See also* Arithmetic mean

Average treatment effect 276, 294

Bayes' law: for simple events 114; for composite events 116; in subjective approach 129

Bernoulli, J. 210

Bias 328, 329, 335; of purposive sampling 40; in using random numbers 86; of estimates 245, 284; in measuring 269; in assessing a treatment effect 279, 327; correction for 250; selection 336; assessment 336, 355; assignment 329, 353, 363; self-selection 355. *See also* Unbiased estimate

Binomial coefficients 47. *See also* Number of samples

Binomial distribution 167; tables of 167, 413, 414; use of Tables B and C for 168; symmetry in 169, 173; expectation of 170; variance of 171; as approximation to hypergeometric 176; standardized 181; normal tendency of 186, 187; normal approximation to 191, 210; Poisson approximation to 195, 198; generalized 200, 211; as approximation to generalized binomial 200; as multinomial marginal 232; in sign test 354, 365. *See also* Multinomial distribution; Quality control; Statistical control; Test of binomial probability

Binomial probabilities; estimation of 243, 245, 247, 250, 255; comparing two 246, 251. *See also* Test of binomial probability

Binomial random variable 167; as sum of indicators 170; expectation of 170; variance of 171; and law of large numbers 209

Binomial theorem 60, 167, 173

Binomial trials 77, 80, 166

Binomial (trials) model 77; realism of 78; relation to sampling model of 78, 176; for defectives in mass production 80, 169; probabilities in 80; generalization to unequal probabilities 81,

INDEX OF EXAMPLES